Advances in Bone Graft Materials

Advances in Bone Graft Materials

Editor
Jung-Bo Huh

MDPI • Basel • Beijing • Wuhan • Barcelona • Belgrade • Manchester • Tokyo • Cluj • Tianjin

Editor
Jung-Bo Huh
Pusan National University
YangSan
Korea

Editorial Office
MDPI
St. Alban-Anlage 66
4052 Basel, Switzerland

This is a reprint of articles from the Special Issue published online in the open access journal *Materials* (ISSN 1996-1944) (available at: https://www.mdpi.com/journal/materials/special_issues/bone_graft_materials).

For citation purposes, cite each article independently as indicated on the article page online and as indicated below:

LastName, A.A.; LastName, B.B.; LastName, C.C. Article Title. *Journal Name* **Year**, *Volume Number*, Page Range.

ISBN 978-3-0365-7022-8 (Hbk)
ISBN 978-3-0365-7023-5 (PDF)

© 2023 by the authors. Articles in this book are Open Access and distributed under the Creative Commons Attribution (CC BY) license, which allows users to download, copy and build upon published articles, as long as the author and publisher are properly credited, which ensures maximum dissemination and a wider impact of our publications.

The book as a whole is distributed by MDPI under the terms and conditions of the Creative Commons license CC BY-NC-ND.

Contents

So-Yeun Kim, Eun-Bin Bae, Jae-Woong Huh, Jong-Ju Ahn, Hyun-Young Bae, Won-Tak Cho and Jung-Bo Huh
Bone Regeneration Using a Three-Dimensional Hexahedron Channeled BCP Block Combined with Bone Morphogenic Protein-2 in Rat Calvarial Defects
Reprinted from: *Materials* **2019**, *12*, 2435, doi:10.3390/ma12152435 1

Eun-Bin Bae, Ha-Jin Kim, Jong-Ju Ahn, Hyun-Young Bae, Hyung-Joon Kim and Jung-Bo Huh
Comparison of Bone Regeneration between Porcine-Derived and Bovine-Derived Xenografts in Rat Calvarial Defects: A Non-Inferiority Study
Reprinted from: *Materials* **2019**, *12*, 3412, doi:10.3390/ma12203412 15

Ji-Youn Hong, Seok-Yeong Ko, Wonsik Lee, Yun-Young Chang, Su-Hwan Kim and Jeong-Ho Yun
Enhancement of Bone Ingrowth into a Porous Titanium Structure to Improve Osseointegration of Dental Implants: A Pilot Study in the Canine Model
Reprinted from: *Materials* **2020**, *13*, 3061, doi:10.3390/ma13143061 31

Bing-Chen Yang, Jing-Wei Lee, Chien-Ping Ju and Jiin-Huey Chern Lin
Physical/Chemical Properties and Resorption Behavior of a Newly Developed Ca/P/S-Based Bone Substitute Material
Reprinted from: *Materials* **2020**, *13*, 3458, doi:10.3390/ma13163458 45

Janine Waletzko-Hellwig, Michael Saemann, Marko Schulze, Bernhard Frerich, Rainer Bader and Michael Dau
Mechanical Characterization of Human Trabecular and Formed Granulate Bone Cylinders Processed by High Hydrostatic Pressure
Reprinted from: *Materials* **2021**, *14*, 1069, doi:10.3390/ma14051069 73

Solomiya Kyyak, Andreas Pabst, Diana Heimes and Peer W. Kämmerer
The Influence of Hyaluronic Acid Biofunctionalization of a Bovine Bone Substitute on Osteoblast Activity In Vitro
Reprinted from: *Materials* **2021**, *14*, 2885, doi:10.3390/ma14112885 85

Won-Tak Cho, So-Yeun Kim, Sung-In Jung, Seong-Soo Kang, Se-Eun Kim, Su-Hyun Hwang, Chang-Mo Jeong and Jung-Bo Huh
Effects of Gamma Radiation-Induced Crosslinking of Collagen Type I Coated Dental Titanium Implants on Osseointegration and Bone Regeneration
Reprinted from: *Materials* **2021**, *14*, 3268, doi:10.3390/ma14123268 99

Lauren A. Boller, Madison A.P. McGough, Stefanie M. Shiels, Craig L. Duvall, Joseph C. Wenke and Scott A. Guelcher
Settable Polymeric Autograft Extenders in a Rabbit Radius Model of Bone Formation
Reprinted from: *Materials* **2021**, *14*, 3960, doi:10.3390/ma14143960 113

Da-Seul Kim, Jun-Kyu Lee, Ji-Won Jung, Seung-Woon Baek, Jun Hyuk Kim, Yun Heo, et al.
Promotion of Bone Regeneration Using Bioinspired PLGA/MH/ECM Scaffold Combined with Bioactive PDRN
Reprinted from: *Materials* **2021**, *14*, 4149, doi:10.3390/ma14154149 129

So-Yeun Kim, You-Jin Lee, Won-Tak Cho, Su-Hyun Hwang, Soon-Chul Heo, Hyung-Joon Kim and Jung-Bo Huh
Preliminary Animal Study on Bone Formation Ability of Commercialized Particle-Type Bone Graft with Increased Operability by Hydrogel
Reprinted from: *Materials* **2021**, *14*, 4464, doi:10.3390/ma14164464 . . . . **141**

Thomas Frankenberger, Constantin Leon Graw, Nadja Engel, Thomas Gerber, Bernhard Frerich and Michael Dau
Sustainable Surface Modification of Polyetheretherketone (PEEK) Implants by Hydroxyapatite/Silica Coating—An In Vivo Animal Study
Reprinted from: *Materials* **2021**, *14*, 4589, doi:10.3390/ma14164589 . . . . **155**

Chun-Sik Bae, Seung-Hyun Kim, Taeho Ahn, Yeonji Kim, Se-Eun Kim, Seong-Soo Kang, et al.
Multiple Porous Synthetic Bone Graft Comprising EngineeredMicro-Channel for Drug Carrier and Bone Regeneration
Reprinted from: *Materials* **2021**, *14*, 5320, doi:10.3390/ma14185320 . . . . **169**

Article

Bone Regeneration Using a Three-Dimensional Hexahedron Channeled BCP Block Combined with Bone Morphogenic Protein-2 in Rat Calvarial Defects

So-Yeun Kim [1,†], Eun-Bin Bae [2,†], Jae-Woong Huh [2,3], Jong-Ju Ahn [2], Hyun-Young Bae [2], Won-Tak Cho [2] and Jung-Bo Huh [2,*]

1 Department of Prosthodontics, Biomedical Research Institute, Pusan National University Hospital, Busan 49241, Korea
2 Department of Prosthodontics, Dental Research Institute, Dental and Life Science Institute, BK21 PLUS Project, School of Dentistry, Pusan National University, Yangsan 50612, Korea
3 Seroun Dental Clinic, Suyeong-ro, Nam-gu, Busan 48445, Korea
* Correspondence: neoplasia96@hanmail.net; Tel.: +82-10-8007-9099; Fax: +82-55-360-5134
† These authors contributed equally to this work.

Received: 3 July 2019; Accepted: 29 July 2019; Published: 31 July 2019

Abstract: It is important to obtain sufficient bone mass before implant placement on alveolar bone, and synthetic bone such as biphasic calcium phosphate (BCP) has been studied to secure this. This study used a BCP block bone with a specific structure of the three-dimensional (3D) hexahedron channel and coating with recombinant human bone morphogenetic protein-2 (rhBMP-2) impregnated carboxymethyl cellulose (CMC) was used to examine the enhancement of bone regeneration of this biomaterial in rat calvarial defect. After the preparation of critical-size calvarial defects in fifteen rats, defects were divided into three groups and were implanted with the assigned specimen (n = 5): Boneplant (untreated 3D hexahedron channeled BCP block), Boneplant/CMC (3D hexahedron channeled BCP block coated with CMC), and Boneplant/CMC/BMP (3D hexahedron channeled BCP block coated with CMC containing rhBMP-2). After 4 weeks, the volumetric, histologic, and histometric analyses were conducted to measure the newly formed bone. Histologically, defects in the Boneplant/CMC/BMP group were almost completely filled with new bone compared to the Boneplant and Boneplant/CMC groups. The new bone volume ($P < 0.05$) and area ($P < 0.001$) in the Boneplant/CMC/BMP group (20.12% ± 2.17, 33.79% ± 3.66) were much greater than those in the Boneplant (10.77% ± 4.8, 16.48% ± 9.11) and Boneplant/CMC (10.72% ± 3.29, 16.57% ± 8.94) groups, respectively. In conclusion, the 3D hexahedron channeled BCP block adapted rhBMP-2 with carrier CMC showed high possibility as an effective bone graft material.

Keywords: biphasic calcium phosphate; bone regeneration; bone substitute; recombinant human bone morphogenetic protein-2; carboxymethyl cellulose

1. Introduction

In dental clinics, implant procedures have become routine, and most often, and almost uniquely, the most limiting factor is when a large amount of bone is lost. If enough bone is available, the difficulty of the implant procedure will be greatly reduced [1]. In reality, however, the bone volume is often small-scale due to various complications, such as periodontitis, trauma, and alveolar bone resorption due to a long edentulous period. What is required as an ideal bone graft material for this purpose is a scaffold that promotes new bone formation, acts as a pathway for bone-forming material, can be optimally degraded and replaced with new bone, and has sufficient strength and stability [2,3]. Examples of reconstructed alveolar bone include using a bone graft or guided bone regeneration [4], using frozen and radiation sterilized allogenic bone grafting [5], use of graft materials from animals or synthetic production [6], recent use of marine collagen and substitutes [7], and so on. Among them, until now, the most promising method for gaining the alveolar bone has been ideally presented with an autologous bone graft with osteogenic potential. However, there are concerns about the availability and morbidity of the donor, difficulty in shaping, and complexity of the surgery [8,9]. On the other hand, the application of allografts or xenografts is also limited by the drawbacks of immune reaction complications, disease transmission, and the absence of osteogenetic properties that could lead to bone resorption and nonunion [10,11].

New bone graft materials such as 3D scaffolds have been recently studied in the dental field as alternatives to overcome the weak points of previous graft materials. As an alternative to bone grafts of sufficient size and diversiform shape, synthetic bone grafts are becoming increasingly popular because they are capable of sufficient production and have no immune reactions [12–15]. Block structures instead of powders contribute to stability by themselves. Depending on the size and shape, as well as the architecture and geometry design, they may promote the clustering of stem cells or osteoprogenitor cells and induce them to function as attachment elements within the scaffold [16]. The internal porous structure helps the circulation of tissue ingrowth, body fluids, and nutrients, and the 3D environmental factors, such as pore size and orientation, also influence the tissue regeneration broadly [17]. At this time, aligned pores are more advantageous for cell migration than irregular pores [18].

Meanwhile, biphasic calcium phosphate (BCP), a synthetic bone graft material, is a mixture of stable phase hydroxyapatite (HA) and soluble phase beta tricalcium phosphate (β-TCP) that is widely used because of its biocompatible, osteoconductive, bioactive, safe, and predictable properties [19,20]. BCP ceramic is considered a promising scaffold for use with tissue engineering strategies for large bone augmentation. In dentistry, BCP bioceramic is also recommended as an alternative or supplementary material for autogenous bone [21]. The biomaterial should have a clinically manageable macrostructure, and a microstructure that induces cell adhesion and proliferation [22]. Because their chemical properties, size, and shape can be easily controlled, they have become a multipurpose matrix for bone formation and development [19].

However, to date, the osteoinductivity of biomaterials is still limited and shows a large difference compared to the one with rhBMP-2 [23]. The recombinant human bone morphogenetic protein-2 (rhBMP-2), an osteoinductive growth factor, is sometimes used in the dental field [24,25]. It has a high bone regeneration capability, which differentiates mesenchymal stem cells (MSCs), transforms pre-osteoblasts into osteoblasts, and acts as a trigger for migration of osteoblasts [26,27]. This leads to early stage osteogenic differentiation of stem cells and osteoprogenitor cells to initiate bone formation [28,29]. Meanwhile, the uncontrolled release of rhBMP-2 causes numerous complications, such as heterotopic ossification, osteolysis, or cancer [30,31]. Despite these problems, rhBMP-2 has outstanding effects on bone defect reconstruction and implant osseointegration acceleration, so research is continuing into developing an effective delivery system for its release without adverse effects. [8,32]. A carrier maintains the release at the local site for the time required to induce bone regeneration [8]. A number of delivery systems composed of various materials have been studied for controlled sustained release of rhBMP-2 [33]. Carboxymethyl cellulose (CMC), which is used as a carrier in this study due to its non-toxicity, biodegradability, and hydrophilic characteristics, is a polysaccharide additive used recently in various industries [34–38]. CMC itself has the ability to contribute directly to the synthesis process of calcium phosphate and can be an effective template for biomimetic mineralization. In studies evaluating CMC as a carrier for rhBMP-2, the results showed that alkaline phosphatase activity of fibroblast cells was increased when rhBMP-2 and CMC were used separately, but CMC combined with rhBMP-2 exhibited much better effects [39,40]. By adding CMC to recombinant human osteogenic protein, the quality of union, total histologic appearance, handling characteristics, and stability are reported to be improved in a previous study [41]. According to Yoo et al. [42], in micro-computed tomography and tissue morphometry analysis, BCP with cross-linked CMC promoted new bone formation and increased new bone area ratio. In addition to maintaining the graft volume, CMC has been shown to promote bone metabolism, resulting in the formation of thick new bone around the graft materials in BCP mixed with cross-linked CMC.

The 3D scaffold synthetic block bone can be made with the desired structure and porosity, and can contain the desired biological materials [43]. Pae et al. [44] reported favorable osteogenetic activity of a 3D hexahedron channeled BCP block bone, the same material with used in this study. In comparison with the particle bone, it showed unremarkable new bone formation similar to the experimental group. This block bone is three-dimensionally connected with a tubule with a concave inner surface, made in line with regularity, and has a porosity of 95%. Even when broken, due to the structural characteristics, it was not completely flawed and had the strength to maintain the volume. Three factors affecting the osteoinduction of biomaterials are chemical composition, macrostructure, and surface structure, and this graft material used in this study was expected to meet these requirements [45]. Another previous study that used this 3D hexahedron channeled synthetic bone block obtained favorable but not particularly high amounts of new bone volume in a rabbit calvarial defect model [44]. Therefore, in addition to the volume maintenance ability of this scaffold, we planned to gain more new bone through delivering the CMC and rhBMP-2, and develop a better new bone graft material. Therefore, using this 3D block bone graft material, in this study, we compared the bone formation capacity of in vivo rat calvarial defect model by combining CMC and rhBMP-2 to increase the bone regeneration ability of this 3D hexahedron channeled synthetic BCP block bone.

2. Materials and Methods

2.1. Preparation of 3D Hexahedron Channeled Synthetic Block Bone

All the specimens of 3D Hexahedron Channeled synthetic block bone were supplied (BoneplantTM, Ezekiel, Chungchung-Namdo, Korea) after sterilization using gamma irradiation. The 3D hexahedron channeled bone block was made of BCP microporous ceramics, with a mixture of 60% HA and 40% β-TCP. The pore size of the disc-shaped bone block (diameter: 8 mm, thickness: 2 mm) was 400–700 μm with 95% porosity (Figure 1). Prior to the animal experiment, the bone blocks were coated with 100 μL of CMC (1.5%, CowellMedi, Busan, Korea) or CMC containing 5 μg of rhBMP-2 (CowellMedi, Busan, Korea) in a completely sterile laboratory [39,46]. The low-level sonication was performed to uniformly coat CMC on the Boneplant surface. The experimental groups of this study were as below:

- Boneplant group: Untreated 3D hexahedron channeled BCP block;
- Boneplant/CMC group: 3D Hexahedron channeled BCP block coated with CMC;
- Boneplant/CMC/BMP group: 3D Hexahedron channeled BCP block coated with CMC containing rhBMP-2.

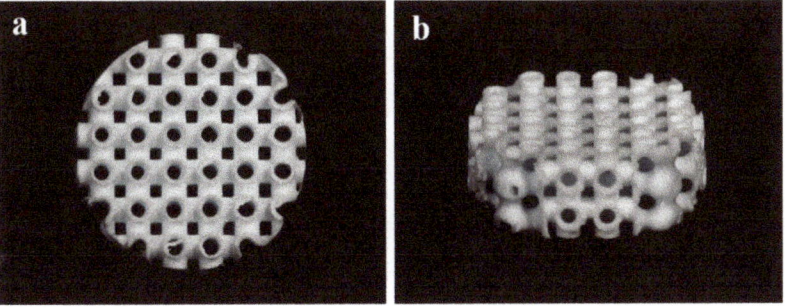

Figure 1. The images of BCP block bone with a 3D hexahedron channel. (**a**) Top view; (**b**) top-side view.

2.2. Scanning Electron Microscope Surface Analysis

The specimens were observed using a Field Emission Scanning Electron Microscope (FE-SEM, S-4700, Hitachi, Tokyo, Japan) to assess morphologies of surface microstructures at magnifications of ×40, ×250, and ×1000. The samples of each group were coated with platinum in ion sputter (E1010, HITACHI, Tokyo, Japan) and were observed using FE-SEM at accelerating voltage of 5 kV.

2.3. In Vivo Animal Study

2.3.1. Experimental Animals

Fifteen 12-week-old male Sprague-Dawley rats (5 per group) (Koatech, Pyeongtaek, Korea) weighing 250–300 g were used for this animal experiment. The animals adapted for at least 7 days before surgery and were caged individually with rodent pellets and water supplied ad-libitum. Animal management and surgical procedures were performed at the Pusan National University Laboratory Animal Resource Center and all the experiments followed routines approved by the Institutional Animal Care and Use Committee of Pusan National University (PNU-2018-2101).

2.3.2. Surgical Procedures

All the rats were anesthetized by general intramuscular injection of a mixture (0.1 mL/10 g) of Tiletamine-zolazepam (Zoletil50, Virbac Korea, Seoul, Korea) and xylazine (Rompun lnj, Bayer Korea, Seoul, Korea) during the surgical procedures (Figure 2). The surgical sites were clearly shaved and scrubbed with Betadine® (povidone-iodine) for disinfection. The local anesthesia was applied using lidocaine (2% Lidocaine HCl and Epinephrine Injection (1:100,000), Yuhan, Seoul, Korea). After making a sagittal incision across the middle of the cranium, the full-thickness flap was elevated to expose the calvarial bone. A critical-sized circular bony defect (8 mm in diameter) was created in the middle of the cranium using a saline-cooled trephine bur (Osung, Kimpo, Korea). Each defect was randomly implanted with an assigned bone block, and covered with a membrane (10 × 10 mm, collagen membrane, GENOSS, Suwon, Korea). The skin and periosteum were respectively closed using absorbable sutures (4-0, Vicryl®, Ethicon, NJ, USA). After 4 weeks, all if the rats were sacrificed by CO_2 asphyxiation and the surgical sites with surrounding bone were carefully harvested. The obtained tissue samples were placed in 10% neutral-buffered formalin (Sigma Aldrich, St. Louis, MO, USA) for 14 days.

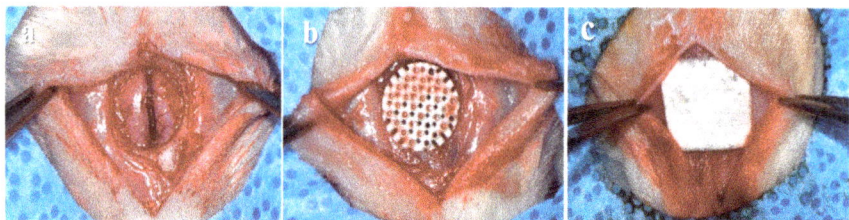

Figure 2. Surgical procedures of this animal experiment. (**a**) Bone defect formation on the cranium. (**b**) Implantation of randomized block bone. (**c**) Placement of collagen membrane.

2.3.3. Micro-Computed Tomography (μCT) Analysis

To measure the new bone volume within the defect area, the specimens were imaged by micro-computed tomography (SMX-90CT, Shimadzu, Kyoto, Japan) at 90 kV, at an intensity of 109 μA. The region of interest (ROI) was set equal to bony defect size (diameter of 8 mm, height of 1.5 mm) (Figure 3). The images distinguished mineralized bone, soft tissues, and scaffold by adjusting the threshold. The percentages of new bone volume were calculated by customized program corded by cording software (MATLAB 2018a, MathWorks, Natick, MA, USA).

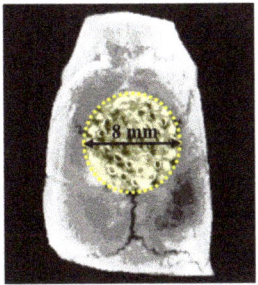

Figure 3. Region of interest (ROI) for micro-computed tomography (µCT) analysis.

2.3.4. Histologic and Histometric Procedures

The harvested tissue samples were sequentially dehydrated with ethanol (70, 80, 90, and 100%) and infiltrated with embedding methylacrylate-based resin (Technovit 7200 VLC, Heraeus Kulzer, Dormagen, Germany) and then hardened in an ultraviolet polymerization unit (EXAKT 520, Exakt-Apparatebau, Norderstedt, Germany). The final tissue slides with a thickness of 30 µm were prepared from initial sections with 400 µm thickness by cutting and grinding procedures using a microtome (KULZER EXAKT 300, Exakt-Apparatebau, Norderstedt, Germany) and grinding machine (KULZER EXAKT 400CS, Exakt-Apparatebau, Norderstedt, Germany). After staining the specimens with hematoxylin and eosin (H&E) solution, all of the sections were photographed using a microscope (BX51, Olympus, Tokyo, Japan) with a digital CCD camera (Polaroid, MA, USA) For histometic analysis, the newly formed bones within defects were measured using an i-solution image analysis program (IMT, Daejeon, Korea) by a single investigator (Figure 4).

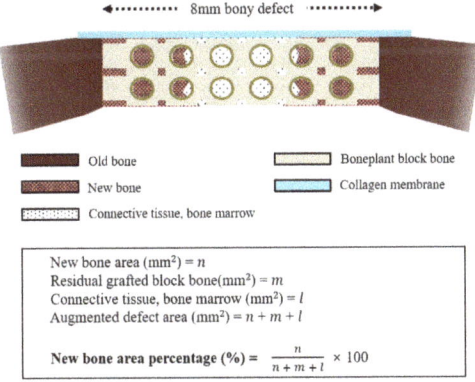

Figure 4. Schematic design for histometric analysis.

2.4. Statistical Analysis

Resulting data of each analysis are expressed as mean ± standard deviation (SD). The in vivo results were analyzed by Kruskal-Wallis test with post-hoc Mann-Whitney test using statistical program (SPSS 25.0, SPSS, IL, USA). P-values < 0.05 were considered statistically significant.

3. Results

3.1. Observations of Surface Morphology

SEM analysis was used to examine the surface morphologies of Boneplant group and CMC coated groups (Boneplant/CMC and Boneplant/CMC/BMP group). Boneplants showed the 3D hexahedron channel (pore size of 400–700 μm; Figure 5a,b). The rough surfaces were observed in the Boneplant group via aggregation of BCP particles (Figure 5c,e). In CMC coated groups, smooth surface were observed covered with CMC compared to the Boneplant group (Figure 5d,f).

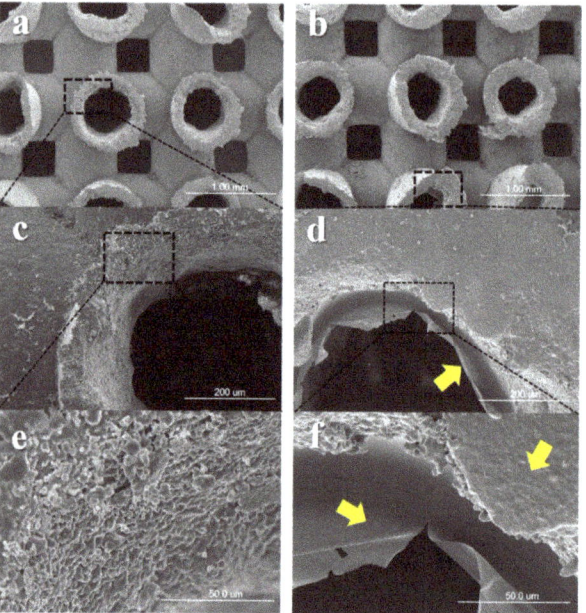

Figure 5. Captured images using scanning electron microscopy. (**a,c,e**) Boneplant group. (**b,d,f**) CMC coated groups, Boneplant/CMC, and Boneplant/CMC/BMP group. Yellow arrow: coated CMC on Boneplant surfaces [Original magnifications: ×40 (**a,b**) ×250 (**c,d**), and ×1000 (**e,f**)].

3.2. In Vivo Animal Study

3.2.1. Clinical Findings

All of the animals survived during the healing periods without adverse effects, such as infection, inflammation, and specimen exposure at surgical sites.

3.2.2. Volumetric Findings

The volumetric results of new bone in the Boneplant, Boneplant/CMC, and Boneplant/CMC/BMP groups were 10.77% ± 4.87, 10.72% ± 3.29, and 20.12% ± 2.17, respectively (Table 1, Figure 6g). The newly formed bone was detected between 3D channel pores, especially in the Boneplant/CMC/BMP group (Figure 6a–f). The Boneplant/CMC/BMP groups showed higher new bone volume than the others ($P < 0.05$). However, there was no significant difference between Boneplant and Boneplant/CMC groups ($P > 0.05$).

Table 1. New bone volume within regions of interest (n = 5 per group).

	Groups	Mean	SD	P-Value
New Bone Volume (%)	Boneplant	10.77	4.87	0.013 *
	Boneplant/CMC	10.72	3.29	
	Boneplant/CMC/BMP	20.12	2.17	

The symbol * indicates statistical significance ($P < 0.05$).

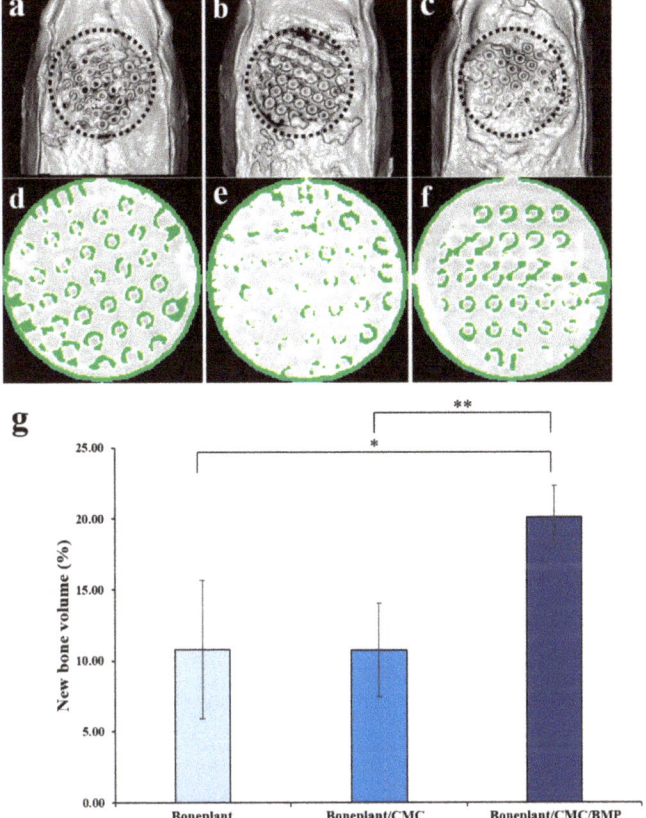

Figure 6. Micro-computed tomography (μCT) findings. (**a–f**) Obtained μCT images of each group. (**a,d**) Boneplant; (**b,e**) Boneplant/CMC; (**c,f**) Boneplant/CMC/BMP groups. (**a–c**) 3D-reconstructed μCT images. (**d–f**) colored μCT images. Green colored area: newly formed bone. Gray colored area: grafted block bone. (**g**) New bone volume within region of interest. The symbol * indicates statistical significance ($P < 0.05$). The symbol ** indicates statistical significance ($P < 0.01$).

3.2.3. Histological Findings

The histological abnormal findings such as inflammatory response were not observed in any tissue samples, and implanted bone graft materials were not collapsed and showed good space maintenance in bony defects (Figure 7). In all three groups, the newly regenerated bones infiltrated into the pores of the 3D hexahedron channel structure. In Boneplant and Boneplant/CMC groups, bone marrow and connective tissues were detected around the new bone. The defects of the Boneplant/CMC/BMP group were almost filled with new bone.

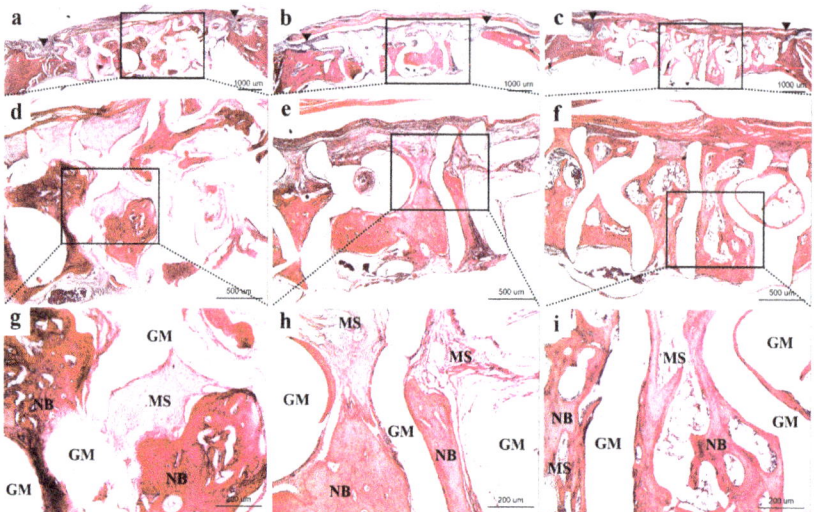

Figure 7. Representative tissue slides at 4-weeks post-surgery: (**a**,**d**,**g**) Boneplant; (**b**,**e**,**h**) Bonplant/CMC; and (**c**,**f**,**i**) Boneplant/CMC/BMP groups. Note: NB = new bone; MS = marrow space; GM = bone graft material. Arrow head = margin of defect (H&E stain, original magnification: (**a**,**b**,**c**) ×12.5, (**d**,**e**,**f**) ×40, (**g**,**h**,**i**) ×100).

3.2.4. Histometric Findings

The histometric results of measuring the new bone area at 4-weeks post-surgery are shown in Table 2 and Figure 8. The Boneplant/CMC/BMP group exhibited more than two-times more new bone area percentage compared with the other groups, indicating statistically different significance ($P < 0.001$).

Table 2. New bone area percentages within the region of interest (n = 5 per group).

	Groups	Mean	SD	P-Value
New Bone Area (%)	Boneplant	16.48	9.11	0.000 ***
	Boneplant/CMC	16.57	8.94	
	Boneplant/CMC/BMP	33.79	3.66	

The symbol *** indicates statistical significance ($P < 0.001$).

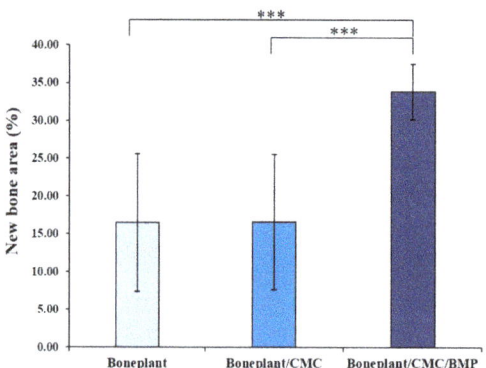

Figure 8. New histometrical bone area percentages of each group. The symbol *** indicates statistical significance ($P < 0.001$).

4. Discussion

Bone grafting is a difficult task in dental practice, but it is an essential goal to be solved. Various methods and materials for bone grafting have been studied, but finally in this study the focus is on the use of originated bone graft materials in the laboratory, i.e. synthetic bone [47]. Therefore, in this study, we tried to study more stable and predictive bone graft materials and methods by evaluating new bone formation ability using synthetic bone.

Synthetic bone can be made in the form of a block bone, which is advantageous for volume maintenance, and there is no need to worry about additional surgery or lack of bone quantity. In addition, structural morphology, such as porosity, pore size, interconnectivity, and orientation of synthetic bone, affect bone regeneration [48,49]. According to Frame et al. [50], during the alveolar bone augmentation, grafting of compact solid HA alloplastic bone showed less bone formation than that of porous HA alloplastic bone graft. The scaffold-type synthetic bone produced in 3D can be given a variety of internal structures, such as tube, pore, etc. In the pore structure, fluids and nutrients flow into it and circulate [51]. Micropores permit the capture and concentration of proteins that induce differentiation upon contact with undifferentiated cells [52,53]. The pore enables molecular transport necessary for bone regeneration, and an interconnected internal structure increases cell attachment rate [54]. Studies have shown that compared to the convex surface, the concave surface promotes cell attachment and proliferation and is in charge of the initiation of the bone formation process [52,55]. The pipe-shaped concave surface in the scaffold used in this study is a favorable form for cell attachment and proliferation. The regular hexahedron is connected with empty pipes and the pipe channels are connected in and out. The blood circulates in and out of the six directions of the pipe structure. Bone materials do not roll down because of the matrix structure of the block, which has a sealing effect when it breaks. When it is pushed, highly reactive surfaces are attached to each other to maintain volume.

Micro CT analysis showed no difference between Boneplant and Boneplant/CMC groups, but Boneplant/CMC/BMP group showed differences in new bone volume production from the other two groups. Boneplant alone showed favorable new bone formation when used alone, but rhBMP-2 using CMC as a carrier showed more bone formation. This was also evident in the area value assessed by histometric finding. The difference between the Boneplant and Boneplant/CMC groups was not found, but the new bone area of Boneplant/CMC/BMP group was twice as large as the other two groups. This suggests that a larger area of new bone was obtained when rhBMP-2 was used, which is consistent with the volume results from the micro CT analysis. This volumetric analysis has the limitation that it could not distinguish the tissues perfectly, but it is considered that a relative comparison could be conducted because we used a single set of equipment and the same analysis level for each tissue in order to distinguish the tissues as clearly as possible.

The use of Boneplant in histological findings was found to be superior to the maintenance of space and volume at defect sites. There was hardly any depression on the superior or inferior sides of the defect site. In all three groups, new bone appeared inside the pore, but Boneplant/CMC/BMP groups were filled with new bone, while Boneplant and Boneplant/CMC groups were found to have more connective tissue around the new bone. The pore size of the scaffolds has been shown to affect nutrient transport during tissue regeneration, as well as cell adhesion and migration [56,57]. The appropriate pore size for bone tissue engineering varies slightly depending on conditions, such as 200–400 μm [14], interconnected pore of 300–500 μm [48], 100–300 μm, 600 μm [58], and aligned channel of 270 μm [18], and a clear consensus has yet to be reached. However, if the pore is too large, the blood flow is excessive and it is difficult for the cell differentiation to mature, resulting in a lot of fibrous tissue. On the other hand, if the pore is too small, the pore is clogged before the bone regeneration is completed [18]. The pore size of the block bone used in our study ranged from 400 to 700 μm. Since the channel is not one-way but is six-directional, it is expected that the blood flow was not excessively fast, but it is estimated that some connective tissues were formed inside the pore because of its large pore size. In the Boneplant/CMC/BMP group, rapid differentiation of osteogenic cells was induced by rhBMP-2, suggesting less formation of connective tissue and better quality of direct contact in the pores of new bone.

Among the various biomaterials, the ceramics, especially calcium phosphate, have the advantages of mechanical strength, high affinity for proteins, lower degradation rate, and biocompatibility, so this is the most interesting bone substitute nowadays [59]. Kruyt et al. [60] compared the BCP, β-TCP, and HA with the bone growth ability, and the result showed higher bone growth in BCP and β-TCP. Lim et al. [61] compared the loading of rhBMP-2 or platelet-rich plasma (PRP) in the β-TCP scaffold and the bone regeneration effect of rhBMP-2 was much higher. Magne et al. [62] reported that the BCP loaded directly with human growth hormone showed non-constant hormone release, which was rapid in the first 48 hours and slowed down after this. Calcium phosphate can be 3D printed to have pores and this porosity can be a stimulant of osteogenetic behavior. In the previous study, osteogenic and volume abilities were favorable for bone regeneration in BCP block bone having a 3D hexahedron channel structure [44]. In this study, CMC and rhBMP-2 were loaded for faster bone regeneration, and bone maturity and bone mass were significantly increased when rhBMP-2 was added. Releasing a substantial portion of rhBMP-2 and discharging it in a sustained release pattern is helpful for osseointegration and regeneration of the bone defect around the implant [63]. Herford et al. [32] evaluated the effect of rfBMP-2 in the distraction osteogenesis procedure. They concluded that the addition of rhBMP-2 in the distraction osteogenesis technique induces rapid bone regeneration and soft tissue healing in the bone defect site. On the other hand, locally high concentrations due to short half time induce osteoclasts [64]. Therefore, it is important to maintain sustained release at a constant rate using the carrier for release [65]. The effects of diverse delivery systems on growth factors have been studied. In particular, hydrogel [66], collagen sponge [67], and an altered scaffold surface were presented to prolong growth factor release while avoiding primary massive release [68]. In this study, CMC was applied as a carrier for the sustained release. The addition of rhBMP-2 with CMC resulted in a better quality of new bone distribution with less formation of connective tissue in the graft site.

As a limit in this study, firstly, in the animal experiment, the number of the experimental group is the minimum number, and long-term observation could not be performed because it takes time to use large animals and wait to sacrifice them. In animal studies, systematization of overall animal testing, including appropriate drug use and sacrifice timing, customized for each clinical setting will further improve the predictability of animal clinical trials. In addition, rhBMP-2 showed the potential of added autoinducing in this study and proved its potential, but no resolution of carcinogenesis, dosage, long-term results, and suitable carriers has been reported. We did not investigate the kinetic release of the rhBMP-2 in this study and instead referred to the dose concentration used in the previous article [69]. It would be meaningful to measure the releasing profile of rhBMP-2 in the Boneplant/CMC/BMP group, but this study excluded this test because the difference between fast release and slow release of rhBMP-2 was not the main evaluation variable. Eventually, this is the fundamental problem of exposing too much rhBMP-2 to the local area, and additional and continuing research on the appropriate dose and delivery system of rhBMP-2 to biomaterial is needed.

5. Conclusions

Bone formation was assessed by grafting using a 3D BCP synthetic block bone with an internal hexahedron channel structure. When the 3D hexahedron channeled BCP block bone was implanted, the volume of the bone was secured to 50% or more of the transplantation site, and BCP itself was effective as a transplanted bone at the defect site to maintain volume and produce new bone. The addition of rhBMP-2 with CMC showed significant volume and a large area of new bone. The addition of rhBMP-2 reduced the formation of connective tissue inside the pores and improved the quality of the new bone. Therefore, 3D hexahedron channeled BCP block combined with rhBMP-2 is expected to have a better effect for bone regeneration during bone grafting, and further studies are required for other types of 3D BCP block bone with rhBMP-2.

Author Contributions: Conceptualization, J.-B.H. and J.-W.H.; formal analysis, E.-B.B. and J.-J.A.; investigation, E.-B.B., J.-J.A., H.-Y.B., and W.-T.C.; data curation, E.-B.B., J.-J.A, H.-Y.B., and W.-T.C.; writing—original draft preparation, S.-Y.K. and E.-B.B.; writing—review and editing, S.-Y.K., J.-W.H., J.-B.H., and E.-B.B.

Acknowledgments: This study was supported by Dental Research Institute (PNUDH DRI-2017-01), Pusan National University Dental Hospital.

Conflicts of Interest: The authors declare no conflict of interest.

References

1. Javed, F.; Ahmed, H.B.; Crespi, R.; Romanos, G.E. Role of primary stability for successful osseointegration of dental implants: Factors of influence and evaluation. *Interv. Med. Appl. Sci.* **2013**, *5*, 162–167. [CrossRef] [PubMed]
2. Del Fabbro, M.; Rosano, G.; Taschieri, S. Implant survival rates after maxillary sinus augmentation. *Eur. J. Oral Sci.* **2008**, *116*, 497–506. [CrossRef] [PubMed]
3. Guarino, V.; Causa, F.; Ambrosio, L. Bioactive scaffolds for bone and ligament tissue. *Expert Rev. Med Devices* **2007**, *4*, 405–418. [CrossRef] [PubMed]
4. Gultekin, B.A.; Bedeloglu, E.; Kose, T.E.; Mijiritsky, E. Comparison of Bone Resorption Rates after Intraoral Block Bone and Guided Bone Regeneration Augmentation for the Reconstruction of Horizontally Deficient Maxillary Alveolar Ridges. *BioMed Res. Int.* **2016**, *2016*, 1–9. [CrossRef] [PubMed]
5. Krasny, M.; Krasny, K.; Fiedor, P.; Zadurska, M.; Kamiński, A. Long-term outcomes of the use of allogeneic, radiation-sterilised bone blocks in reconstruction of the atrophied alveolar ridge in the maxilla and mandible. *Cell Tissue Bank.* **2015**, *16*, 631–638. [CrossRef] [PubMed]
6. Le, B.; Borzabadi-Farahani, A.; Nielsen, B. Treatment of labial soft tissue recession around dental implants in the esthetic zone using guided bone regeneration with mineralized allograft: A retrospective clinical case series. *J. Oral Maxillofac. Surg.* **2016**, *74*, 1552–1561. [CrossRef] [PubMed]
7. Zhang, X.; Vecchio, K.S. Conversion of natural marine skeletons as scaffolds for bone tissue engineering. *Front. Mater. Sci.* **2013**, *7*, 103–117. [CrossRef]
8. Herford, A.S.; Tandon, R.; Stevens, T.W.; Stoffella, E.; Cicciu, M. Immediate distraction osteogenesis: The sandwich technique in combination with rhBMP-2 for anterior maxillary and mandibular defects. *J. Craniofacial Surg.* **2013**, *24*, 1383–1387. [CrossRef]
9. Cicciù, M.; Herford, A.; Stoffella, E.; Cervino, G.; Cicciù, D. Protein-Signaled Guided Bone Regeneration Using Titanium Mesh and Rh-BMP2 in Oral Surgery: A Case Report Involving Left Mandibular Reconstruction after Tumor Resection. *Open Dent. J.* **2012**, *6*, 51–55. [CrossRef]
10. Athanasiou, V.T.; Papachristou, D.J.; Panagopoulos, A.; Saridis, A.; Scopa, C.D.; Megas, P. Histological comparison of autograft, allograft-DBM, xenograft, and synthetic grafts in a trabecular bone defect: An experimental study in rabbits. *Med. Sci. Monit.* **2009**, *16*, BR24–BR31.
11. Dimitriou, R.; Jones, E.; McGonagle, D.; Giannoudis, P.V. Bone regeneration: Current concepts and future directions. *BMC Med.* **2011**, *9*, 66. [CrossRef] [PubMed]
12. Asa'Ad, F.; Pagni, G.; Pilipchuk, S.P.; Giannì, A.B.; Giannobile, W.V.; Rasperini, G. 3D-printed scaffolds and biomaterials: Review of alveolar bone augmentation and periodontal regeneration applications. *Int. J. Dent.* **2016**, *2016*, 1–15. [CrossRef] [PubMed]
13. Hong, P.; Boyd, D.; Beyea, S.D.; Bezuhly, M. Enhancement of bone consolidation in mandibular distraction osteogenesis: A contemporary review of experimental studies involving adjuvant therapies. *J. Plast. Reconstr. Aesthet. Surg.* **2013**, *66*, 883–895. [CrossRef] [PubMed]
14. Shim, J.H.; Moon, T.S.; Yun, M.J.; Jeon, Y.C.; Jeong, C.M.; Cho, D.W.; Huh, J.B. Stimulation of healing within a rabbit calvarial defect by a PCL/PLGA scaffold blended with TCP using solid freeform fabrication technology. *J. Mater. Sci. Mater. Med.* **2012**, *23*, 2993–3002. [CrossRef] [PubMed]
15. Lee, M.H.; You, C.; Kim, K.H. Combined effect of a microporous layer and type I collagen coating on a biphasic calcium phosphate scaffold for bone tissue engineering. *Materials* **2015**, *8*, 1150–1161. [CrossRef] [PubMed]
16. Kinoshita, Y.; Maeda, H. Recent developments of functional scaffolds for craniomaxillofacial bone tissue engineering applications. *Sci. World J.* **2013**, *2013*, 1–21. [CrossRef]

17. Bružauskaitė, I.; Bironaitė, D.; Bagdonas, E.; Bernotienė, E. Scaffolds and cells for tissue regeneration: Different scaffold pore sizes-different cell effects. *Cytotechnology* **2016**, *68*, 355–369. [CrossRef] [PubMed]
18. Seong, Y.J.; Kang, I.G.; Song, E.H.; Kim, H.E.; Jeong, S.H.; Seong, Y.; Kang, I.; Song, E.; Kim, H.; Jeong, S. Calcium phosphate-collagen scaffold with aligned pore channels for enhanced osteochondral regeneration. *Adv. Heal. Mater.* **2017**, *6*, 1700966. [CrossRef]
19. Lobo, S.E.; Arinzeh, T.L. Biphasic Calcium Phosphate Ceramics for Bone Regeneration and Tissue Engineering Applications. *Materials* **2010**, *3*, 815–826. [CrossRef]
20. Lim, H.C.; Zhang, M.L.; Lee, J.S.; Jung, U.W.; Choi, S.H. Effect of different hydroxyapatite: β–Tricalcium phosphate ratios on the osteoconductivity of biphasic calcium phosphate in the rabbit sinus model. *Int. J. Oral Maxillofac. Implant.* **2015**, *30*, 56–72. [CrossRef]
21. LeGeros, R.Z.; Lin, S.; Rohanizadeh, R.; Mijares, D.; LeGeros, J.P. Biphasic calcium phosphate bioceramics: Preparation, properties and applications. *J. Mater. Sci. Mater. Electron.* **2003**, *14*, 201–209. [CrossRef]
22. Herford, A.S.; Lü, M.; Akin, L.; Cicciù, M. Evaluation of a porcine matrix with and without platelet-derived growth factor for bone graft coverage in pigs. *Int. J. Oral Maxillofac. Implant.* **2012**, *27*, 1351–1358.
23. Chan, O.; Coathup, M.; Nesbitt, A.; Ho, C.Y.; Hing, K.; Buckland, T.; Campion, C.; Blunn, G. The effects of microporosity on osteoinduction of calcium phosphate bone graft substitute biomaterials. *Acta Biomater.* **2012**, *8*, 2788–2794. [CrossRef] [PubMed]
24. Solofomalala, G.D.; Guery, M.; Lesiourd, A.; Le Huec, J.C.; Chauveaux, D.; Laffenetre, O. Bone morphogenetic proteins: From their discoveries till their clinical applications. *Eur. J. Orthop. Surg. Traumatol.* **2007**, *17*, 609–615. [CrossRef]
25. Kim, H.C.; Kim, S.N.; Lee, J.Y.; Kim, U.C. Successful strategy of treatment using rhBMP-2 for maxillary sinus graft. *J. Korean Dent. Assoc.* **2015**, *53*, 14–27.
26. Liu, S.; Liu, Y.; Jiang, L.; Li, Z.; Lee, S.; Liu, C.; Wang, J.; Zhang, J. Recombinant human BMP-2 accelerates the migration of bone marrow mesenchymal stem cells via the CDC42/PAK1/LIMK1 pathway in vitro and in vivo. *Biomater. Sci.* **2018**, *7*, 362–372. [CrossRef] [PubMed]
27. Ikeuchi, M.; Dohi, Y.; Horiuchi, K.; Ohgushi, H.; Noshi, T.; Yoshikawa, T.; Yamamoto, K.; Sugimura, M. Recombinant human bone morphogenetic protein-2 promotes osteogenesis within atelopeptide type I collagen solution by combination with rat cultured marrow cells. *J. Biomed. Mater. Res.* **2002**, *60*, 61–69. [CrossRef] [PubMed]
28. Jansen, J.; Vehof, J.; Ruhe, P.; Kroeze-Deutman, H.; Kuboki, Y.; Takita, H.; Hedberg, E.; Mikos, A. Growth factor-loaded scaffolds for bone engineering. *J. Controll. Release* **2005**, *101*, 127–136. [CrossRef]
29. Babensee, J.E.; McIntire, L.V.; Mikos, A.G. Growth Factor Delivery for Tissue Engineering. *Pharm. Res.* **2000**, *17*, 497–504. [CrossRef]
30. Epstein, N.E. Pros, cons, and costs of INFUSE in spinal surgery. *Surg. Neurol. Int.* **2011**, *2*, 10. [CrossRef]
31. Tannoury, C.A.; An, H.S. Complications with the use of bone morphogenetic protein 2 (BMP-2) in spine surgery. *Spine J.* **2014**, *14*, 552–559. [CrossRef] [PubMed]
32. Herford, A.S.; Cicciù, M.; Eftimie, L.F.; Miller, M.; Signorino, F.; Famà, F.; Cervino, G.; Lo Giudice, G.; Bramanti, E.; Lauritano, F. rhBMP-2 applied as support of distraction osteogenesis: A split–mouth histological study over nonhuman primates mandibles. *Int. J. Clin. Exp. Med.* **2016**, *9*, 17187–17194.
33. Blokhuis, T.J. Formulations and delivery vehicles for bone morphogenetic proteins: Latest advances and future directions. *Injury* **2009**, *40*, S8–S11. [CrossRef]
34. Salama, A.; El-Sakhawy, M. Preparation of polyelectrolyte/calcium phosphate hybrids for drug delivery application. *Carbohydr. Polym.* **2014**, *113*, 500–506. [CrossRef] [PubMed]
35. Liuyun, J.; Yubao, L.; Chengdong, X. A novel composite membrane of chitosan–carboxymethyl cellulose polyelectrolyte complex membrane filled with nano–hydroxyapatite I. Preparation and properties. *J. Mater. Sci. Mater. Med.* **2009**, *20*, 1645–1652. [CrossRef]
36. Stelzer, G.I.; Klug, E. Carboxymethylcellulose. In *Handbook of Watersoluble Gums and Resins*; McGraw-Hill: New York, NY, USA, 1980; pp. 1–4.
37. Barbucci, R.; Magnani, A.; Consumi, M. swelling behavior of carboxymethylcellulose hydrogels in relation to cross-linking, pH, and charge density. *Macromolecules* **2000**, *33*, 7475–7480. [CrossRef]
38. Schweizer, S.; Taubert, A. Polymer-controlled, bio-inspired calcium phosphate mineralization from aqueous solution. *Macromol. Biosci.* **2007**, *7*, 1085–1099. [CrossRef]

39. Salama, A.; Abou-Zeid, R.E.; El-Sakhawy, M.; El-Gendy, A.A. Carboxymethyl cellulose/silica hybrids as templates for calcium phosphate biomimetic mineralization. *Int. J. Boil. Macromol.* **2015**, *74*, 155–161. [CrossRef]
40. Santa-Comba, A.; Pereira, A.; Lemos, R.; Santos, D.; Amarante, J.; Pinto, M.; Tavares, P.; Bahia, F. Evaluation of carboxymethylcellulose, hydroxypropylmethylcellulose, and aluminum hydroxide as potential carriers for rhBMP-2. *J. Biomed. Mater. Res.* **2001**, *55*, 396–400. [CrossRef]
41. Cook, S.D.; Salkeld, S.L.; Patron, L.P. Bone defect healing with an osteogenic protein-1 device combined with carboxymethylcellulose. *J. Biomed. Mater. Res. Part B Appl. Biomater.* **2005**, *75*, 137–145. [CrossRef]
42. Yoo, H.S.; Bae, J.H.; Kim, S.E.; Bae, E.B.; Kim, S.Y.; Choi, K.H.; Moon, K.O.; Jeong, C.M.; Huh, J.B. The effect of bisphasic calcium phosphate block bone graft materials with polysaccharides on bone regeneration. *Materials* **2017**, *10*, 17. [CrossRef] [PubMed]
43. Hollister, S.J. Porous scaffold design for tissue engineering. *Nat. Mater.* **2005**, *4*, 518–524. [CrossRef] [PubMed]
44. Pae, H.C.; Kang, J.H.; Cha, J.K.; Lee, J.S.; Paik, J.W.; Jung, U.W.; Choi, S.H. Bone regeneration using three-dimensional hexahedron channel structured BCP block in rabbit calvarial defects. *J. Biomed. Mater. Res. Part B Appl. Biomater.* **2019**. [CrossRef] [PubMed]
45. García-Gareta, E.; Coathup, M.J.; Blunn, G.W. Osteoinduction of bone grafting materials for bone repair and regeneration. *Bone* **2015**, *81*, 112–121. [CrossRef] [PubMed]
46. Lee, J.H.; Jeong, B. The effect of hyaluronate-carboxymethyl cellulose on bone graft substitute healing in a rat spinal fusion model. *J. Korean Neurosurg. Soc.* **2011**, *50*, 409–414. [CrossRef] [PubMed]
47. Cicciù, M.; Cervino, G.; Herford, A.S.; Famà, F.; Bramanti, E.; Fiorillo, L.; Lauritano, F.; Sambataro, S.; Troiano, G.; Laino, L. Facial bone reconstruction using both marine or non-marine bone substitutes: Evaluation of current outcomes in a systematic literature review. *Mar. Drugs* **2018**, *16*, 27. [CrossRef] [PubMed]
48. Karageorgiou, V.; Kaplan, D. Porosity of 3D biomaterial scaffolds and osteogenesis. *Biomater.* **2005**, *26*, 5474–5491. [CrossRef]
49. Bai, F.; Wang, Z.; Lu, J.; Liu, J.; Chen, G.; Lv, R.; Wang, J.; Lin, K.; Zhang, J.; Huang, X. The Correlation Between the Internal Structure and Vascularization of Controllable Porous Bioceramic Materials In Vivo: A Quantitative Study. *Tissue Eng. Part A* **2010**, *16*, 3791–3803. [CrossRef]
50. Frame, J.W.; Rout, P.; Browne, R. Ridge augmentation using solid and porous hydroxylapatite particles with and without autogenous bone or plaster. *J. Oral Maxillofac. Surg.* **1987**, *45*, 771–777. [CrossRef]
51. Wei, M.Z.; Da, P.W.; Jian, Y.X. Biological characteristics and clinical application of scaffold materials for bone tissue engineering. *J. Clin. Rehabil. Tissue Eng. Res.* **2007**, *11*, 9781.
52. Graziano, A.; d'Aquino, R.; Angelis, M.G.C.D.; De Francesco, F.; Giordano, A.; Laino, G.; Piattelli, A.; Traini, T.; De Rosa, A.; Papaccio, G. Scaffold's surface geometry significantly affects human stem cell bone tissue engineering. *J. Cell. Phys.* **2008**, *214*, 166–172. [CrossRef] [PubMed]
53. Ripamonti, U. Osteoinduction in porous hydroxyapatite implanted in heterotopic sites of different animal models. *Biomaterials* **1996**, *17*, 31–35. [CrossRef]
54. Lundgren, A.; Sennerby, L.; Lundgren, D. Guided jaw-bone regeneration using an experimental rabbit model. *Int. J. Oral Maxillofac. Surg.* **1998**, *27*, 135–140. [CrossRef]
55. Ripamonti, U.; Richter, P.W.; Thomas, M.E. Self-inducing shape memory geometric cues embedded within smart hydroxyapatite-based biomimetic matrices. *Plast. Reconstr. Surg.* **2007**, *120*, 1796–1807. [CrossRef] [PubMed]
56. Lien, S.M.; Ko, L.Y.; Huang, T.J. Effect of pore size on ECM secretion and cell growth in gelatin scaffold for articular cartilage tissue engineering. *Acta Biomater.* **2009**, *5*, 670–679. [CrossRef] [PubMed]
57. Harley, B.A.; Kim, H.D.; Zaman, M.H.; Yannas, I.V.; Lauffenburger, D.A.; Gibson, L.J. Microarchitecture of three-dimensional scaffolds influences cell migration behavior via junction interactions. *Biophys. J.* **2008**, *95*, 4013–4024. [CrossRef] [PubMed]
58. Chiu, Y.C.; Fang, H.Y.; Hsu, T.T.; Lin, C.Y.; Shie, M.Y. The characteristics of mineral trioxide aggregate/polycaprolactone 3-dimensional scaffold with osteogenesis properties for tissue regeneration. *J. Endod.* **2017**, *43*, 923–929. [CrossRef]
59. Agrawal, V.; Sinha, M. A review on carrier systems for bone morphogenetic protein-2. *J. Biomed. Mater. Res. Part B Appl. Biomater.* **2017**, *105*, 904–925. [CrossRef]

60. Kruyt, M.C.; Wilson, C.E.; De Bruijn, J.D.; Van Blitterswijk, C.A.; Oner, C.F.; Verbout, A.J.; Dhert, W.J. The effect of cell-based bone tissue engineering in a goat transverse process model. *Biomaterials.* **2006**, *27*, 5099–5106. [CrossRef]
61. Lim, H.P.; Mercado-Pagan, A.E.; Yun, K.D.; Kang, S.S.; Choi, T.H.; Bishop, J.; Koh, J.T.; Maloney, W.; Lee, K.M.; Yang, Y.P. The effect of rhBMP-2 and PRP delivery by biodegradable β-tricalcium phosphate scaffolds on new bone formation in a non-through rabbit cranial defect model. *J. Mater. Sci. Mater. Med.* **2013**, *24*, 1895–1903. [CrossRef]
62. Magne, D.; Bluteau, G.; Lopez-Cazaux, S.; Weiss, P.; Pilet, P.; Ritchie, H.H.; Daculsi, G.; Guicheux, J. Development of an odontoblast in vitro model to study dentin mineralization. *Connect. Tissue Res.* **2004**, *45*, 101–108. [CrossRef] [PubMed]
63. Jo, J.H.; Choi, S.W.; Choi, J.W.; Paik, D.H.; Kang, S.S.; Kim, S.E.; Jeon, Y.C.; Huh, J.B. Effects of different rhBMP-2 release profiles in defect areas around dental implants on bone regeneration. *Biomed. Mater.* **2015**, *10*, 45007. [CrossRef] [PubMed]
64. Kaneko, H.; Arakawa, T.; Mano, H.; Kaneda, T.; Ogasawara, A.; Nakagawa, M.; Toyama, Y.; Yabe, Y.; Kumegawa, M.; Hakeda, Y. Direct stimulation of osteoclastic bone resorption by bone morphogenetic protein (BMP)-2 and expression of BMP receptors in mature osteoclasts. *Bone* **2000**, *27*, 479–486. [CrossRef]
65. Takaoka, K.; Saito, N. New synthetic biodegradable polymers as BMP carriers for bone tissue engineering. *Biomaterials* **2003**, *24*, 2287–2293.
66. Fukui, T.; Ii, M.; Shoji, T.; Matsumoto, T.; Mifune, Y.; Kawakami, Y.; Akimaru, H.; Kawamoto, A.; Kuroda, T.; Saito, T.; et al. Therapeutic effect of local administration of low-dose simvastatin-conjugated gelatin hydrogel for fracture healing. *J. Bone Miner. Res.* **2012**, *27*, 1118–1131. [CrossRef] [PubMed]
67. Monjo, M.; Rubert, M.; Wohlfahrt, J.C.; Rønold, H.J.; Ellingsen, J.E.; Lyngstadaas, S.P. In vivo performance of absorbable collagen sponges with rosuvastatin in critical-size cortical bone defects. *Acta Biomater.* **2010**, *6*, 1405–1412. [CrossRef]
68. Walter, M.S.; Frank, M.J.; Rubert, M.; Monjo, M.; Lyngstadaas, S.P.; Haugen, H.J. Simvastatin-activated implant surface promotes osteoblast differentiation in vitro. *J. Biomater. Appl.* **2014**, *28*, 897–908. [CrossRef]
69. Park, J.C.; Bae, E.B.; Kim, S.E.; Kim, S.Y.; Choi, K.H.; Choi, J.W.; Bae, J.H.; Ryu, J.J.; Huh, J.B. Effects of BMP-2 delivery in calcium phosphate bone graft materials with different compositions on bone regeneration. *Materials* **2016**, *9*, 954. [CrossRef]

© 2019 by the authors. Licensee MDPI, Basel, Switzerland. This article is an open access article distributed under the terms and conditions of the Creative Commons Attribution (CC BY) license (http://creativecommons.org/licenses/by/4.0/).

Article

Comparison of Bone Regeneration between Porcine-Derived and Bovine-Derived Xenografts in Rat Calvarial Defects: A Non-Inferiority Study

Eun-Bin Bae [1,†], Ha-Jin Kim [2,†], Jong-Ju Ahn [1], Hyun-Young Bae [1], Hyung-Joon Kim [2,*] and Jung-Bo Huh [1,*]

1. Department of Prosthodontics, Dental Research Institute, Dental and Life Science Institute, BK21 PLUS Project, School of Dentistry, Pusan National University, Yangsan 50612, Korea; 0228dmqls@hanmail.net (E.-B.B.); tarov0414@daum.net (J.-J.A.); h.02@hanmail.net (H.-Y.B.)
2. Department of Oral Physiology, Dental Research Institute, Dental and Life Science Institute, School of Dentistry, Pusan National University, Yangsan 50612, Korea; ya120010@naver.com
* Correspondence: hjoonkim@pusan.ac.kr (H.-J.K.); neoplasia96@hanmail.net (J.-B.H.); Tel.: +82-10-6326-4189 (H.-J.K.); Tel.: +82-10-8007-9099 (J.-B.H.); Fax: +82-55-510-8208 (H.-J.K.); Fax: +82-55-360-5134 (J.-B.H.)
† These authors contributed equally to this work.

Received: 26 September 2019; Accepted: 16 October 2019; Published: 18 October 2019

Abstract: The present study aimed to compare the bone-regeneration capacity of porcine-derived xenografts to bovine-derived xenografts in the rat calvarial defect model. The observation of surface morphology and in vitro cell studies were conducted prior to the animal study. Defects with a diameter of 8 mm were created in calvaria of 20 rats. The rats were randomly treated with porcine-derived (Bone-XP group) or bovine-derived xenografts (Bio-Oss group) and sacrificed at 4 and 8 weeks after surgery. The new bone regeneration was evaluated by micro-computed tomography (μCT) and histomorphometric analyses. In the cell study, the extracts of Bone-XP and Bio-Oss showed a positive effect on the regulation of osteogenic differentiation of human mesenchymal stem cells (hMSCs) without cytotoxicity. The new bone volume of Bone-XP (17.52 ± 3.78% at 4 weeks and 32.09 ± 3.51% at 8 weeks) was similar to that of Bio-Oss (11.6 ± 3.88% at 4 weeks and 25.89 ± 7.43% at 8 weeks) ($p > 0.05$). In the results of new bone area, there was no significant difference between Bone-XP (9.08 ± 5.47% at 4 weeks and 25.22 ± 13.56% at 8 weeks) and Bio-Oss groups (5.83 ± 2.56% at 4 weeks and 21.68 ± 11.11% at 8 weeks) ($p > 0.05$). It can be concluded that the porcine-derived bone substitute may offer a favorable cell response and bone regeneration similar to those of commercial bovine bone mineral.

Keywords: bone regeneration; bone substitute; xenograft; porcine bone

1. Introduction

An insufficient alveolar bone volume can produce the problems of implant insertion and prognosis, therefore, bone graft materials have been commonly used to reconstruct the osseous defects in the implant and periodontic surgeries. Bone grafting not only fills the boneless space but also provides structural stability and facilitates bone tissue growth [1]. For successful bone regeneration, numerous bone substitutes have been developed and introduced using autogenous bone, allografts, synthetic bone, and xenografts [2,3].

Among them, autogenous bone has been considered a gold standard for bone regeneration from biological and histological vantage points [4]. Nonetheless, autografts have not frequently been applied in clinical application due to concerns about additional injury, donor site limitations, and morbidity from the bone harvest [5,6]. On the contrary, xenogeneic bone substitutes derived from bones of other species also a have sufficient osteoconductivity and biocompatibility and have generally been used in the dental field [7]. Besides, unlike autogenous bone, xenografts circumvent a second operative site and have no limit in terms of the available bone amount [1,8].

Currently, the deproteinized cancellous bovine bone matrix (DBBM) is mainly transplanted in alveolar bone defects and in sinus floor augmentation [9,10]. The most well-known DBBM in dentistry is Bio-Oss® with successful preclinical and clinical results, which has similar porous structure to human bone and shows high biocompatibility with oral hard tissues, and also meets the criteria of bone conductivity [1,11–13]. Such a graft consists of hydroxyapatite prepared by alkaline and heat treatment (300 °C) for eliminating the organics of the bone. However, in using the bovine-derived bone substitute, the fear of bovine spongiform encephalopathy (BSE) transmission is still potentially inherent [14]. BSE is a kind of transmissible spongiform encephalopathies (TSE), a fatal neurodegenerative disorder that can be transmitted to humans [15].

Porcine bone also shares a similar physiological, anatomic, and genetic makeup to human [16–18]. Hence the porcine-derived xenografts with relatively low zoonosis risk have been recently developed and made commercially available as an alternative to bovine-derived xenografts [16,19,20]. Porcine xenogeneic bone substitute has crystal structures similar to the human-derived bones like bovine bone [18,21]. Moreover, the stiffness and Ca/P ratio of porcine xenograft are closer to human trabecular bone than bovine bone [22]. Despite the porcine-derived xenografts having crystalline structures resembling human osseous tissue seem to be a profitable bone graft, the researches and data on porcine-derived xenografts still lack compared to bovine-derived grafts.

In the present study, we expected that the porcine-derived xenograft (Bone-XP®) recently introduced in the dental field may serve either non-inferior or superior effects on bone regeneration compare to the well-known commercially available bovine-derived bone substitute (Bio-Oss®). Thus, this study was undertaken to compare the bone regeneration capacity of porcine-derived xenografts to bovine-derived xenografts in the rat calvarial defect model.

2. Materials and Methods

2.1. Experimental Xenogenetic Bone Substitues

Two kinds of commercially available xenogeneic bone substitutes derived from different species (porcine and bovine) were used in this study. For porcine-derived xenograft (Bone-XP group), Bone-XP® (particle size of 0.2 mm–1.0 mm) was purchased from Medpark (Busan, Korea). Bovine bone graft material (Bio-Oss group), Bio-Oss® (particle size of 0.25–1.0 mm), was purchased from Geistlich Biomaterials (Wolhusen, Switzerland).

2.2. In Vitro Study

2.2.1. Scanning Electron Microscopy (SEM)

To compare the surface morphologies of porcine and bovine xenogeneic bone substitute, scanning electron microscopy (SEM) observation was conducted. After coating the Bio-Oss and Bone-XP with Au using a sputter coater (SCD 005, BAL-TEC, Balzers, Liechtenstein), the SEM (Hitachi S3500N, Hitachi, Tokyo, Japan) was operated at 15 kV.

2.2.2. Preparation of Extracts

We mixed 1 g of each xenogeneic bone substitute (Bio-Oss and Bone-XP) with 10 mL of alpha-modification of Eagle's medium (α-MEM; Welgene, Deagu, Korea) and stored at 37 °C and 5% CO_2 for 1 day. Each suspension was centrifuged once, for 5 min at 1200× g. These suspensions were filtered through the membrane (pore size: 0.2 µm) and stored 4 °C before use. The concentration of extract solutions in culture media ranging 20% was treated.

2.2.3. Culture of Human Bone Marrow Mesenchymal Stem Cells (hMSCs)

Human mesenchymal stem cells (hMSCs) (LONZA, Walkersville, MD, USA) were used for this in vitro cell study. hMSCs culture medium consisted of alpha-modification of Eagle's medium (α-MEM; Welgene Inc., Deagu, Korea) supplemented with 10% fetal bovine serum (FBS; Gibco BRL, Carlsbad, CA, USA), 100 U/mL penicillin and 100 μg/mL streptomycin (Gibco BRL), at 37 °C, 5% CO_2. The cultured medium was changed regularly every three days. The cells were detached with 0.25% trypsin/1 mM EDTA (Gibco BRL) and passaged after reaching 80%–90% confluence. hMSCs between Passages 4 and 5 were used for all the experiments.

2.2.4. Differentiation toward Osteoblasts

For osteogenic differentiation, hMSCs were seeded into a 48-well plate (3×10^4 cells/mL) in α-MEM and incubated at 37 °C and 5% CO_2 for 24 h. The media was replaced to osteogenic differentiation media, i.e., 10% α-MEM supplemented with 50 μg/mL ascorbic acid-2-phosphate (Sigma-Aldrich, Milan, Italy) and 10 mM β-glycerophosphate (Sigma-Aldrich), which function as a positive control. Media was changed every two days.

2.2.5. Cell Viability and Proliferation Assay

The cell viability and proliferation were assessed using the CCK-8 assay kit (Dojindo, Rockville, MD, USA) at 0, 1, 2, and 3 days. hMSCs were seeded into 48-well plates (Nunc, Roskilde, Denmark) at a density of 1×10^4 cells/mL. Each plate was pre-incubated in a humidified incubator with 5% CO_2 at 37 °C, and 20 μL of the CCK-8 solution was added to each well and then incubated for 2 h. 100 μL/well aliquots were transferred to a 96-well plate (Nunc) and absorbance was measured at 450 nm using an Opsys MR micro-plate reader (DYNEX Technologies Inc., Denkendorf, Germany). The cell viability was determined in 24 h absorbance data. The cell proliferation rate was assessed in optical density (OD) units.

2.2.6. Alkaline Phosphatase (ALP) Activity Assay

To evaluate the osteogenic differentiation, we used the Leukocyte Alkaline Phosphatase Kit (Sigma-Aldrich) according to the manufacturer's protocol. hMSCs were plated into 48-well plates (Nunc) at a density of 3×10^4 cells/mL and cultured for 3 and 8 days in the osteogenic media. Quantification of the staining images were done using ImageJ software program (U.S. National Institutes of Health, Bethesda, MD, USA).

2.2.7. Real-Time Polymerase Chain Reaction (PCR) Analysis

The real-time polymerase chain reaction (PCR) analysis was performed to examine the gene level of osteogenic differentiation markers. hMSCs were seeded into a 6-well plate (2×10^4 cells/mL). The cells cultured under the basal α-MEM, and the osteogenic α-MEM media were used as controls. Five days after the induction of osteogenesis, total RNA was extracted from different treated cells by using TRIzol (Life Technologies, Grand Island, NY, USA), and the total RNA concentration was measured by using a NanoDrop ND-1000 spectrophotometer (Technologies Inc., Wilmington, DE, USA). Complementary DNA (cDNA) was synthesized from 1.5 μg of the RNA. Amplification was done by using SYBR Green Master Mix reagents (Kapa Biosystems, Woburn, MA, USA) and an ABI 7500 instrument (Applied Biosystems, Carlsbad, CA, USA) according to the manufacturer's protocols. The relative gene expressions (*ALP*, Osteopontin (*OPN*), and Runt-related transcription factor 2 (*RUNX2*)) were calculated using the relative 2-ΔΔCt method. The control gene, *Actin*, was used to normalize the target genes. All reactions were performed in three samples. The primers were synthesized and provided by Bionics (Daejeon, Korea) (Table 1).

Table 1. Primer sequences used for real-time polymerase chain reaction (PCR) analysis.

Target Genes	Sequences
ALP	F: 5′-ATTTCTCTTGGGCAGGCAGAGAGT-3′ R: 5′-ATCCAGAATGTTCCACGGAGGCTT-3′
OPN	F: 5′-AGACACATATGATGGCCGAGG-3′ R: 5′-GGCCTTGTATGCACCATTCAA-3′
Runx2	F: 5′-CTCTACTATGGCACTTCGTCAGG-3′ R: 5′-GCTTCCATCAGCGTCAACAC-3′
Actin	F: 5′-ACTCTTCCAGCCTTCCTTCC-3′ R: 5′-TGTTGGCGTACAGGTCTTTG-3′

2.3. In Vivo Animal Study

2.3.1. Experimental Animal and Operative Procedures

Twenty Sprague-Dawley rats (male, 13-week-old, Koatech, Pyeongtaek, Korea) were used in this animal experiment. The individually caged rats were fed with rodent pellets and water and adapted to the laboratory for a week prior to surgery. The room conditions were kept constant at a temperature of 25 ± 1 °C and at a humidity of $55 \pm 7\%$. All the animal care and surgical procedures were done at the Laboratory Animal Resource Center of Pusan National University and were approved by the Institutional Animal Care and Use Committee of Pusan National University (PNU-2018-2101). During the surgical procedure, the rats were anesthetized using intramuscular injection of a combination of xylazine (Rompun, Bayer Korea, Seoul, Korea) and tiletamine zolazepam (Zoletil50, Virbac Korea, Seoul, Korea) for general anesthesia. The surgical sites of rats were shaved and then disinfected with povidone-iodine (Betadine, Korea Pharma, Seoul, Korea). The local anesthesia was performed using 2% lidocaine (Huons, Soengnam, Korea). After making a sagittal incision across the center of the skull using surgical scalpel blade (No. 15, Swann-Morton Ltd., Sheffield, UK), the cranium was exposed by raising the full thickness flap. Using a saline-cooled trephine bur (Osung, Kimpo, Korea), a critical-sized osseous defect (8 mm in diameter) was made in the center of the skull (Figure 1a). The quantified xenogeneic bone grafts (0.03 ± 0.002 g) were randomly implanted in each defect (Figure 1b). The defect was covered with a resorbable collagen membrane (10 × 10 mm, Cola-D, Medpark, Seoul, Korea) (Figure 1c). Skin and periosteum were sutured using 4-0 Vicryl suture (Ethicon, Livingston, UK). At 4 and 8 weeks after surgery, the experimental rats were sacrificed by CO_2 inhalation. The tissue samples were carefully harvested and immersed in 10% neutral buffered formalin (Sigma-Aldrich) for 2 weeks.

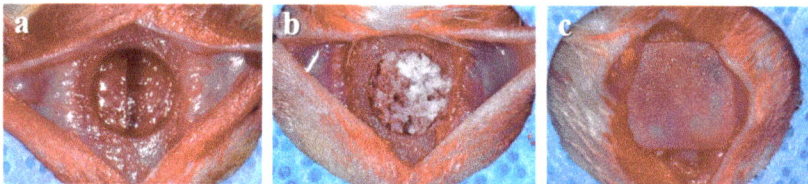

Figure 1. Operative procedures using rat calvarial defect model. (**a**) Created calvarial defect, (**b**) insertion of bone grafts, (**c**) placement of collagen membrane.

2.3.2. Micro-Computed Tomography (µCT) Analysis

All the harvested samples were scanned using µCT (SMX-90CT, Shimadzu, Kyoto, Japan) at 90 kV, intensity of 109 µA for measuring the percentage of new bone volume. The new bone volume was calculated using a customized program corded by cording software (MATLAB 2018b, MathWorks, Natick, MA, USA). The area of interest (AOI) was set as the same for all the specimens (diameter of

8 mm, height of 1.5 mm) (Figure 2). The images were divided into soft tissue, bone graft material and new bone and were classified by threshold values.

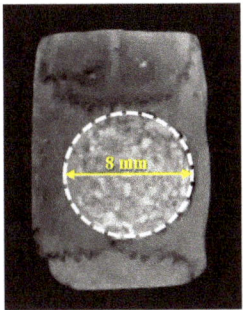

Figure 2. Area of interest (AOI) for volumetric analysis.

2.3.3. Histological and Histomorphometric Analysis

To prepare the decalcified histological sections, fixed-tissue specimens were soaked in 2.5% sodium hypochlorite/17% Ethylenediaminetetraacetic acid (EDTA) solution, and the solution was replaced daily for 2 weeks. Afterwards, specimens were dehydrated in a graded ethanol series and xylene and were embedded in paraffin. The paraffin blocks were cut into 3–4 μm thickness using a microtome (Microm HM 325, Waltham, MA, USA) and then attached to poly-L-lysine-coated slides. The tissue specimens were stained with hematoxylin-eosin (H&E) and Masson's trichrome (MT) staining solutions and photographed by optical microscope (Olympus BX, Tokyo, Japan) attached CCD camera (Polaroid DMC2 digital Microscope Camera, Polaroid, Cambridge, MA, USA). The newly formed bone area at the defect sites were consistently measured using a i-solution (IMT, Daejeon, Korea) by a single investigator (Figure 3).

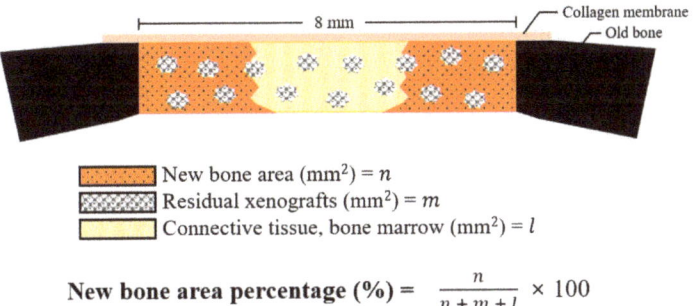

Figure 3. Schematic diagram of histometric analysis.

2.4. Statistical Analysis

All the statistical analyses were performed using SPSS statistical analysis software (ver. 24.0, IBM, Armonk, NY, USA) with a confidence level of 95% ($p < 0.05$). The in vitro and in vivo results were analyzed by using the independent student's t-test to determine the significance of differences between groups.

3. Results

3.1. In Vitro Findings

3.1.1. Observations of Surface Morphology

To investigate the surface morphologies of bone grafts, the SEM images were captured at magnification of ×60, ×500, and ×2000 (Figure 4). The xenografts of both groups showed similar particle size (Figure 4a,b) and macro-porous structure (Figure 4c–f) exhibiting the rough surfaces.

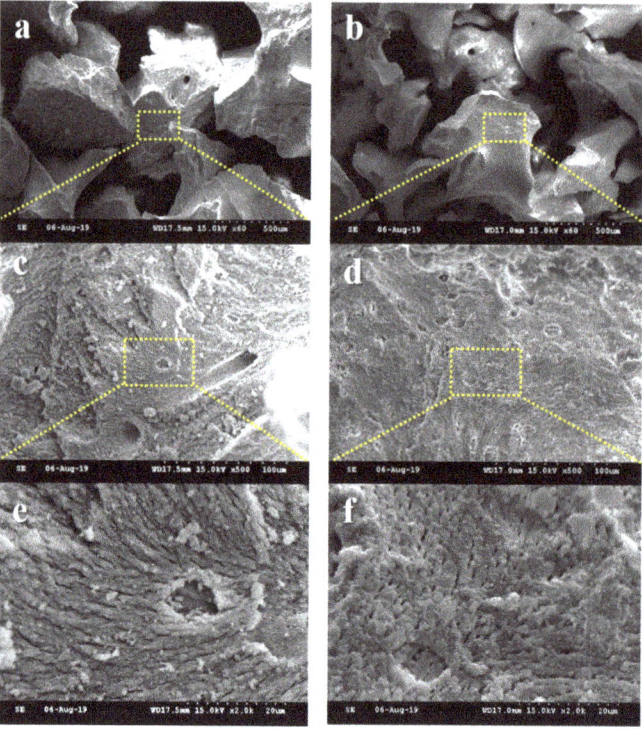

Figure 4. Comparative scanning electron microscope (SEM) images of each group. (**a,c,e**) Bio-Oss and (**b,d,f**) Bone-XP groups. [Original magnification: ×60 (**a,b**), ×500 (**c,d**), ×2000 (**e,f**)].

3.1.2. Cell Viability and Proliferation

The viability and proliferation of hMSCs on the two different extracts were analyzed using the CCK-8 assay, 0, 1, 2, and 3 days after cell seeding. At each time point, cell numbers were calculated following the normalization of the absorbance units to that obtained for the cells cultured on the Control group (Figure 5). At 1 day, the number of viable cells was similar for the extracts of Bio-Oss group and Bone-XP group and no significant differences were observed (Figure 5a). After 1, 2, and 3 days of culture, the cell proliferation of hMSCs on the two extracts was no different to that of the Control group (Figure 5b). These results indicate that the hMSCs cultured on two extractsf had no affect toxicity, and the proliferation rate was quite modest.

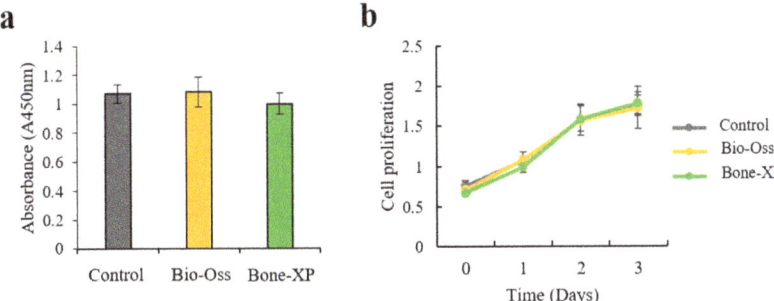

Figure 5. (**a**) Cell viability and (**b**) proliferation of extracts on human mesenchymal stem cells (hMSCs).

3.1.3. Alkaline Phosphatase (ALP) Staining

The alkaline phosphatase staining was conducted to analyze the effect of Control untreated hMSCs and those treated either with Bio-Oss group or Bone-XP group extracts after incubation for 3 and 8 days (Figure 6). On day 3, ALP staining slightly increased in the cells cultured on both the Bio-Oss group and Bone-XP group compared to the Control group. In addition, there was no obvious staining difference between Bio-Oss group and Bone-XP group-treated hMSCs. On day 8, both Bio-Oss group and Bone-XP group treated hMSCs, which were not substantially different from each other, showed higher staining than the Control, untreated cells. These data suggest that the extracts of Bio-Oss and Bone-XP had a positive effect on the regulation of osteogenic differentiation of hMSCs.

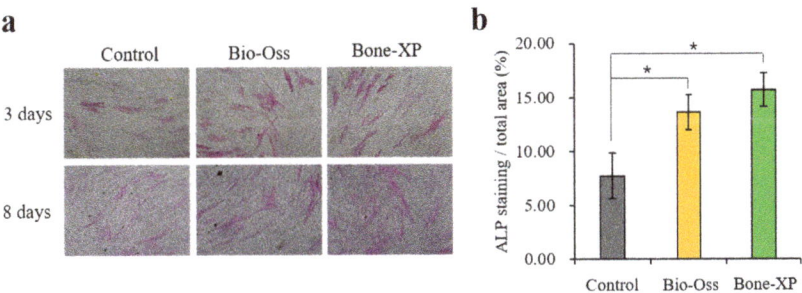

Figure 6. Cell osteogenic differentiation assay. (**a**) Alkaline phosphatase (ALP) staining and (**b**) the quantitative analysis. The symbol * indicates statistical significance compare to control (* $p < 0.05$).

3.1.4. Analysis of Real-time Polymerase Chain Reaction (PCR)

It is important to note that *ALP* and *RUNX2* were an early marker of osteoblastic lineage, and to evaluate the effects of Bio-Oss and Bone-XP extracts on osteoblast differentiation. The levels of osteogenesis-related genes were examined by real-time PCR (Figure 7). hMSCs were cultured in medium supplemented with osteogenic factors and 20% extracts for 5 days. Compared with control, untreated hMSCs, the up-regulated genes by Bio-Oss and Bone-XP extracts induced the factors involved in osteoblast differentiation and matrix mineralization, *ALP* and *OPN* (Figure 7a,b) and key osteogenic transcriptional factor; *RUNX2* (Figure 7c). Additionally, there was no statistically significant difference between Bio-Oss and Bone-XP-treated hMSCs (Figure 7a,b).

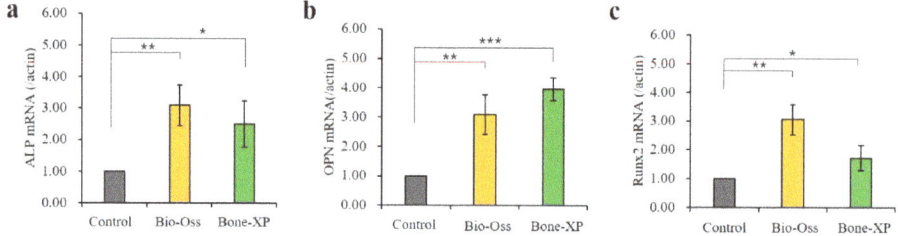

Figure 7. Real-time polymerase chain reaction (PCR) analysis of hMSCs on extracts. (**a**) *ALP*, (**b**) *OPN*, and (**c**) *RUNX2* were selected as the osteogenic differentiation related genes. The symbol * indicates statistical significance compare to control group (* $p < 0.05$, ** $p < 0.01$, *** $p < 0.001$).

3.2. In Vivo Findings

3.2.1. Clinical Findings

All the rats survived during the operative procedures and healing periods. In all the surgical sites, any side effects such as inflammation, swelling, or exposure of the specimens were not observed.

3.2.2. Volumetric Findings

On μCT images, bone grafts were well-positioned in the AOI without scattering of particles (Figure 8a–d). Within the AOI, the similar amount of new bone areas classified by threshold were observed in both experimental groups at 4 weeks and 8 weeks, respectively (Figure 8i–l). At 4 weeks, the new bone volumes (%) of Bio-Oss and Bone-XP groups were 11.6 ± 3.88% and 17.52 ± 3.78%, respectively (Table 2 and Figure 9). At 8 weeks, new bone volumes of Bio-Oss group and Bone-XP group were 25.89 ± 7.43% and 32.09 ± 3.51%, respectively. There was no significant difference in new bone volume between groups at both 4 and 8 weeks ($p > 0.05$).

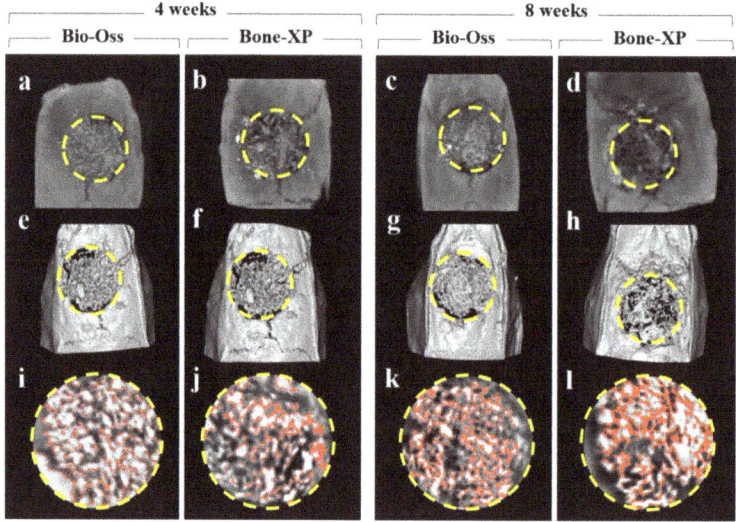

Figure 8. Micro-computed tomography (μCT) analysis images. (**a–d**) μCT images. (**e–h**) 3D reconstructed μCT images. (**i–l**) Classified new bone in area of interest (AOI). Yellow circle: AOI, Red colored area: newly formed bone.

Table 2. New bone volume within area of interest (AOI). (n = 5).

		Groups	Mean	SD	p-Value
New bone volume (%)	4 weeks	Bio-Oss	11.6	3.88	0.092
		Bone-XP	17.52	3.78	
	8 weeks	Bio-Oss	25.89	7.43	0.38
		Bone-XP	32.09	3.51	

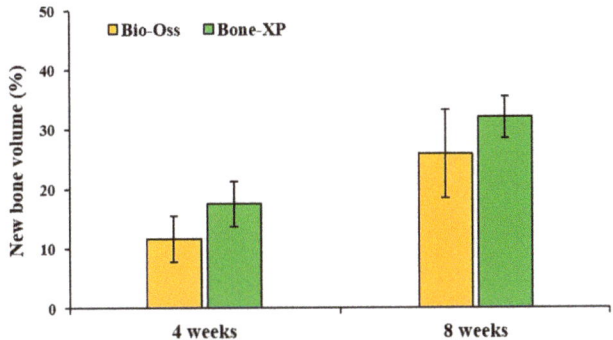

Figure 9. New bone volume percentages within area of interest (AOI).

3.2.3. Histological and Histomorphometric Findings

In all the histologic sections, the implanted bone-graft materials were holding a stable position within the defective site and showed good space-maintenance for bone reconstruction (Figures 10 and 11). At all the time point of 4 and 8 weeks, there were no remarkable inflammatory reactions such as presence of inflammatory cells or hematoma forms, and a number of active osteoblasts' lines could be easily found surrounding the newly formed bone (Figures 10i–l and 11i–l). In addition, connective tissues and bone marrow were observed around the residual bone graft materials. At 4 weeks after surgery, the new bones were minimally regenerated from the defect margin in both groups (Figure 4). The bone rebuilding was time-dependent, the mineralization and amount of new bone were increased at 8 weeks compared to those of 4 weeks in both groups (Figure 11).

The histomeric results of new bone area at 4 and 8 weeks after surgery are shown in Table 3 and Figure 12. At 4 weeks post-surgery, the mean ± standard deviation (SD) of new bone areas (%) in Bio-Oss and Bone XP groups were 5.83 ± 2.56%, 9.08 ± 5.47%, respectively. At 8 weeks, the new bone areas of Bio-Oss and Bone XP groups were 21.68 ± 11.11% and 25.22 ± 13.56%, respectively. There was no significant difference in new bone volume between groups at both 4 and 8 weeks ($p > 0.05$).

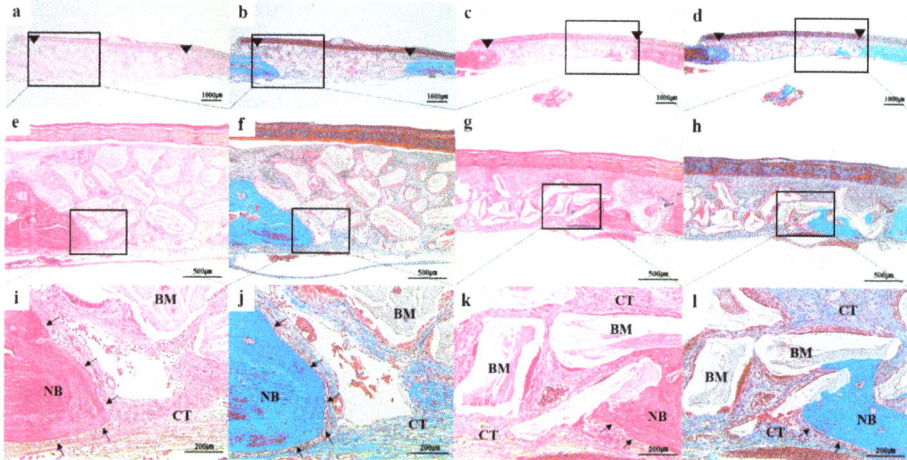

Figure 10. Histologic sections of (**a,b,e,f,i,j**) Bio-Oss and (**c,d,g,h,k,l**) Bone-XP groups at 4 weeks post-surgery. (**a,c,e,g,i,k**) haematoxylin and eosin (H&E) stained slides; (**b,d,f,h,j,l**) Masson's trichrome (MT) stained slides; Arrowhead: original defect edge; Arrow: lines of osteoblasts NB: newly generated bone; CT: connective tissue; BM: residual bone grafts. [Original magnification: (**a–d**) ×12.5, (**e–h**) ×40, (**i–l**) ×100].

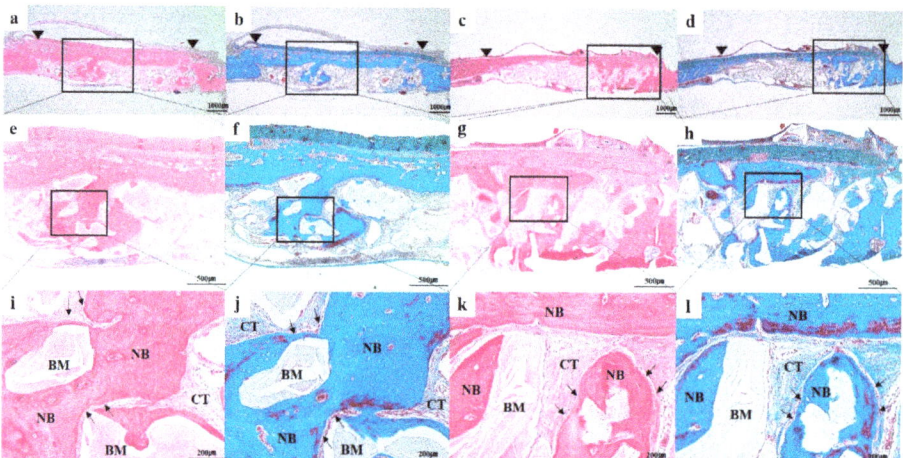

Figure 11. Histologic sections of (**a,b,e,f,i,j**) Bio-Oss and (**c,d,g,h,k,l**) Bone-XP groups at 8 weeks post-surgery. (**a,c,e,g,i,k**) haematoxylin and eosin (H&E) stained slides; (**b,d,f,h,j,l**) Masson's trichrome (MT) stained slides; Arrowhead: original defect edge; Arrow: lines of osteoblasts NB: newly generated bone; CT: connective tissue; BM: residual bone grafts. [Original magnification: (**a–d**) ×12.5, (**e–h**) ×40, (**i–l**) ×100].

Table 3. New bone area within area of interest (AOI). (n = 5).

		Groups	Mean	SD	*p*-Value
New bone area (%)	4 weeks	Bio-Oss	5.83	2.56	0.139
		Bone-XP	9.08	5.47	
	8 weeks	Bio-Oss	21.68	11.11	0.273
		Bone-XP	25.22	13.56	

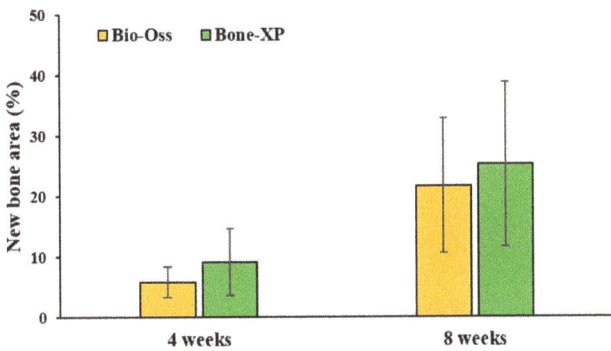

Figure 12. New bone area percentages within area of interest (AOI).

4. Discussion

Porcine-derived xenografts are a biocompatible material and have similar structures to that of human bone [23]. Bone-XP® used in our study is a heat-treated mineralized porcine bone, commercially available in dentistry. To remove the residual organic components, Bone-XP® was produced by thermal-treatment at high temperature. The high-temperature thermal treatment gives a higher crystalline structure and longer hydroxyapatite crystal [24,25]. While the heat-treatment may influence the Ca/P ratio, it is reported that the Ca/P ratio of Bone-XP (1.65–1.66) is closer to the Ca/P ratio of human bone (1.68–1.71) than that of bovine bone (1.92) [22]. An et al. [26] conducted animal experiments using the porcine-derived xenogeneic bone substitutes and reported the uniform new bone formation in rabbit tibia at 16 weeks post-surgery. They also conducted the multicenter clinical trial in orthopedics and proved the safety, stability, and appropriate absorption rate of porcine bone in the ilium. In addition, in the case report for alveolar ridge preservation using porcine bone mineral, the smooth new bone formation could be observed around the grafted xenografts without any adverse events such as inflammatory reaction or fibrous film formation [27]. The measured new bone rate (37.4%) was not inferior to the similar previous studies on ridge preservation using allogeneic bone and DBBM [27–29]. Even though there are lots of advantageous points of thermal-treated mineralized porcine xenografts, since most of the previous works on porcine bone grafts have been as to the collagenized porcine xenografts (CPX) [30–33], it is a lack of relevant research on anorganic porcine xenograft materials compared to CPX or DBBM. To expand the choice of bone grafts in dental applications, therefore, we conducted the in vitro and in vivo evaluation of porcine-derived xenografts in comparison with bovine-derived xenografts.

In our SEM observation, both Bone-XP and Bio-Oss groups exhibited similar naturally rough surfaces. The surface roughness of material influences cell morphology, behavior, and adhesion [34]. Hatano et al. [35] reported that osteogenic cells grown on the micro-rough surface showed higher proliferation, ALP activity, and osteogenic gene expression in comparison with cells on the smooth surface. Therefore, the micro-rough surfaces of Bone-XP and Bio-Oss are considered to be favorable to adherence and proliferation of osteogenic cells.

MSCs can differentiate into cells of the mesodermal lineage, such as adipocyte, osteocyte, and chondrocyte, etc. The proliferative and osteogenic characteristics of MSCs are used for reconstructing bone defects in clinical conditions [36,37]. Previous studies demonstrated that Bio-Oss has a positive effect on the osteogenic ability of the hMSCs [38]. In this study, hMSCs were used to evaluate the osteogenic activity of Bone-XP in comparison with the Bio-Oss. The cells cultured on the two different extracts of bone grafts showed no significant effect in terms of their cytotoxicity and proliferation rate. Upon osteogenic differentiation, both investigated extracts showed significant increases compared to control untreated cells. To determine the mechanism of osteogenic induction by the investigated extracts, we examined the related gene expressions by quantitative PCR (qPCR). In bone, *ALP* is localized on the entire cell surface of pre-osteoblasts and has long been used as an osteoblastic marker [39,40]. *OPN*, a secreted matrix glycoprotein, is biosynthesized by preosteoblast, osteoblast, osteocytes and plays a significant role in the bone remodeling procedure [41,42]. *RUNX2*, a key transcription modulator, is the main transcription factor for osteoblast differentiation and bone formation. In our study, both extracts of experimental xenografts induced ALP, OPN, and RUNX2 expressions in comparison to the control, but there was no significant difference between investigated extracts. The results of osteogenic differentiation and gene expression suggest that the Bone-XP is an osteoinductive material and can be used as a replacement for the Bio-Oss.

From a surgical standpoint of our in vivo experiment, Bone-XP gave favorable handling properties similar to Bio-Oss and could be easily implanted into the bone defects. In the study on the physicochemical property of porcine-derived bone which is essentially identical grafts with Bone-XP [1], the porcine-derived xenografts showed higher wettability than that of bovine-derived xenografts (Bio-Oss). High wettability of biomaterial can beneficially effect protein adsorption and cellular behavior [1,43]. In our surgical procedure, Bone-XP was easily wetted by body fluid after transplantation as well and it can be considered that Bone-XP is a relatively hydrophilic bone graft material.

This study conducted µCT, histologic, and histometric evaluations to compare the new bone formation between the porcine-derived and bovine-derived xenografts in a critical-sized rat calvarial defect. Osteoblast from the defect margin utilizes the grafted bone substitute as a frame upon which to regenerate bone [44]. The newly formed bone should replace bone grafts as the latter resorbs during the healing period. In our histological analysis, the lines of osteoblast were easily observed surrounding the grafted xenografts in all groups, and this indicates that both xenografts used in this study have an excellent osteoconductivity [45]. Furthermore, in the result of the volumetric and histometric analysis, Bone-XP and Bio-Oss group induced comparable proportions of new bone regeneration at all the time points of 4 and 8 weeks. These results demonstrated that Bone-XP has the non-inferior capacity of new bone regeneration compared to that of Bio-Oss.

Based on the results obtained in the present study, Bone-XP®, recently commercially available porcine xenograft, is a biocompatible, osteoinductive, and osteoconductive bone substitute in comparison with Bio-Oss®. However, this in vivo study was conducted using limited sample size and observation time points, and therefore future preclinical trials using large animal and clinical trials are needed to verify these results.

5. Conclusions

Within the limitations of the present study, it can be concluded that the newly investigated porcine-derived bone substitute, Bone-XP®, may offer a favorable cell response and bone regeneration similar to those of commercial bovine bone mineral.

Author Contributions: Conceptualization, J.-B.H.; Data curation, E.-B.B., H.-J.K. (Ha-Jin Kim), J.-J.A. and H.-Y.B.; Formal analysis, E.-B.B., H.-J.K. (Ha-Jin Kim), J.-J.A. and H.-Y.B.; Investigation, E.-B.B., H.-J.K. (Ha-Jin Kim), J.-J.A. and H.-Y.B.; Methodology, H.-J.K. (Hyung-Joon Kim) and J.-B.H.; Supervision, H.-J.K. (Hyung-Joon Kim) and J.-B.H.; Writing-original draft, E.-B.B., H.-J.K. (Ha-Jin Kim) and J.-J.A.; Writing-review and editing, E.-B.B., H.-J.K. (Hyung-Joon Kim) and J.-B.H.

Funding: This study was supported by the National Research Foundation of Korea (NRF) grant funded by the Korean government (MSIP) (2017R1A2B4005820) and was supported by a grant of the Korea Health Technology R&D Project through the Korea Health Industry Development Institute (KHIDI), funded by the Ministry of Health & Welfare, Republic of Korea (grant number: HI17C2397).

Conflicts of Interest: The authors declare no conflict of interest.

References

1. Lee, J.H.; Yi, G.S.; Lee, J.W.; Kim, D.J. Physicochemical characterization of porcine bone-derived grafting material and comparison with bovine xenografts for dental applications. *J. Periodontal Implant Sci.* **2017**, *47*, 388–401. [CrossRef] [PubMed]
2. Venkataraman, N.; Bansal, S.; Bansal, P.; Narayan, S. Dynamics of bone graft healing around implants. *J. Int. Clin. Dent. Res. Organ.* **2015**, *7*, 40.
3. Bauer, T.W.; Muschler, G.F. Bone graft materials: An overview of the basic science. *Clin. Orthop. Relat. Res.* **2000**, *371*, 10–27. [CrossRef]
4. Misch, C.M. Autogenous bone: Is it still the gold standard? *Implant Dent.* **2010**, *19*, 361. [CrossRef]
5. van den Bergh, J.P.; ten Bruggenkate, C.M.; Krekeler, G.; Tuinzing, D.B. Sinus floor elevation and grafting with autogenous iliac crest bone. *Clin. Oral Implant. Res.* **1998**, *9*, 429–435. [CrossRef]
6. Cypher, T.J.; Grossman, J.P. Biological principles of bone graft healing. *J. Foot Ankle Surg.* **1996**, *35*, 413–417. [CrossRef]
7. Jensen, S.S.; Terheyden, H. Bone augmentation procedures in localized defects in the alveolar ridge: Clinical results with different bone grafts and bone-substitute materials. In *Database of Abstracts of Reviews of Effects (DARE): Quality-Assessed Reviews [Internet]*; Centre for Reviews and Dissemination (UK): York, UK, 2009.
8. Jensen, S.; Bosshardt, D.; Buser, D. Bone grafts and bone substitute materials. *Buser D Ed.* **2009**, *20*, 1–96.
9. Jensen, T.; Schou, S.; Stavropoulos, A.; Terheyden, H.; Holmstrup, P. Maxillary sinus floor augmentation with Bio-Oss or Bio-Oss mixed with autogenous bone as graft: A systematic review. *Clin. Oral Implant. Res.* **2012**, *23*, 263–273. [CrossRef]
10. Pang, K.M.; Um, I.W.; Kim, Y.K.; Woo, J.M.; Kim, S.M.; Lee, J.H. Autogenous demineralized dentin matrix from extracted tooth for the augmentation of alveolar bone defect: A prospective randomized clinical trial in comparison with anorganic bovine bone. *Clin. Oral Implant. Res.* **2017**, *28*, 809–815. [CrossRef]
11. Pinholt, E.M.; Bang, G.; Haanaes, H.R. Alveolar ridge augmentation in rats by Bio-Oss. *Eur. J. Oral Sci.* **1991**, *99*, 154–161. [CrossRef]
12. Haas, R.; Mailath, G.; Dörtbudak, O.; Watzek, G. Bovine hydroxyapatite for maxillary sinus augmentation: Analysis of interfacial bond strength of dental implants using pull-out tests. *Clin. Oral Implant. Res.* **1998**, *9*, 117–122. [CrossRef]
13. Berglundh, T.; Lindhe, J. Healing around implants placed in bone defects treated with Bio-Oss®. An experimental study in the dog. *Clin. Oral Implant. Res.* **1997**, *8*, 117–124. [CrossRef]
14. Noumbissi, S.S.; Lozada, J.L.; Boyne, P.J.; Rohrer, M.D.; Clem, D.; Kim, J.S.; Prasad, H. Clinical, histologic, and histomorphometric evaluation of mineralized solvent-dehydrated bone allograft (Puros) in human maxillary sinus grafts. *J. Oral Implantol.* **2005**, *31*, 171–179. [CrossRef]
15. Gill, D.S.; Tredwin, C.J.; Gill, S.K.; Ironside, J.W. The transmissible spongiform encephalopathies (prion diseases): A review for dental surgeons. *Int. Dent. J.* **2001**, *51*, 439–446. [CrossRef] [PubMed]
16. Salamanca, E.; Lee, W.-F.; Lin, C.-Y.; Huang, H.-M.; Lin, C.-T.; Feng, S.-W.; Chang, W.-J. A novel porcine graft for regeneration of bone defects. *Materials* **2015**, *8*, 2523–2536. [CrossRef]
17. Feng, W.; Fu, L.; Liu, J.; Li, D. The expression and distribution of xenogeneic targeted antigens on porcine bone tissue. *Transplant. Proc.* **2012**, *44*, 1419–1422. [CrossRef] [PubMed]
18. Bracey, D.; Seyler, T.; Jinnah, A.; Lively, M.; Willey, J.; Smith, T.; Van Dyke, M.; Whitlock, P. A Decellularized Porcine Xenograft-Derived Bone Scaffold for Clinical Use as a Bone Graft Substitute: A Critical Evaluation of Processing and Structure. *J. Funct. Biomater.* **2018**, *9*, 45. [CrossRef]
19. Ramírez-Fernández, M.; Calvo-Guirado, J.L.; Delgado-Ruiz, R.A.; Maté-Sánchez del Val, J.E.; Vicente-Ortega, V.; Meseguer-Olmos, L. Retracted: Bone response to hydroxyapatites with open porosity of animal origin (porcine [OsteoBiol®mp3] and bovine [Endobon®]): A radiological and histomorphometric study. *Clin. Oral Implant. Res.* **2011**, *22*, 767–773. [CrossRef]

20. Go, A.; Eun Kim, S.; Mi Shim, K.; Lee, S.M.; Hwa Choi, S.; Sik Son, J.; Soo Kang, S. Osteogenic effect of low-temperature-heated porcine bone particles in a rat calvarial defect model. *J. Biomed. Mater. Res. Part A* **2014**, *102*, 3609–3617. [CrossRef]
21. Yung, G.P.; Schneider, M.K.; Seebach, J.D. Immune responses to α1, 3 galactosyltransferase knockout pigs. *Curr. Opin. Organ Transplant.* **2009**, *14*, 154–160. [CrossRef]
22. Park, S.-A.; Shin, J.-W.; Yang, Y.-I.; Kim, Y.-K.; Park, K.-D.; Lee, J.-W.; Jo, I.-H.; Kim, Y.-J. In vitro study of osteogenic differentiation of bone marrow stromal cells on heat-treated porcine trabecular bone blocks. *Biomaterials* **2004**, *25*, 527–535. [CrossRef]
23. Linde, F.; Hvid, I.; Pongsoipetch, B. Energy absorptive properties of human trabecular bone specimens during axial compression. *J. Orthop. Res.* **1989**, *7*, 432–439. [CrossRef] [PubMed]
24. Gao, Y.; Cao, W.-L.; Wang, X.-Y.; Gong, Y.-D.; Tian, J.-M.; Zhao, N.-M.; Zhang, X.-F. Characterization and osteoblast-like cell compatibility of porous scaffolds: Bovine hydroxyapatite and novel hydroxyapatite artificial bone. *J. Mater. Sci. Mater. Med.* **2006**, *17*, 815–823. [CrossRef] [PubMed]
25. Nazirkar, G.; Singh, S.; Dole, V.; Nikam, A. Effortless effort in bone regeneration: A review. *J. Int. Oral Health* **2014**, *6*, 120. [PubMed]
26. An, K.C.; Choi, J.S.; Kim, T.H.; Lee, Y.J.; Yoon, T.L.; Shin, J.W.; Chung, H.J.; Kim, S.W. Biocompatibility Evaluation of Heat-treated Mineralized Porcine Cancellous Bone: Using Animal & Clinical Study. *J. Korean Orthop. Res. Soc.* **2009**, *12*, 33.
27. Park, E.S.; Yu, J.A.; Choi, S.H.; Lee, D.W. Ridge preservation using porcine bone mineral and cross-linked collagen membrane in damaged socket: A case report. *J. Korean Acad. Osseointegr.* **2017**, *9*, 1–6.
28. Lee, D.-W.; Pi, S.-H.; Lee, S.-K.; Kim, E.-C. Comparative histomorphometric analysis of extraction sockets healing implanted with bovine xenografts, irradiated cancellous allografts, and solvent-dehydrated allografts in humans. *Int. J. Oral Maxillofac. Implant.* **2009**, *24*, 609–615.
29. Wood, R.A.; Mealey, B.L. Histologic comparison of healing after tooth extraction with ridge preservation using mineralized versus demineralized freeze-dried bone allograft. *J. Periodontol.* **2012**, *83*, 329–336. [CrossRef]
30. Scarano, A.; Lorusso, F.; Ravera, L.; Mortellaro, C.; Piattelli, A. Bone regeneration in iliac crestal defects: An experimental study on sheep. *Biomed Res. Int.* **2016**, *2016*. [CrossRef]
31. Calvo-Guirado, J.L.; Gómez-Moreno, G.; Guardia, J.; Ortiz-Ruiz, A.; Piatelli, A.; Barone, A.; Martínez-González, J.M.; Meseguer-Olmo, L.; López-Marí, L.; Dorado, C.B. Biological response to porcine xenograft implants: An experimental study in rabbits. *Implant Dent.* **2012**, *21*, 112–117. [CrossRef]
32. Iezzi, G.; Degidi, M.; Piattelli, A.; Mangano, C.; Scarano, A.; Shibli, J.A.; Perrotti, V. Comparative histological results of different biomaterials used in sinus augmentation procedures: A human study at 6 months. *Clin. Oral Implant. Res.* **2012**, *23*, 1369–1376. [CrossRef] [PubMed]
33. Scarano, A.; Piattelli, A.; Assenza, B.; Quaranta, A.; Perrotti, V.; Piattelli, M.; Iezzi, G. Porcine bone used in sinus augmentation procedures: A 5-year retrospective clinical evaluation. *J. Oral Maxillofac. Surg.* **2010**, *68*, 1869–1873. [CrossRef] [PubMed]
34. Chang, H.-I.; Wang, Y. Cell responses to surface and architecture of tissue engineering scaffolds. In *Regenerative Medicine and Tissue Engineering-Cells and Biomaterials*; InTechOpen: London, UK, 2011.
35. Hatano, K.; Inoue, H.; Kojo, T.; Matsunaga, T.; Tsujisawa, T.; Uchiyama, C.; Uchida, Y. Effect of surface roughness on proliferation and alkaline phosphatase expression of rat calvarial cells cultured on polystyrene. *Bone* **1999**, *25*, 439–445. [CrossRef]
36. Cooper, L.; Harris, C.; Bruder, S.; Kowalski, R.; Kadiyala, S. Incipient analysis of mesenchymal stem-cell-derived osteogenesis. *J. Dent. Res.* **2001**, *80*, 314–320. [CrossRef]
37. Jafarian, M.; Eslaminejad, M.B.; Khojasteh, A.; Abbas, F.M.; Dehghan, M.M.; Hassanizadeh, R.; Houshmand, B. Marrow-derived mesenchymal stem cells-directed bone regeneration in the dog mandible: A comparison between biphasic calcium phosphate and natural bone mineral. *Oral Surg. Oral Med. Oral Pathol. Oral Radiol. Endodontol.* **2008**, *105*, e14–e24. [CrossRef]
38. Yu, B.-H.; Zhou, Q.; Wang, Z.-L. Periodontal ligament versus bone marrow mesenchymal stem cells in combination with Bio-Oss scaffolds for ectopic and in situ bone formation: A comparative study in the rat. *J. Biomater. Appl.* **2014**, *29*, 243–253. [CrossRef]
39. Liu, S.; Liu, D.; Chen, C.; Hamamura, K.; Moshaverinia, A.; Yang, R.; Liu, Y.; Jin, Y.; Shi, S. MSC transplantation improves osteopenia via epigenetic regulation of notch signaling in lupus. *Cell Metab.* **2015**, *22*, 606–618. [CrossRef]

40. Liu, W.; Liu, Y.; Guo, T.; Hu, C.; Luo, H.; Zhang, L.; Shi, S.; Cai, T.; Ding, Y.; Jin, Y. TCF3, a novel positive regulator of osteogenesis, plays a crucial role in miR-17 modulating the diverse effect of canonical Wnt signaling in different microenvironments. *Cell Death Dis.* **2013**, *4*, e539. [CrossRef]
41. Nilsson, S.K.; Johnston, H.M.; Whitty, G.A.; Williams, B.; Webb, R.J.; Denhardt, D.T.; Bertoncello, I.; Bendall, L.J.; Simmons, P.J.; Haylock, D.N. Osteopontin, a key component of the hematopoietic stem cell niche and regulator of primitive hematopoietic progenitor cells. *Blood* **2005**, *106*, 1232–1239. [CrossRef]
42. Grassinger, J.; Haylock, D.N.; Storan, M.J.; Haines, G.O.; Williams, B.; Whitty, G.A.; Vinson, A.R.; Be, C.L.; Li, S.; Sørensen, E.S. Thrombin-cleaved osteopontin regulates hemopoietic stem and progenitor cell functions through interactions with α9β1 and α4β1 integrins. *Blood* **2009**, *114*, 49–59. [CrossRef]
43. Kubies, D.; Himmlová, L.; Riedel, T.; Chánová, E.; Balík, K.; Douderova, M.; Bártová, J.; Pesakova, V. The interaction of osteoblasts with bone-implant materials: 1. The effect of physicochemical surface properties of implant materials. *Physiol. Res.* **2011**, *60*, 95. [PubMed]
44. Laurencin, C.; Khan, Y.; El-Amin, S.F. Bone graft substitutes. *Expert Rev. Med Devices* **2006**, *3*, 49–57. [CrossRef] [PubMed]
45. Kumar, P.; Vinitha, B.; Fathima, G. Bone grafts in dentistry. *J. Pharm. Bioallied Sci.* **2013**, *5*, S125. [CrossRef] [PubMed]

© 2019 by the authors. Licensee MDPI, Basel, Switzerland. This article is an open access article distributed under the terms and conditions of the Creative Commons Attribution (CC BY) license (http://creativecommons.org/licenses/by/4.0/).

Article

Enhancement of Bone Ingrowth into a Porous Titanium Structure to Improve Osseointegration of Dental Implants: A Pilot Study in the Canine Model

Ji-Youn Hong [1,†], Seok-Yeong Ko [2,†], Wonsik Lee [3], Yun-Young Chang [4], Su-Hwan Kim [5,6] and Jeong-Ho Yun [2,7,*]

1. Department of Periodontology, Periodontal-Implant Clinical Research Institute, School of Dentistry, Kyung Hee University, 26, Kyungheedae-ro, Dongdaemun-gu, Seoul 02447, Korea; jkama7@gmail.com
2. Department of Periodontology, College of Dentistry and Institute of Oral Bioscience, Jeonbuk National University, 567, Baekje-daero, Deokjin-gu, Jeonju-si, Jeollabuk-do 54896, Korea; dentquartz@naver.com
3. Advanced Process and Materials R&D Group, Korea Institute of Industrial Technology, 7-47 Songdo-dong, Yeonsu-gu, Incheon 406-840, Korea; wonslee@kitech.re.kr
4. Department of Dentistry, Inha International Medical Center, 424, Gonghang-ro, 84-gil, Unseo-dong, Jung-gu, Incheon 22382, Korea; bewitme@naver.com
5. Department of Periodontics, Asan Medical Center, 88, Olympic-ro 43-gil, Songpa-gu, Seoul 05505, Korea; suhwank@gmail.com
6. Department of Dentistry, University of Ulsan College of Medicine, 88, Olympic-ro 43-gil, Songpa-gu, Seoul 05505, Korea
7. Research Institute of Clinical Medicine of Jeonbuk National University-Biomedical Research Institute of Jeonbuk National University Hospital, 20, Geonjiro, Deokjin-gu, Jeonju-si, Jeollabuk-do 54907, Korea
* Correspondence: grayheron@hanmail.net; Tel.: +82-63-250-2289
† These authors contributed equally to this study.

Received: 8 May 2020; Accepted: 6 July 2020; Published: 8 July 2020

Abstract: A porous titanium structure was suggested to improve implant stability in the early healing period or in poor bone quality. This study investigated the effect of a porous structure on the osseointegration of dental implants. A total of 28 implants (14 implants in each group) were placed in the posterior mandibles of four beagle dogs at 3 months after extraction. The control group included machined surface implants with an external implant–abutment connection, whereas test group implants had a porous titanium structure added to the apical portion. Resonance frequency analysis (RFA); removal torque values (RTV); and surface topographic and histometric parameters including bone-to-implant contact length and ratio, inter-thread bone area and ratio in total, and the coronal and apical parts of the implants were measured after 4 weeks of healing. RTV showed a significant difference between the groups after 4 weeks of healing ($p = 0.032$), whereas no difference was observed in RFA. In the test group, surface topography showed bone tissue integrated into the porous structures. In the apical part of the test group, all the histometric parameters exhibited significant increases compared to the control group. Within the limitations of this study, enhanced bone growth into the porous structure was achieved, which consequently improved osseointegration of the implant.

Keywords: porosity; dental implant; osseointegration; bone formation; titanium

1. Introduction

A dental implant has been accepted as a reliable treatment modality for edentulous ridge with high long-term survival [1], and improvements in implant design, surface treatment, and surgical technique led to a marked increase in implant stability [2,3]. However, the results are mostly based on the selection of subjects with the exclusion of any clinical conditions that might have a negative effect on the healing around the implants. There are several possible risk factors associated with early implant failure or impaired healing, including smoking, head and neck radiation [4,5], bone quality and osteoporosis [6]. Osseointegration was defined as a direct and functional connection between bone and an artificial implant. However, the macroscopic (body structure and thread geometry) and microscopic (chemical composition and surface treatment) characteristics of dental implants could influence the success of the osseointegration [7].

The topographical features in an implant surface can be defined in terms of their scales, which were produced by surface modification treatments such as titanium plasma-spraying, grit-blasting, acid-etching, or combinations [2,8]. Apart from the macro-level that is related to the threaded screw of implant geometry and macroporous surface, mechanical interlocking is maximized by microtopographic roughness [9]. In addition, nanotopography is associated with the biological activities of cells to stimulate bone formation on an implant surface [10]. However, the majority of current surface treatments are unreliable to achieve reproducible nanoscale features as they range randomly from nanometers to millimeters.

Another approach in surface modification was the production of porous bodies of titanium metal and its alloys, and sintering of metal powders onto the surface was commonly used for porous coatings [11,12]. Advantages of porous surface-enhanced implant include the induction of new bone tissue ingrowth and neovascularization into the porous scaffold in three-dimensional (3D) aspects [13], elastic modulus closer to the cancellous bone that allows load distribution [14], and enhanced transport of metabolites and space for new bone through substantial porous interconnectivity [15]. To maximize the potential benefits from the porous structures, precise control of overall porosity and pore size was identified as important, although the optimal ranges were yet to be determined [16]. However, conventional methods had limitations of low volumetric porosity, irregular dimensions of pores, and poor interconnectivity [17]. Furthermore, possible mechanical failures related to a lack of yield strength and separation of coating materials led to soft tissue encapsulation and loosening of implants [18].

Recent approaches have utilized methods such as selective melting with laser or electron beam, 3D printing, casting or vapor deposition to control the internal pore geometry and distribution [12,19]. The porous scaffolds were sometimes combined with threaded implants for additional advantages in terms of primary mechanical stability and removability. One of the products that had been widely studied to adapt the porous structure to the root form implant was the porous tantalum trabecular metal (PTTM) enhanced implant [18,20,21]. PTTM was fabricated by foam-like vitreous carbon scaffold that resulted in the open-cell structure similar to the trabecular bone [18]. The PTTM part was added to the middle portion of the implant by laser welding and was combined with the screw-type design of titanium alloy surface at the cervical and apical portions, which were microtextured by grit-blasting with hydroxyapatite particles. From the animal studies, histomorphometric evaluations have demonstrated more new bone growth at the PTTM occupying the middle portion compared to the conventional surface within the 12-week study periods, and suggested the potential benefits of the porous structures in the compromised bone quality. However, there were limited biomechanical improvements assessed by the resonance frequency analysis in PTTM and the implant stability appeared comparable to the conventional microtextured surface.

In the present study, a novel method of utilizing the powder injection molding technique has been employed to form a porous titanium structure, which was fabricated on the apical portion of the machined screw-type implant. The effect of the newly developed porous structure on osseointegration was compared to the smooth-surfaced implant in the canine model.

2. Materials and Methods

2.1. Design of the Implants

A threaded machined surface implant (c.p. titanium grade 4) with an external-type abutment connection (MegaGen Implant Co., Ltd., Gyeongsan, Korea) was used in the control group. The implant measured 4.1 mm in diameter, 8.1 mm in length, and had a straight configuration of the implant body (core diameter of 3.25 mm) with a homogenous thread height of 0.35 mm (Figure 1a). The test group implant had a porous titanium structure fabricated on the implant core at the apical portion; the core had a reduced diameter of 1.25 mm and a thread height of 1.35 mm to afford the space for the porous scaffold (1 mm in depth and 0.83 mm in width) (Figure 1b). The resulting profile had 3-mm spiral shape threads in the apical part, a regular pitch distance of 1.25 mm and a thread angle of 45°.

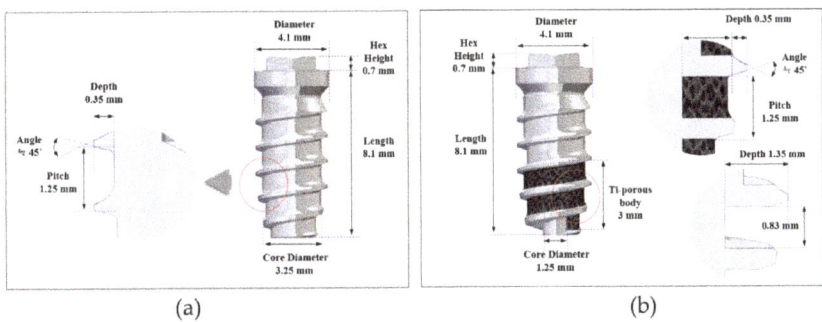

Figure 1. Design of the implants: (**a**) overall profiles and dimensions of the implant in the control group; (**b**) overall profiles and dimensions of the implant in the test group.

2.2. Fabrication of Porous Titanium Structure

The porous titanium structure on the implant core was fabricated at the Korea Institute of Industrial Technology (KITECH) using insert powder injection molding technology (Figure 2a). Briefly, a feedstock, which was a mixture of titanium hydride (TiH$_2$) powder, space holder and some polymeric binders, was prepared as a material for powder injection molding. Expandable polystyrene (EPS) beads with an average diameter of 325 µm were selected as space holders to form open-pore structure made from the contact between the beads during the expansion that occurred above 80 °C. The feedstock was injected into the narrow cavity between the threads at the apical third portion of the implant insert. The molded implants were inserted again into a mold designed for expansion of the EPS beads and kept for 20 min in an oven at 110 °C. The expanded beads were removed in a solvent, resulting in an open-pore structure consisting of TiH$_2$ powder and binders (Figure 2b). The polymeric binders were removed completely during the thermal debinding process, slowly increasing the temperature up to 700 °C under argon atmosphere. During this process, TiH$_2$ powder was also transformed to Ti powder by dehydrogenation reaction that occurred in the temperature range of from 350 to 500 °C. Finally, the open-pore scaffold of titanium powder between the threads was sintered for 3 h at 1100 °C in high vacuum (Figure 2c). During sintering, the titanium scaffold and implant core were combined into a single body by interdiffusion of titanium atoms at the interface, and titanium fixtures with open-pore titanium structure between the implant threads were formed. The porous structure had an average porosity of 68.1%, an average strut thickness of 61.4 µm (range: 30 to 100 µm), and an average pore size of 243 µm (range: 200 to 350 µm). The interconnected area between the pores that acts as the path for ingrowth of bone was measured to be 122.0 µm (range: 50 to 235 µm) in mean diameter from 2-dimensional image analysis.

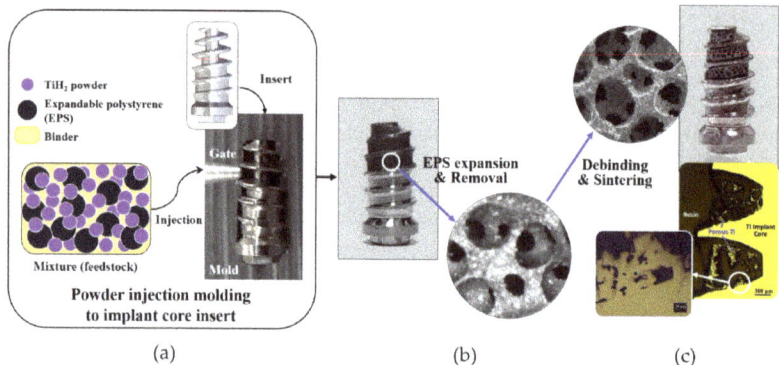

Figure 2. Insert powder injection molding process to fabricate the porous titanium fixture: (**a**) after a titanium implant with a narrow cavity in the deep inter-thread area was inserted into a mold, a mixture of titanium hydride (TiH_2) powder, expandable polystyrene (EPS) beads and some polymeric binders were injected into the narrow cavity between the threads of the implant; (**b**) the EPS beads within the molded area were removed in a solvent after the expansion of EPS; (**c**) subsequently, through the debinding process, binders were removed and after dehydrogenation of TiH_2 to Ti powder and the sintering process, the porous titanium fixtures with an open-pored titanium structure between the implant threads were produced and combined into a single body.

2.3. Animal Experiment

Four 12-month-old male beagle dogs weighing 12.0 to 17.0 kg were used in this study. The dogs were kept in separate cages under standard laboratory conditions. Animal selection, management, surgical procedures, and preparations were performed according to the protocols approved by the Institutional Animal Care and Use Committee at Korea Animal Medical Science Institute, Guri, Korea (Approval No. 16-KE-234). The study was conducted following the Animal Research: Reporting In Vivo Experiments (ARRIVE) guidelines [22].

Surgical procedures were performed under general anesthesia with intravenous injection of a solution (0.1 mL/kg) containing 1:1 ratio of tiletamine/zolazepam (Zoletil 50, Virbac S.A., Virbac Laboratories 06516, Carros, France) and xylazine hydrochloride (Rumpun, Bayer, Seoul, Korea). Infiltration anesthesia with 2% lidocaine HCl with 1:100,000 epinephrine (Huons, Seoul, Korea) was used at the surgical sites. The premolars (P1–P4) and the first molar (M1) in both the mandibles were carefully extracted and a total of 28 implants (14 implants for each group) were placed after 12 weeks. In each quadrant, three or four implants from the control or test group were randomly allocated. After sequential osteotomies, implants were installed under 40 Ncm (newton centimeter) of torque and submerged (Figure 3a–c). Antibiotics and nonsteroidal anti-inflammatory drugs were administered for 5 days. Sutures were removed after one week and animals were sacrificed after 4 weeks of healing by intravenous injection of 1 mL of suxamethonium chloride (50 mg/mL).

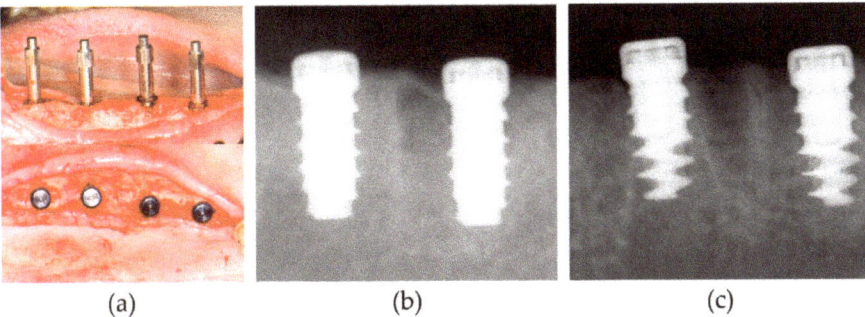

Figure 3. Surgical procedure of implant placement: (**a**) Smartpeg connection to measure implant stability quotient (ISQ) values using the resonance frequency analysis (RFA) device (Osstell Mentor) immediately after the implant placement (upper), and coverscrew connection (lower); (**b**) periapical X-ray images were taken in the control group; (**c**) periapical X-ray images were taken in the test group. Radiolucency near the implant surface was shown at the apical part of the test group compared to the control group.

2.4. Resonance Frequency Analysis (RFA), Removal Torque Test and Topographical Analysis

Implant stability quotient (ISQ) value was measured immediately after the implant placement and at the time of animal sacrifice. A SmartPeg (Type 04, REF 100350, Osstell AB, Gothenburg, Sweden) was connected to each implant and a commercially available RFA equipment (Osstell Mentor, Osstell AB, Gothenburg, Sweden) was adjusted at the mesial and buccal direction of the implant (Figure 3a). The mean ISQ values in the mesial and buccal direction were recorded.

In each group, 7 implants were randomly selected and removal torque values (RTV) were measured using torque meter (MARK-10 torque gauge, MARK-10 Corporation, NY, USA) on the day of sacrifice. Removed implant specimens were dehydrated in graded ethanol series and sputter-coated with platinum (LEICA EM ACE200, sputter current 40 mA, Leica Microsystems, Wetzlar, Germany). Surface topography was examined under a field-emission scanning electron microscope (FE-SEM, SUPRA 40VP, Carl Zeiss, Oberkochen, Germany) and photograph images were taken at the magnifications of 20× and 100× with 5.0 kV.

2.5. Histologic and Histometric Analysis

Among the 14 implants allocated for each group, 7 specimens were processed for histologic and histometric analysis, as the rest of the 7 implants were tested for RTVs described in Section 2.4. Specimens of the implants and surrounding tissues were dissected into blocks and fixed in 10% buffered formalin solution. After sequential ethanol dehydration, nondecalcified specimens were embedded in methylmethacrylate (Technovit 7200, Kulzer GmbH, Hanau, Germany) and sectioned along the implant axis in the bucco-lingual plane using a diamond saw with 30–50 μm thickness. Hematoxylin and eosin (H&E)-stained sections were evaluated under a light microscope fitted with a camera and histometric measurements were completed using an automated image analysis program (Image-Pro Plus, Media Cybernetics, Rockville, MD, USA).

Parameters were measured from two parts (apical and coronal) of the implant, which was transversely divided along its long axis. The apical part included the area between the most apical border of the fixture and 3 mm above and coronal part was from the coronal border of the apical part to the most coronal endpoint thread of the fixture (Figure 4a). The following parameters were measured: (a) the bone-to-implant contact length (BICL, in mm), which was the sum of length of the implant surface in direct contact with surrounding bone; (b) the bone-to-implant contact ratio (BICR, in %), which was the percentage of BICL out of the length measured for the implant surface outline; (c) the inter-thread bone area (BA, in mm^2), which was the sum of bone area observed between the threads;

and (d) the inter-thread bone area ratio (BAR, in %), which was the percentage of BA in the region of interest (ROI). ROI in the control group and coronal part of the test group was determined by outlining the space between the threads (inter-thread space). To determine the corresponding ROI in the apical part of the test group, superimposition of the counterpart in the control implant was performed to outline the virtual boundary of the original shape of the inter-thread area (Figure 4b).

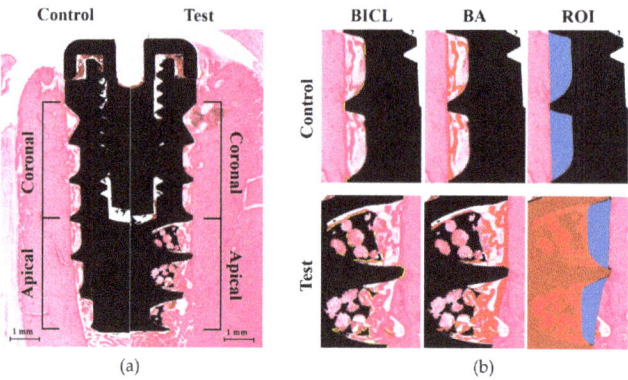

Figure 4. Schematic images of the histometric measurements: (**a**) Parameters were measured within the coronal and apical part, which were transversely divided along the long axis of the implant in both the control (left) and test groups (right) (magnification of 20×). (**b**) In each part of the control or test group (magnification of 50×), the bone-to-implant contact length (BICL) was determined by the total length of the implant surface in contact with the surrounding bone (yellow outlines in BICL). The inter-thread bone area (BA) was measured by the total bone area between the threads (red-colored area in BA). The region of interest (ROI) of the control group was outlined for the inter-thread space (blue colored area in ROI at the upper line). The ROI of the apical part in the test group (blue-colored area in ROI at the lower line) was determined by the virtual boundary, which was made from the superimposed counterpart of the control implant (red-colored implant shape in ROI at the lower line). Consequently, the original shape of the inter-thread space was outlined and the BA ratio (BAR) was calculated from the percentage of BA within each ROI.

2.6. Statistical Analysis

Statistical analyses were performed using SPSS Ver. 12.0 (SPSS, Chicago, IL, USA). Normality of data distribution was determined by Shapiro–Wilk test. For ISQ and RTV in both groups, paired t-test was used to compare the differences of the parameters between the two groups at each time period and the differences in ISQ between the baseline and 4 weeks in each group. Regarding histometric parameters, comparisons between the groups in each part and between the two parts (coronal and apical parts) in each group were performed using Student's t-test (in the data of BICL and BICR in the apical part, BA and BAR in the coronal, apical and total area), or Wilcoxon's rank-sum test (in the data of BICL and BICR in apical part and total area). The level of statistical significance was set at $p < 0.05$.

3. Results

3.1. Clinical Findings

The experimental sites in all the animals demonstrated uneventful healing and did not exhibit any adverse reaction throughout the postoperative healing period.

3.2. Resonance Frequency Analysis and Removal Torque Value

No significant differences were observed in ISQ values between the two groups at each time period and between the baseline and 4 weeks in each group. However, RTV in the test group (20.5 ± 6.8) was higher than that of the control group (8.0 ± 3.6) with a significant difference at the 4-week healing period ($p = 0.03$) (Table 1).

Table 1. Implant stability quotient (ISQ) value and removal torque value (RTV) after 4 weeks of healing in the test and control groups (mean ± SD).

		Test Group (n = 14)	Control Group (n = 14)
ISQ Value	Baseline	66.7 ± 4.0	69.5 ± 8.3
	4 weeks	67.5 ± 5.0	68.4 ± 6.3
		Test Group (n = 7)	Control Group (n = 7)
RTV (Ncm)	4 weeks	20.5 ± 6.8 [1]	8.0 ± 3.6

[1] Statistically significant difference between the two groups in the paired t-test ($p < 0.05$).

3.3. Surface Topography from FE-SEM Images

The original surface topography of the test group implant showed a titanium structure with regular distribution of pores in similar size ranging from 200 to 350 μm in the apical part (Figure 5a,b), whereas the control group showed a smooth texture of the machined surface (Figure 5c,d). After the removal torque test, the porous body in the test group exhibited structural destruction with a few integrated bone tissues in close proximity (Figure 5e,f). However, the interface between the porous structure and core of the fixture was maintained. The control group implant did not show any specific destruction, although there were some traces of scrapes on the surface (Figure 5g,h).

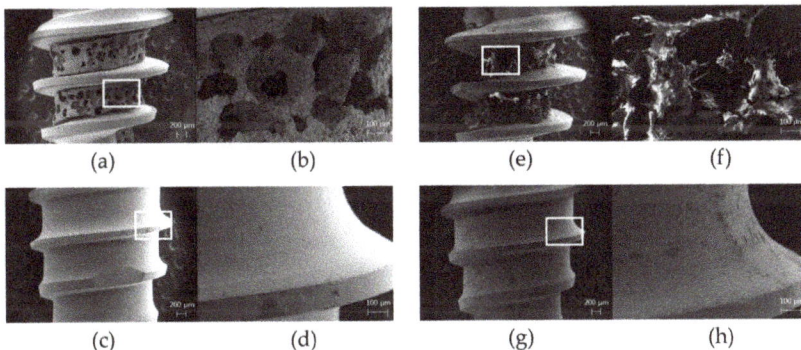

Figure 5. Field-emission scanning electron microscope (FE-SEM) images before the implant placement and after the removal: (**a**) overall image of the test group before the placement (magnification of 20×); (**b**) magnified view of the white box in (**a**) showing a porous titanium structure in the apical part of the test group implant with regularly distributed pores (magnification of 100×); (**c**) overall image of the control group before the placement (magnification of 20×); (**d**) magnified view of the white box in (**c**) showing a smooth machined surface in the control group implant (magnification of 100×); (**e**) overall image of the test group after the removal of implants at 4-week healing periods (magnification of 20×); (**f**) magnified view of the white box in (**e**) showing the destruction of the porous structures and some bone tissues at the porous structure after the removal (magnification of 100×); (**g**) overall image of the control group after the removal of implants at 4-week healing periods (magnification of 20×); (**h**) magnified view of the white box in (**g**) showing no specific destruction in the surface profile except for some scrapes after the removal (magnification of 100×).

3.4. Histologic Analysis

New bone (NB) formation and BIC were observed at entire length of the implants in both the test and control groups (Figure 6a,f). In the coronal part of the test group, newly formed hard tissues projected from the parent bone (PB) surface towards the drilled osteotomy sites along the threads of the implant (Figure 6b,c). Osteoid and NB lined with osteoblast-like cells were found in the space between the threads, which were in direct contact with the implant surface and sometimes bridged PB with the implant surface. Some part of the PB surface underwent a remodeling process with reversal lines parallel to the long axis of the bone wall. These histologic features were also observed at both the coronal (Figure 6g,h) and apical parts (Figure 6i,j) of the control group which had the same surface topography of the coronal part in the test group. In the apical part of the test group, newly formed woven bone with osteoblast-like cells on its surface projected from the PB wall into the drilled space, and ingrowth of NB into the porous structure exhibited direct contact with the implant surface (Figure 6d). NB shown in the porous structure was integrated with the pore entrances and in direct contact with the inner surfaces of regularly distributed porous scaffolds (Figure 6e). NB was bridged between one another or to the PB surface and surrounded by the densely packed connective tissue matrix.

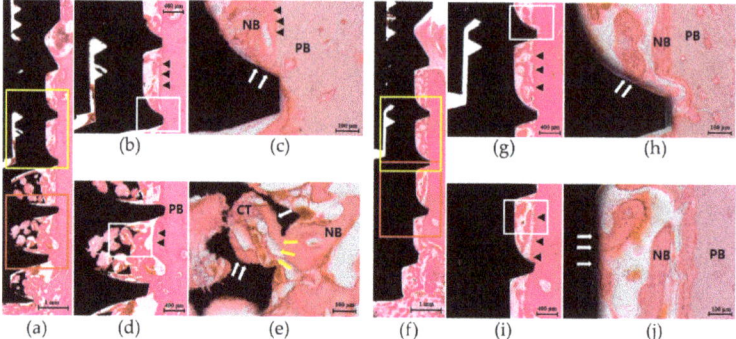

Figure 6. Representative histologic images of the implant in the test group and control group at 4 weeks of healing (H&E staining): (**a**) Overall view of the whole length of the test group implant showing a smooth surface profile in the coronal part (yellow box) and the porous structure in the apical part (red box) (magnification of 20×). (**b**) In the coronal part, new bone (NB) projection from the parent bone (PB) into the drilled osteotomy sites along the implant threads was observed (magnification of 50×). (**c**) Magnified view of the white box in (**b**) showing NB in direct contact with the implant surface (white arrows) (magnification of 200×). (**d**) In the apical part, NB connected with the PB surface and in contact with the surface of the porous structure was shown (magnification of 50×). (**e**) Magnified view of the white box in (**d**) showing NB ingrowth through the pore entrance (yellow arrows) and in direct contact with the surface of the porous scaffold (white arrows) (magnification of 200×). NB was lined with osteoblast-like cells on its surface and surrounded by the densely packed connective tissue matrix (CT). Reversal lines (black arrowheads) at the PB surface were found along the bony wall. (**f**) Overall view of the whole length of the control group implant showing a smooth surface profile in both the coronal (yellow box) and apical parts (red box) (magnification of 20×). (**g**) In the coronal part, new bone (NB) projected from the parent bone (PB) surface and towards the inter-thread space was shown (magnification of 50×). (**h**) Magnified view of the white box in (**g**) showing NB in direct contact with the implant surface (white arrows) and lined with osteoblast-like cells on its surface (magnification of 200×). (**i**) In the apical part, histologic appearance similar to that of the coronal part was shown (magnification of 50×). (**j**) Magnified view of the white box in (**i**) showing NB directly bridged to the implant surface (magnification of 200×). Lamellated reversal lines (black arrowheads) can be seen at the PB surface along the bony wall.

3.5. Histometric Analysis

Significant increases in BICL ($p < 0.007$), BICR ($p = 0.011$), BA ($p = 0.014$) and BAR ($p = 0.028$) of the total area were shown in the test group compared to the control group. The apical part of the test group presented significant increases in BICL ($p < 0.001$), BICR ($p = 0.001$), BA ($p = 0.011$) and BAR ($p = 0.020$) compared to the control group; all the parameters in the coronal part were similar in both the groups. The test group also showed significant differences between the coronal and apical part in BICL ($p = 0.005$), BICR ($p = 0.010$), BA ($p = 0.009$), and BAR ($p = 0.049$), whereas the control group showed no differences between the two parts (Table 2).

Table 2. Histometric parameters after 4 weeks of healing in the test and control groups (mean ± SD).

		Test Group (n = 7)	Control Group (n = 7)
	Coronal part	1.88 ± 0.98	1.31 ± 0.76
BICL (mm)	Apical part	3.43 ± 0.68 [1,2]	1.30 ± 0.78
	Total	5.31 ± 1.28 [1]	2.60 ± 1.21
	Coronal part	17.83 ± 9.53	12.56 ± 7.26
BICR (%)	Apical part	31.70 ± 7.53 [1,2]	12.89 ± 7.57
	Total	24.83 ± 6.44 [1]	12.69 ± 5.74
	Coronal part	0.28 ± 0.11	0.21 ± 0.08
BA (mm^2)	Apical part	0.48 ± 0.12 [1,2]	0.25 ± 0.12
	Total	0.77 ± 0.24 [1]	0.45 ± 0.16
	Coronal part	24.9 ± 10.84	20.7 ± 7.74
BAR (%)	Apical part	37.2 ± 10.04 [1,2]	23.0 ± 9.75
	Total	31.5 ± 8.03 [1]	21.8 ± 6.36

BICL, the bone-to-implant contact length; BICR, the bone-to-implant contact ratio; BA, the inter-thread bone area; BAR, the inter-thread bone area ratio. [1] Statistically significant difference from the control group in Wilcoxon's rank-sum test (used for BICL and BICR in total area) and in Student's t-test (used for BICL, BICR, BA, and BAR in apical part, BA and BAR in total area) ($p < 0.05$). [2] Statistically significant difference from the coronal part in Student's t-test ($p < 0.05$).

4. Discussion

In the present study, the porous titanium structure fabricated at the apical portion of the implant resulted in the interconnected open-pore structure, which led to an increase in the implant surface and its osteoconductivity to enhance bone ingrowth into the porous scaffold, thereby improving osseointegration of the implant.

The percentage of porosity on the overall surface and the size of pores are known as the determining factors in bone ingrowth [23]. Conventional methods like the sintering of beads on the titanium alloys have been reported to have limited degree of porosity (around 35%) and exhibit difficulty in controlling the profile of the topography [11]. As per the recent approaches, the PTTM- enhanced titanium implant could exhibit an increased percentage of the porosity (up to 70–80%) owing to the open-cell structure of dodecahedral repeats resembling trabecular bone [17,24]. In animal studies, the porous tantalum implants showed greater bone-to-implant contact with increased osteogenic activity compared to the solid titanium [20,25]. The clinical benefits of porous tantalum implants were also reported in the retrospective studies as there were high survival rates and less peri-implant bone loss [26], and a pilot study of failed implants immediately replaced by using the porous tantalum implants showed successful outcomes in 5 years of follow-up as well although the sample size was limited [27]. However, there still existed difficulty in manipulation of pure tantalum and high costs for purification [28].

The powder injection molding technique was introduced to process the fine ceramics in the past two decades and could offer the reproducible mass production of complicated structures like near-net-shapes, even in hard materials like ceramics [29]. In dental fields, powder injection molding zirconia implants were tested in the animal model, and this technique was suggested to provide an enhanced tissue response to the roughened surface when compared with that of a machined titanium surface [30]. In the present study, porous titanium structure produced by the powder injection molding technique resulted in the formation of interconnected open pores with an average porosity of about 70%. The structure was similar to the alveolar bone of the human mandible with D1 type bone (showing primarily of homogenous dense cortical bone) as evaluated through micro-computed tomography study, thus providing a negative template with a natural trabecular pattern with higher bone density [31]. Potential benefits of rapid vascularization, osteoblastic differentiation, and new bone ingrowth could be expected, as they mimic the natural trabecular bone. As the porosity could be controlled by adjusting the amount of polymers or the temperature of EPS expansion, it might also be possible to alter the characteristics of porous structures to match with the natural bone, or customized as per the need of individual patients.

The multithreaded root-form implant has clinical benefits of simple osteotomy, implant placement, close mechanical proximity to the bone to increase primary stability, and less traumatic retrieval under conditions of failure [3]. When combined with the adequately controlled porous structure, additional effects of enhanced neovascularization and new bone formation inside the porous scaffold termed as osseoincorporation can be expected [32,33]. However, the porous structure in the present study has possible problems, such as a higher risk of bacterial plaque accumulation, mucosal and peri-implant diseases when compared to the machined surface upon exposure to complex oral environments [34]. Hybrid surface implants have been suggested to reduce the prevalence of peri-implantitis by including the machined surface or less roughened texture in the coronal part of the implant together with the rough surface treatments in the apical part, which played important role in healing between the bone and the surface [35]. In correspondence with the rationale of the hybrid implant, the test group implant was expected to have both advantages in terms of accelerated healing and increased bone-to-implant contact by the porous structure at the apical part, and less biological complications related to the inflammation at the coronal smooth surface area. In addition, an external connection was utilized to minimize the risk of fracture from the thinned lateral wall of the body after the decrease in core diameter, while providing space for the porous structure. The crestal bone–implant interface is an important area in stress distribution during load transmission [36], and there exists a lack of data regarding mechanical failure in the porous structure with reference to stress and fatigue when loading, thus stating inappropriateness of this structure in the coronal portion of the implant body.

Implant stability measured using RFA showed ISQ values over 60 in both the groups and at both the observation periods with no significant differences. The ISQ value over 60 was reported to demonstrate clinical stability of the implant despite the differences in implant designs, surgical models, and devices used, and the factors affecting RFA include stiffness of the implant-bone interface, distance to first bone contact, and marginal bone loss [37]. It can be assumed that relatively standardized bone density among the surgical sites, fully engaged implant surface within the surrounding bone, and the same topographic aspects at the coronal portion contribute to the similar outcome of ISQ at baseline and after 4 weeks of healing between the two groups. The presence of a porous titanium structure in the apical portion and its loss of close proximity to the bone wall cut by the twist drill may not have a great influence on the ISQ value. On the other hand, the removal torque test is an indirect method to measure the shear strength that ruptures the bone–implant interface and gives information about the bone growth aspects at the surface [38]. Hence, a significant increase in RTV of the test group implant at 4 weeks might be explained by the increased secondary stability at the porous structure at the apical portion. The fractured bone compartment together with the porous surface in FE-SEM images after the reverse torque rotation also supported the findings of bone ingrowth at the apical portion. However, some of the destructions were observed in the center portion of the porous body, whereas the interface at the attachment to the implant core still remained. The destroyed sites might suggest the weak points in the mechanical properties of the implants, which should further be improved by the development of the manufacturing techniques.

Significant increases in total BICL, BICR, BA, and BAR in the test group implants were attributed to the apical portion, and histologic findings supported the results by showing enhanced NB formation in direct contact with the interconnected open pores with an increased surface area at the apical portion and improved BIC at 4 weeks. Healing in the trabecular compartment relies on the process of osteoconduction and de novo bone formation at the implant surface, resulting in contact osteogenesis [33], and porous configuration could promote osteoconductivity by increasing space for blood clot stabilization and recruitment of various cells involved in biologic cascades of peri-implant healing [39]. A porous structure designed to resemble the trabecular patterns of highest bone quality might permit an adequate osteoconductive scaffold for bone ingrowth, which is in accordance with other studies [40,41]. Apparently, it also might help in providing enhanced micro-mechanical interdigitation with the parent bone and increase primary stability irrespective of the condition of the recipient site, including compromised bone quality such as D4 type bone (showing the fine trabecular bone composing almost the total volume) where the cortical stiffness could not be anticipated and the posterior maxilla that lacks bone height.

In the present study, a porous titanium structure fabricated by the powder injection molding technique was able to provide three-dimensional interconnected porosity on the implant surface and thereby enhanced the new bone ingrowth at the surface. The histometric findings, including the fact that the bone-to-implant contact and new bone formation inside the porous material demonstrated the improvements in osseointegration by the porous structure in the early healing dynamics, were in accordance with some other studies utilizing trabecular-like scaffolds to the dental implants [20,21,25]. However, the test implant used in this study was designed to combine the porous structure with the machined surface implant for a pilot approach to focus on the efficacy of the newly developed structure at the apical portion, and other variables, including overall macrogeometry and topography of the implants, were intended to be controlled. For the clinical applications, various modifications in the microstructures other than the machined surface at the coronal aspect should be considered, and it might also be necessary to develop the macrostructural designs of the implant–abutment connection and the platform that can show more favorable outcomes in the biomechanical aspects. Improvements in the fabrication technologies to standardize porosity and increase the mechanical strength of the porous structures are necessary to obtain reliable clinical outcomes. In addition, the healing events around the porous implants in the long-term period and the tissue dynamics after the loading should further be observed. Finally, the effects of the porous titanium structure on the compromised bone conditions in clinical situations such as osteoporosis, grafted bone or simultaneous sinus floor elevation should be further investigated.

5. Conclusions

The findings of the study suggest that the porous titanium structure might increase apical bone-to-implant contact due to the increased surface area and enhance new bone formation with increased osteoconductivity in the early healing period, thereby leading to improvements in the osseointegration of the implants.

Author Contributions: Conceptualization, S.-Y.K., W.L., and J.-H.Y.; data curation, J.-Y.H., S.-Y.K., and J.-H.Y.; formal analysis, J.-Y.H., S.-Y.K., and J.-H.Y.; funding acquisition, W.L. and J.-H.Y.; methodology, J.-Y.H., S.-Y.K., Y.-Y.C., S.-H.K., and J.-H.Y.; supervision, W.L. and J.-H.Y.; writing—original draft, J.-Y.H. and S.-Y.K.; writing—review and editing, J.-Y.H., W.L., Y.-Y.C., S.-H.K. and J.-H.Y. All authors have read and agreed to the published version of the manuscript.

Funding: This work was supported by the Technology Innovation Program (10049237, Development of rapid mold manufacturing technology for mass customized medical devices with SLS hybrid 3D printing technology) funded by the Ministry of Trade, Industry & Energy (MI, Korea), the fund of Biomedical Research Institute, Jeonbuk National University Hospital and the research funds of Jeonbuk National University in 2018.

Conflicts of Interest: The authors declare no conflict of interest. The funders had no role in the design of the study; in the collection, analyses, or interpretation of data; in the writing of the manuscript, or in the decision to publish the results.

References

1. Boioli, L.T.; Penaud, J.; Miller, N. A meta-analytic, quantitative assessment of osseointegration establishment and evolution of submerged and non-submerged endosseous titanium oral implants. *Clin. Oral Implants Res.* **2001**, *12*, 579–588. [CrossRef]
2. Le Guéhennec, L.; Soueidan, A.; Layrolle, P.; Amouriq, Y. Surface treatments of titanium dental implants for rapid osseointegration. *Dent. Mater.* **2007**, *23*, 844–854. [CrossRef]
3. Javed, F.; Ahmed, H.B.; Crespi, R.; Romanos, G.E. Role of primary stability for successful osseointegration of dental implants: Factors of influence and evaluation. *Interv. Med. Appl. Sci.* **2013**, *5*, 162–167. [CrossRef]
4. Moy, P.K.; Medina, D.; Shetty, V.; Aghaloo, T.L. Dental implant failure rates and associated risk factors. *Int. J. Oral Maxillofac. Implants* **2005**, *20*, 569–577. [PubMed]
5. van Steenberghe, D.; Jacobs, R.; Desnyder, M.; Maffei, G.; Quirynen, M. The relative impact of local and endogenous patient-related factors on implant failure up to the abutment stage. *Clin. Oral Implants Res.* **2002**, *13*, 617–622. [CrossRef]
6. Beikler, T.; Flemmig, T.F. Implants in the medically compromised patient. *Crit. Rev. Oral Biol. Med.* **2003**, *14*, 305–316. [CrossRef] [PubMed]
7. Giudice, A.; Bennardo, F.; Antonelli, A.; Barone, S.; Wagner, F.; Fortunato, L.; Traxler, H. Influence of clinician's skill on primary implant stability with conventional and piezoelectric preparation techniques: An ex-vivo study. *J. Biol. Regul. Homeost. Agents* **2020**, *34*, 739–745.
8. Shibata, Y.; Tanimoto, Y. A review of improved fixation methods for dental implants. Part I: Surface optimization for rapid osseointegration. *J. Prosthodont. Res.* **2015**, *59*, 20–33. [CrossRef] [PubMed]
9. Buser, D.; Schenk, R.K.; Steinemann, S.; Fiorellini, J.P.; Fox, C.H.; Stich, H. Influence of surface characteristics on bone integration of titanium implants. A histomorphometric study in miniature pigs. *J. Biomed. Mater. Res.* **1991**, *25*, 889–902. [CrossRef] [PubMed]
10. Scopelliti, P.E.; Borgonovo, A.; Indrieri, M.; Giorgetti, L.; Bongiorno, G.; Carbone, R.; Podestà, A.; Milani, P. The effect of surface nanometre-scale morphology on protein adsorption. *PLoS ONE* **2010**, *5*, e11862. [CrossRef]
11. Pilliar, R.M. Overview of surface variability of metallic endosseous dental implants: Textured and porous surface-structured designs. *Imp. Dent.* **1998**, *7*, 305–314. [CrossRef]
12. Ryan, G.; Pandit, A.; Apatsidis, D.P. Fabrication methods of porous metals for use in orthopaedic applications. *Biomaterials* **2006**, *27*, 2651–2670. [CrossRef] [PubMed]
13. Dabrowski, B.; Swieszkowski, W.; Godlinski, D.; Kurzydlowski, K.J. Highly porous titanium scaffolds for orthopaedic applications. *J. Biomed. Mater. Res. B Appl. Biomater.* **2010**, *95*, 53–61. [CrossRef] [PubMed]
14. Miyazaki, T.; Kim, H.M.; Kokubo, T.; Ohtsuki, C.; Kato, H.; Nakamura, T. Mechanism of bonelike apatite formation on bioactive tantalum metal in a simulated body fluid. *Biomaterials* **2002**, *23*, 827–832. [CrossRef]
15. Yoo, D. New paradigms in hierarchical porous scaffold design for tissue engineering. *Mater. Sci. Eng. C Mater. Biol. Appl.* **2013**, *33*, 1759–1772. [CrossRef]
16. Karageorgio, V.; Kaplan, D. Porosity of 3D biomaterial scaffolds and osteogenesis. *Biomaterials* **2005**, *26*, 5474–5491. [CrossRef]
17. Fujibayashi, S.; Neo, M.; Kim, H.M.; Kokubo, T.; Nakamura, T. Osteoinduction of porous bioactive titanium metal. *Biomaterials* **2004**, *25*, 443–450. [CrossRef]
18. Bencharit, S.; Byrd, W.C.; Altarawneh, S.; Hosseini, B.; Leong, A.; Reside, G.; Morelli, T.; Offenbacher, S. Development and applications of porous tantalum trabecular metal-enhanced titanium dental implants. *Clin. Implant. Dent. Relat. Res.* **2014**, *16*, 817–826. [CrossRef]
19. Pattanayak, D.K.; Fukuda, A.; Matsushita, T.; Takemoto, M.; Fujibayashi, S.; Sasaki, K.; Nishida, N.; Nakamura, T.; Kokubo, T. Bioactive Ti metal analogous to human cancellous bone: Fabrication by selective laser melting and chemical treatments. *Acta Biomater.* **2011**, *7*, 1398–1406. [CrossRef]
20. Lee, J.W.; Wen, H.B.; Gubbi, P.; Romanos, G.E. New bone formation and trabecular bone microarchitecture of highly porous tantalum compared to titanium implant threads: A pilot canine study. *Clin. Oral Implants Res.* **2018**, *29*, 164–174. [CrossRef]
21. Fraser, D.; Funkenbusch, P.; Ercoli, C.; Meirelles, L. Biomechanical analysis of the osseointegration of porous tantalum implants. *J. Prosthet. Dent.* **2020**, *123*, 811–820. [CrossRef] [PubMed]

22. Kilkenny, C.; Altman, D.G. Improving bioscience research reporting: ARRIVE-ing at a solution. *Lab. Anim.* **2010**, *44*, 377–378. [CrossRef]
23. Cornell, C.N.; Lane, J.M. Current understanding of osteoconduction in bone regeneration. *Clin. Orthop. Relat. Res.* **1998**, *355*, S267–S273. [CrossRef]
24. Levine, B.R.; Sporer, S.; Poggie, R.A.; Della Valle, C.J.; Jacobs, J.J. Experimental and clinical performance of porous tantalum in orthopedic surgery. *Biomaterials* **2006**, *27*, 4671–4681. [CrossRef] [PubMed]
25. Fraser, D.; Mendonca, G.; Sartori, E.; Funkenbusch, P.; Ercoli, C.; Meirelles, L. Bone response to porous tantalum implants in a gap-healing model. *Clin. Oral Implants Res.* **2019**, *30*, 156–168. [CrossRef] [PubMed]
26. Edelmann, A.R.; Patel, D.; Allen, R.K.; Gibson, C.J.; Best, A.M.; Bencharit, S. Retrospective analysis of porous tantalum trabecular metal-enhanced titanium dental implants. *J. Prosthet. Dent.* **2019**, *121*, 404–410. [CrossRef]
27. Dimaira, M. Immediate placement of trabecular implants in sites of failed implants. *Int. J. Oral Maxillofac. Implants* **2019**, *34*, e77–e83. [CrossRef] [PubMed]
28. Liu, Y.; Bao, C.; Wismeijer, D.; Wu, G. The physicochemical/biological properties of porous tantalum and the potential surface modification techniques to improve its clinical application in dental implantology. *Mater. Sci. Eng. C Mater. Biol. Appl.* **2015**, *49*, 323–329. [CrossRef]
29. Lin, S.I.E. Near-net-shape forming of zirconia optical sleeves by ceramics injection molding. *Ceram. Int.* **2001**, *27*, 205–214. [CrossRef]
30. Park, Y.S.; Chung, S.H.; Shon, W.J. Peri-implant bone formation and surface characteristics of rough surface zirconia implants manufactured by powder injection molding technique in rabbit tibiae. *Clin. Oral Implants Res.* **2013**, *24*, 586–591. [CrossRef] [PubMed]
31. Lee, J.H.; Kim, H.J.; Yun, J.H. Three-dimensional microstructure of human alveolar trabecular bone: A micro-computed tomography study. *J. Periodontal Implant. Sci.* **2017**, *47*, 20–29. [CrossRef] [PubMed]
32. Bobyn, J.D.; Stackpool, G.J.; Hacking, S.A.; Tanzer, M.; Krygier, J.J. Characteristics of bone ingrowth and interface mechanics of a new porous tantalum biomaterial. *J. Bone Joint Surg. Br.* **1999**, *81*, 907–914. [CrossRef] [PubMed]
33. Davies, J.E. Understanding peri-implant endosseous healing. *J. Dent. Educ.* **2003**, *67*, 932–949. [CrossRef] [PubMed]
34. Hanisch, O.; Cortella, C.A.; Boskovic, M.M.; James, R.A.; Slots, J.; Wikesjö, U.M. Experimental peri-implant tissue breakdown around hydroxyapatite-coated implants. *J. Periodontol.* **1997**, *68*, 59–66. [CrossRef] [PubMed]
35. Lee, C.T.; Tran, D.; Jeng, M.D.; Shen, Y.T. Survival rates of hybrid rough surface implants and their alveolar bone level alteration. *J. Periodontol.* **2018**, *12*, 1390–1399. [CrossRef] [PubMed]
36. Baggi, L.; Cappelloni, I.; Di Girolamo, M.; Maceri, F.; Vairo, G. The influence of implant diameter and length on stress distribution of osseointegrated implants related to crestal bone geometry: A three-dimensional finite element analysis. *J. Prosthet. Dent.* **2008**, *100*, 422–431. [CrossRef]
37. Sennerby, L.; Meredith, N. Implant stability measurements using resonance frequency analysis: Biological and biomechanical aspects and clinical implications. *Periodontology 2000* **2008**, *47*, 51–66. [CrossRef]
38. Klokkevold, P.R.; Johnson, P.; Dadgostari, S.; Caputo, A.; Davies, J.E.; Nishimura, R.D. Early endosseous integration enhanced by dual acid etching of titanium: A torque removal study in the rabbit. *Clin. Oral Implants Res.* **2001**, *12*, 350–357. [CrossRef] [PubMed]
39. Scaglione, S.; Giannoni, P.; Bianchini, P.; Sandri, M.; Marotta, R.; Firpo, G.; Valbusa, U.; Tampieri, A.; Diaspro, A.; Bianco, P.; et al. Order versus disorder: In vivo bone formation within osteoconductive scaffolds. *Sci. Rep.* **2012**, *2*, 274. [CrossRef]
40. Chang, B.S.; Lee, C.K.; Hong, K.S.; Youn, H.J.; Ryu, H.S.; Chung, S.S.; Park, K.W. Osteoconduction at porous hydroxyapatite with various pore configurations. *Biomaterials* **2000**, *21*, 1291–1298. [CrossRef]
41. Götz, H.E.; Müller, M.; Emmel, A.; Holzwarth, U.; Erben, R.G.; Stangl, R. Effect of surface finish on the osseointegration of laser-treated titanium alloy implants. *Biomaterials* **2004**, *25*, 4057–4064. [CrossRef] [PubMed]

© 2020 by the authors. Licensee MDPI, Basel, Switzerland. This article is an open access article distributed under the terms and conditions of the Creative Commons Attribution (CC BY) license (http://creativecommons.org/licenses/by/4.0/).

Article

Physical/Chemical Properties and Resorption Behavior of a Newly Developed Ca/P/S-Based Bone Substitute Material

Bing-Chen Yang [1], Jing-Wei Lee [2], Chien-Ping Ju [1,*] and Jiin-Huey Chern Lin [1,*]

[1] Department of Materials Science and Engineering, College of Engineering, National Cheng Kung University, Tainan 70101, Taiwan; N58011308@ncku.edu.tw
[2] Division of Plastic and Reconstructive Surgery, National Cheng Kung University Hospital, Department of Surgery, College of Medicine, National Cheng Kung University, Tainan 70403, Taiwan; jwlee@mail.ncku.edu.tw
* Correspondence: cpju@mail.ncku.edu.tw (C.-P.J.); chernlin@mail.ncku.edu.tw (J.-H.C.L.); Tel.: +886-6-2748086 (C.-P.J.); +886-6-2080391 (J.-H.C.L.)

Received: 3 July 2020; Accepted: 2 August 2020; Published: 5 August 2020

Abstract: Properly regulating the resorption rate of a resorbable bone implant has long been a great challenge. This study investigates a series of physical/chemical properties, biocompatibility and the behavior of implant resorption and new bone formation of a newly developed Ca/P/S-based bone substitute material (Ezechbone® Granule CBS-400). Experimental results show that CBS-400 is comprised majorly of HA and CSD, with a Ca/P/S atomic ratio of 54.6/39.2/6.2. After immersion in Hank's solution for 7 days, the overall morphology, shape and integrity of CBS-400 granules remain similar to that of non-immersed samples without showing apparent collapse or disintegration. With immersion time, the pH value continues to increase to 6.55 after 7 days, and 7.08 after 14 days. Cytotoxicity, intracutaneous reactivity and skin sensitization tests demonstrate the good biocompatibility features of CBS-400. Rabbit implantation/histological observations indicate that the implanted granules are intimately bonded to the surrounding new bone at all times. The implant is not merely a degradable bone substitute, but its resorption and the formation of new cancellous bone proceed at the substantially same pace. After implantation for 12 weeks, about 85% of the implant has been resorbed. The newly-formed cancellous bone ratio quickly increases to >40% at 4 weeks, followed by a bone remodeling process toward normal cancellous bone, wherein the new cancellous bone ratio gradually tapers down to about 30% after 12 weeks.

Keywords: Ca-based; bone substitute; resorption; animal study; histology

1. Introduction

According to the reports of Weiser et al. [1] and Hall et al. [2], the number of orthopedic surgeries performed worldwide was approximately 24 million in 2004, and forecast to grow to 31 million by 2012. One of the most challenging issues among these surgeries has been the reconstruction of bone defects caused by trauma, tumor removal, congenital deformity, etc., therein, 20–25% cases would require bone grafting [3]. Bone grafting is also highly demanded in dental industry. Institut Straumann AG (Basel, Switzerland) estimated that the total number of dental implants used in 2018 worldwide could be more than 10 million [4]. Cha et al. [5] estimated that bone grafting is required for about one in every four dental implants, meaning that the demand for bone substitutes would be more than 2.5 million units for dental implant applications alone.

Being osteoconductive/osteoinductive, non-immunogenic and able to improve the healing process, autologous bone grafts have been considered the gold standard for the repair of osseous defects [6].

Autologous bone grafts harvested from the iliac crest are commonly used in reconstructive orthopedic surgeries. However, considerable morbidity has been reported related to iliac crest harvesting, with a high complication rate up to 49%, including damage to blood vessels and nerves, joint disruption, fractures, subluxation, herniation of abdominal contents, and delayed iliac abscess [7]. Another major concern with the iliac crest harvest procedure is that the amount of bone available for autografting is limited [8], which has prompted an increased interest in using bone graft substitutes [6].

Xenografts, allografts and synthetic bone substitutes are some common alternatives to autologous bone. However, diseases carried from human or animal bone-derived grafts always pose a potential risk to the patients receiving such grafts. Transmission of human immunodeficiency virus (HIV), hepatitis C virus (HCV), human T-lymphotropic virus (HTLV), unspecified hepatitis, tuberculosis and other bacteria has been documented to be associated with allografts [9]. Xenografts derived from animal bone also carry the risk of transmission of diseases. Kim et al. [10] reported that anorganic bovine bone, a popularly used xenograft in dentistry, has the risk of transmitting bovine spongiform encephalopathy (BSE) prion PrPSc. Proteins were detected in Bio-Oss® (bovine-bone derived) [11] and tibia samples treated under a similar deproteinization condition [12]. More recently, Kim et al. [13] suggested that humans are not safe from the infection of prion disease of other species, due to its ability to cross the species barrier. These findings indicate that the long-term risks of bovine bone-derived xenografts, which are extensively used in dentistry today, should be seriously considered.

Apparently, the use of synthetic materials as bone grafts can avoid the aforementioned disease transmission risks. An ideal degradable/resorbable bone substitute material should have a degradation/resorption rate comparable to the rate of the host repairing bone to facilitate a complete bone repair [14]. Nevertheless, most of the currently available bone substitute materials resorb either too fast or too slowly [15]. For example, hydroxyapatite (HA), a widely-used calcium phosphate, generally resorbs too slowly, while calcium sulfate, another popularly-used resorbable bone substitute, is often considered to resorb too fast [16]. It was reported that an implant resorbing too slow may hamper bone repair, degrade mechanical properties of the repaired bone, and lead to chronic inflammation [15,17]. On the other hand, an implant resorbing too fast can lead to resorption without sufficient new bone ingrowth [18]. There is no doubt that properly regulating the resorption rate of a resorbable bone implant remains one of the biggest challenges in this field.

To avoid disease transmission risks, optimize resorption rate and osteoconductivity, a synthetic, inorganic and highly porous Ca/P/S-based bone-substituting material (Ezechbone® Granule CBS-400) has been newly developed by a National Cheng-Kung University (NCKU)/ Joy Medical Devices (JMD) joint research project. The purpose of the present study was to investigate a series of physical/chemical properties, biocompatibility, and particularly the behavior of implant resorption and new bone formation of this material in a rabbit implantation study.

2. Materials and Methods

The present Ca/P/S-based material for the study (Ezechbone® Granule CBS-400) was proprietarily manufactured by JMD, an ISO 13485/GMP-certified facility in Kaohsiung, Taiwan. The raw materials used to fabricate CBS-400 were all purchased from accredited manufacturers with US Pharmaceutical (USP) grade. The heavy metal (primarily As, Cd, Hg and Pb) concentrations of the final product determined by inductively coupled plasma-mass spectrometry (ICP-MS) (Element XR™, Thermo Fisher Scientific Inc., Waltham, MA, USA) were all within the cited limitations as set forth in ASTM F1185-03 Sections 4.3 and 4.4. To assess the safety and efficacy of the present bone-substituting material, a series of chemical/physical characterization and biocompatibility tests were conducted. The physical/chemical testing items of the present study included phase identification, Ca/P/S atomic ratio, granule morphology, dimensional stability, porosity volume fraction, pore size, pH value, and solubility tests in buffered citric acid and TRIS-HCl solution. The biocompatibility tests included cytotoxicity, intracutaneous reactivity, skin sensitization and animal implantation tests.

2.1. Phase Identification and Chemical Composition

An X-ray diffraction (XRD) system (D2 PHASER, Bruker, Billerica, MA, USA) was used to scan the samples in the range from 5° to 55° (2 theta), at a scan speed of 1°/min. The XRD patterns were obtained using a Ni-filtered Cu-Kα radiation diffractometer, operated at 30 kV and 10 mA equipped with a diffracted-beam monochromator. For better resolution, the Kα2 signal was stripped and only Kα1 X-ray with a wavelength of 1.54060 Å was used for analysis. The XRD patterns were analyzed by matching each characteristic peak with that compiled in the ICDD/JCPDS cards of hydroxyapatite (HA), calcium sulfate dihydrate (CSD), tetracalcium phosphate (TTCP) and anhydrous dicalcium phosphate (DCPA).

The Ca/P and Ca/P/S ratios of the CBS-400 granules were determined using a scanning electron microscope (SEM) (JSM-6510, JEOL, Akishima, Japan), equipped with an energy dispersive spectrometer (EDS) (X-ACT, Oxford Instruments, Abingdon, UK) operated at 10 kV and 70 mA, with a working distance of 10 mm. The granules were ground into powder and then packed into a 6 mm-diameter cylindrical stainless steel mold under a pressure of 100 kgf to form a cylindrical disc. To avoid electric charging of the non-conducting material, the disc samples were sputter-coated with a thin layer of carbon, using a JEOL JEC-560 AUTO CARBON COATER system.

Prior to a semi-quantitative analysis of the samples, an overall chemical analysis was conducted on large areas. The results indicated that the predominant elements of CBS-400 were Ca, P, S, O and C, wherein C signals primarily came from the sputtering process and O signals were also commonly detected due to the unavoidable contamination from the environment. To exclude these artifact effects, the Ca/P and Ca/P/S ratios were normalized by assuming Ca + P + S = 100%.

2.2. Granule Morphology and Dimensional Stability in Hank's Solution

The granule morphology and dimensional stability of the material immersed in Hank's solution were evaluated using the same JEOL JSM-6510 SEM. Three granular samples were immersed in daily refreshed Hank's solution at 37 °C, with a pH value of 7.4 and a sample/liquid ratio of 1 g/20 mL. One sample was taken out after 1 day; the second sample after 3 days; and the third after 7 days. The immersed granules were dried in an oven at 50 °C for 24 h, then mounted on a specimen stub using a carbon tape. To avoid electric charging during SEM examination, the non-conducting samples were sputter-coated with gold for 120 s at 20 mA, using a vacuum sputter system (E-1045, Hitachi High-Technologies Corp., Tokyo, Japan).

To determine particle size distribution of the granules, the CBS-400 particles were randomly placed on a transparent glass slit with a black matte finish paper underneath, and photographed using a digital camera (D90 DSLR, Nikon, Tokyo, Japan). Particle sizes (Minimum Feret diameter) were determined using ImageJ (ver.1.52a) issued by the National Institutes of Health (NIH). Through the Color Threshold function, the plain-white particles were selected from the matte black background. Using the Convert to Mask function, a binary image with scattered black particles was obtained. The Open function, a binary processing tool, was applied to largely eliminate the noise from photographing. The Watershed function, another binary processing tool, was performed to avoid occasionally aggregated particles. The binary image-processed figures were then analyzed using the Analyze Particles function, generating the needed distinct particles max and min Feret diameter information.

2.3. Porosity Volume Fraction and Pore Size

The porosity value of the granular product was determined according to ASTM C-830-00. The measurement was conducted using the liquid (95% v/v ethanol) intrusion method, wherein the apparent porosity, P, was determined by the equation, P (% v/v) = [(saturated weight-dry weight)/(saturated weight-suspended weight)] × 100%.

The pore sizes of the granules were determined using the same SEM and analyzed through ImageJ (ver.1.49). Then, 3 lots of CBS-400 were randomly selected and tested. One sample was randomly selected from each lot and 10 sites were randomly selected from each sample. For each site, 50 pores were randomly selected for measurement. In so doing, a total of 500 pores were measured for each lot and its pore size distribution was obtained.

2.4. pH Value

The pH values of the daily refreshed Hank's solution at 37 °C with a pH value of 7.4, wherein CBS-400 was immersed with a sample/liquid ratio of 1 g/20 mL for 1, 2, 3, 4, 5, 6, 7 and 14 days, were measured. A pH meter (SP-2300, Suntex Instruments, Taipei, Taiwan) was used for the test. The average pH values at each time point and their changes with immersion time were analyzed.

2.5. Solubility Testing in Buffered Citric Acid and TRIS-HCl Solution

The concentrations of Ca, P and S elements released from CBS-400 immersed in citric acid solution and simulated body fluid were measured and compared to the concentrations of these elements in human blood. According to ISO 10993-14:2001(E), this test consists of two parts: the extreme solution test, wherein a low pH buffered citric acid solution with a pH value of 3 at 37 °C was used as a worst-case low-end service environment, and the simulation solution test, wherein a TRIS-HCl buffer solution with a pH value of 7.4 was used to simulate the body's normal pH level.

For the extreme solution test, four CBS-400 samples were immersed for 5 days in the freshly-prepared buffered citric acid solution, which was agitated longitudinally at 2 Hz at 37 °C, with a sample/liquid ratio of 1 g/20 mL. After 5 days, the samples were removed from the containers via filtration and the filtrate was retained for analysis. The concentrations of Ca, P and S elements in the filtrate and blank buffered citric acid solution were analyzed using an ICP-MS system (7500ce, Agilent Technologies, Inc., Santa Clara, CA, USA).

The TRIS-HCl solution for the simulation solution test was prepared by dissolving a desired amount of tris(hydroxymethyl)aminomethane in water. The pH value of the solution was adjusted by HCl to 7.4 at 37 °C. The CBS-400 samples were immersed in the TRIS-HCl solution, which was agitated longitudinally at 2 Hz at 37 °C, with a sample/liquid ratio of 1 g/20 mL for 1, 3 and 5 days. After each time point, 4 samples were removed from the containers via filtration, and the filtrates were retained for analysis. The concentrations of Ca, P and S elements in the filtrates and blank TRIS-HCl solution were analyzed using the same ICP-MS system.

The conversion between concentrations (ppm) and mass of elements released (mg) was based on the equation, [Element concentration in filtrate (ppm) − Element concentration in blank vehicle (ppm)] × Immersion solution volume (mL) = Mass of element released from CBS-400 (mg). The calculated value represented the mass of the element released from 1 g CBS-400, immersed in 20 mL immersion solution at 37 °C.

2.6. Biocompatibility

To assess the safety of CBS-400, a series of biocompatibility tests, including cytotoxicity, intracutaneous reactivity and skin sensitization, were conducted. For intracutaneous reactivity and skin sensitization tests, the animal test protocols adopted were reviewed and approved by the Institutional Animal Care and Use Committee of NCKU (IACUC Approval Number: 102254 and 103309).

2.6.1. Cytotoxicity

The cytotoxicity test, which was designed to determine the biological response of mammalian cells in vitro, was performed according to ISO 10993-5:2009(E) methods, wherein the extraction method was used for the study. The extract of CBS-400 in culture medium with serum at a ratio of 0.1 g/mL was used as the test sample. The CBS-400 test samples and the negative control, Al_2O_3 particles, were gamma-ray sterilized at a dosage of 25 kGy. Culture medium with serum was used as the blank control. An established cell line (NIH/3T3 cells, BCRC 60008) was used for the cytotoxicity test, due to its sensitivity to chemical-induced cytotoxicity [19].

Cell viability was determined by WST-1 assay, a colorimetric assay for mitochondrial dehydrogenase activity in which the absorbance at 450 nm is proportional to the amount of dehydrogenase activity in the cell [20]. Overall, 6 wells were tested for control groups and 24 wells for CBS-400 extract. According to ISO 10993-5:2009(E), the test material has a cytotoxic potential if its cell viability value is less than 70% that of the blank control.

2.6.2. Intracutaneous Reactivity Test

The potential of CBS-400 to produce irritation was assessed by an intracutaneous reactivity test according to ISO 10993-10:2010(E). Three healthy young adult male New Zealand White (NZW) rabbits weighing >2 kg were used for the test for each positive control and CBS-400 extract group. For the CBS-400 extract group, the extracts were separately prepared by immersing CBS-400 granules in polar (saline) and non-polar (sesame oil) solvents, at a ratio of 0.1 g/mL at 37 °C for 72 h. The arrangement of the injection sites followed that set forth in ISO 10993-10:2010(E) clause 6.4.5. The positive control group followed the same procedure, wherein formalin and histamine were used as polar and non-polar positive controls, respectively. The appearance of each injection site was noted immediately after injection, and at 24, 48 and 72 h after injection. The tissue reactions for erythema and oedema at each injection site and each time interval were graded according to the system described in detail in ISO 10993-10:2010(E), clause 6.4.6.

After the grading at 72 h, all erythema grades and oedema grades at 24, 48 and 72 h were totaled for each CBS-400 extract test sample and blank on each individual animal separately. To calculate the score of a test sample or blank on each individual animal, each of the totals was divided by 15 (3 scoring time points ×5 CBS-400 extract or blank sample injection sites). To determine the overall mean score for each test sample and the corresponding blank, the scores for the three animals were averaged. The final test sample score was obtained by subtracting the score of the blank from that of the test sample. The requirement of the test was met if the final score was 1.0 or less.

2.6.3. Skin Sensitization Test

Skin sensitization was assessed by murine local lymph node assay (LLNA): 2-bromodeoxyuridine-enzyme linked immunosorbent assay (BrdU-ELISA) test according to ISO 10993-10:2010(E), OECD guideline for the testing of chemicals 442B, and a report of Takeyoshi et al. [21]. Overall, 15 healthy, 8–9 week non-pregnant female mice of the CBA/CaJNarl strain were used for the study; 5 mice for each group. The extract of CBS-400 in acetone olive oil (AOO) solution (4:1 v/v) was used as the test sample. According to OECD 442B, AOO solution and 2,4-dinitrochlorobenzene (DNCB) were selected respectively as blank and positive control for this test. The extract was prepared by immersing CBS-400 granules in the solvent, at a ratio of 0.1 g/mL at 37 °C for 72 h.

Following an ISO 10993-10:2010(E)-recommended study of Takeyoshi et al. [21], the control substances or extracts were applied to the dorsum of both ears daily for three consecutive days. A single intraperitoneal injection of BrdU (5 mg per mouse per injection) was given on day 4. Then, 24 h after injection of BrdU, the mice were sacrificed and their auricular lymph nodes were removed, weighed and stored at −20 °C prior to analysis with ELISA to measure the level of BrdU incorporation. Unpaired Student's t-test was used to analyze the differences in body weight change, lymph nodes weight and labeling index. The incorporation of BrdU into lymph node cells (LNC) was determined using a commercial cell proliferation assay kit (BioVision Inc., Milpitas, CA, USA; Cat. No. K306-200).

The absorbance at 450 nm was determined using a microplate reader (Multiskan™, Thermo Fisher Scientific Inc., Waltham, MA, USA), with a reference wavelength of 630 nm. The absorbance was defined as the BrdU labeling index. Means and standard errors for the labeling indices were calculated for each treatment group. The stimulation index (SI) was calculated by dividing the labeling indices in each test group by that in the concurrent blank control group. According to the principle set forth in

OECD 442B, the response in cellular proliferation of 1.6 times or more compared to the activity of the vehicle control is a threshold for designating a test material as a sensitizer.

2.7. Animal Implantation Study and Resorption Rate of CBS-400 Implant

An animal implantation study was conducted to evaluate local tissue responses and the resorption process of the CBS-400 granules in the implantation sites and their adjacent areas after 4, 8 and 12 weeks of implantation in the femur condyle of New Zealand White (NZW) rabbits. The animals were acclimatized and cared for as specified in ISO 10993-2:2006(E). Moreover, 22 adult, healthy, male NZW rabbits weighing 2.5–3.5 kg were used for the study. The rabbits were housed individually in stainless steel cages with free access to food and water. An acclimation period of a minimum of 7 days was allowed between receipt of the animals and the start of the study. The animal study was performed at NCKU Medical College Animal Center (Tainan, Taiwan). The animal test protocols adopted were reviewed and approved by the Institutional Animal Care and Use Committee of NCKU (IACUC Approval Number: 102254).

All the animals were operated on under general anesthesia. Zoletil 50 (0.05 mL/100 g, Virbac, Carros, France) was used as general anesthesia, while xylocaine (AstraZeneca, Cambridge, England, UK) was used as a local anesthesia. The implantation sites were shaved and cleansed with 70% v/v ethanol and Betadine® (povidone iodine 10% w/v, Mundipharma, Frankfurt, Germany). To implant CBS-400 in the medial epicondyle of the femur, a longitudinal incision was made on the anterior surface of the femur. The inner side of the knee joint was cut to expose the femur. After that, the periosteum was reflected, and a 2 mm pilot hole was drilled, which was sequentially widened with drills of increasing size until a final diameter of 5 mm was reached. A 5 mm diameter drill burr was used and a ring was inserted at a depth of 10 mm to ensure the appropriate length (10 mm) of the drill hole. (Figure 1). A cylindrical void of 5 mm in diameter and 10 mm in depth was created in rabbit femoral condyle. Cavities without filling were used as a negative control. After filling, subcutaneous tissues and skin were closed up, layer by layer, with sutures.

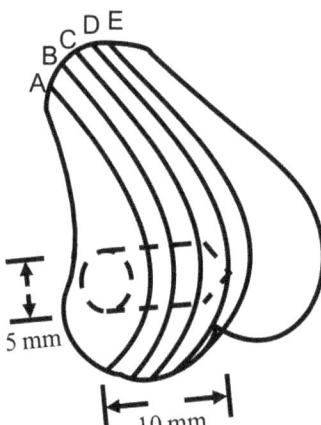

Figure 1. Illustration of 5 mm dia., 10 mm deep cylindrical-shaped CBS-400 implantation site. Lines A, B, C, D and E demonstrate the specific locations of the femur bone being sectioned for histological observation.

The animals were sacrificed at 3 days (3D), 4 weeks (4W), 8 weeks (8W) and 12 weeks (12W) post-operation. The animal numbers were 3 for the 3D group, 1 for the negative control and 6 for the other groups. The harvested samples were divided into two groups: hematoxylin and eosin (H&E)-stained decalcified samples for investigating inflammation/immune responses and toluidine blue

(TB)-stained non-decalcified samples, for investigating implant resorption behavior. TB staining was preferably used for the study of the implant resorption behavior, due to its better overall performance in the quantitative determination of implant resorption and new bone formation rates.

After sacrifice, the femur portions were excised immediately and the excess tissues were removed. The retrieved bone was sectioned using a low-speed diamond blade (IsoMet Blade, Buehler, IL, USA). For consistency, the samples for histological examination were taken from the section located between 2 mm and 6 mm from lateral to medial (section B or C shown in Figure 1). For regular H&E staining, the sectioned samples were fixed in 10% *w/v* neutral buffered formalin (NBF) (pH 7.0) for 3 days, decalcified, dehydrated, and embedded in paraffin. After embedding in paraffin, 5 μm-thick sections of the samples were obtained using a rotary microtome. The slices were then plated, deparaffinized, and stained with H&E. Non-decalcified samples were fixed in the same way, dehydrated in increasing grades of ethanol, then embedded in Buehler EpoxiCure 2 resin. The embedded samples were ground using silicon carbide grit paper, followed by wet-cloth polishing with 1.0, 0.3 and 0.05 μm Al_2O_3 powder, then stained with TB. The samples were examined using a light microscope (DM2500P, Leica, Wetzlar, Germany). For the TB-stained non-decalcified samples, in order to enhance resolution and at the same time to have an overall picture of implant resorption behavior, more than one hundred reflected-light micrographs at a magnification of ×100 were taken sequentially on each section being examined. These 100+ micrographs were then superimposed by Leica Application Suite software, to form a large composite picture covering the entire implant/bone cross-section.

To determine the residual implant ratio, the original 5 mm dia. bone void zone was carefully checked, defined, and artificially contoured on the superimposed composite image being examined. The implant residues appearing on these composite images were distinguished manually. The different regions of new bone, residual implant material, and epoxy that filled the empty space (such as bone marrow) were sketched out according to their respective colors, contrasts and histological features. An image-analyzing software, ImageJ (ver.1.51 g), was used for counting pixels of new bone, bone marrow and residual implant material. The area of bone in the healthy condyle of 12W rabbits without surgery was determined by the same method. The residual implant area ratios were determined according to the equations,

Residual implant ratio (%) = Area of residual implant/Area of original artificially-drawn bone void region

and

New bone formation ratio (%) = Area of new bone/Area of original artificially-drawn bone void region.

Due to the fact that the drawing of an original 5 mm dia. bone void circle could not be absolutely accurate, to assure the entire analyzed area was within the original bone void zone, another 4 mm dia. circle was drawn inside the 5 mm dia. circle. The residual implant and new bone ratios within this smaller circle were also measured and compared with the data from the 5 mm dia. circle.

2.8. Statistical Analysis

The quantitative data are presented as means ± standard deviations. Box plots were used to demonstrate the 5th percentile, 1st quartile (25th percentile), median (50th percentile), 3rd quartile (75th percentile) and 95th percentile, to show the data distribution. One-way ANOVA was used to evaluate the difference in pH values in Hank's solution and the masses of Ca, P and S elements released from CBS-400 in TRIS-HCl group between different immersion days. For the skin sensitization test, unpaired Student's *t*-test was used to analyze the differences in body weight change, lymph nodes weight and labeling index. Unpaired Student's *t*-test was also used to analyze the differences in the masses of Ca, P and S elements released from CBS-400 between TRIS-HCl group and citric acid group, residual implant and new bone ratios between the two circles, as well as the differences in bone area ratio between blank control and implantation groups. Linear regression and polynomial regression (quadratic regression) analyses were used to evaluate the change of residual implant ratio and new bone formation ratio versus implantation time, respectively. To ensure the calculated areas were always within the implantation region, the smaller, 4 mm-dia. circle was chosen for regression analysis. A statistical significance is determined at $p < 0.05$.

3. Results

3.1. Phase Identification and Chemical Composition

As shown in Figure 2A,B, CBS-400 is comprised majorly of HA (JCPDS# 09-0432) and calcium sulfate dihydrate (CSD, JCPDS# 33-0311), with relatively small amounts of tetracalcium phosphate (TTCP, JCPDS# 25-1137) and dicalcium phosphate anhydrate (DCPA, JCPDS# 09-0080). The semi-quantitatively-determined average phase contents of CBS-400 are 40.7 wt % HA, 28.6 wt % CSD, 12.5 wt % TTCP and 18.2 wt % DCPA. The SEM-EDS results shown in Figure 2C indicate that the major detected elements of CBS-400 are Ca (Kα = 3.7 and Kβ = 4.0 KeV), P (Kα = 2.0 KeV) and S (Kα = 2.3 KeV), O (Kα = 0.53 KeV) and C (Kα = 0.28 KeV). The normalized (by assuming Ca + P + S = 100%) Ca/P/S and Ca/P atomic ratios of CBS-400 are 54.61/39.21/6.18 and 1.39, respectively. Raw data of phase contents and SEM-EDS results are provided in Tables S1 and S2, respectively.

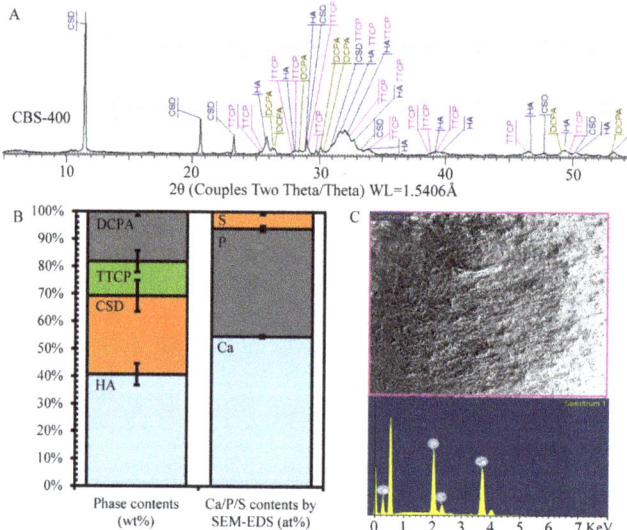

Figure 2. Phase identification and composition analysis of CBS-400. (**A**) XRD pattern; (**B**) XRD-determined phase contents and SEM/EDS-determined elemental contents; (**C**) SEM/EDS spectrum and analyzed region.

3.2. Granule Morphology and Dimensional Stability in Hank's Solution

The CBS-400 granules designed to have particle sizes ranging from about 400 µm to 1200 µm were sieved between 16 mesh (1.18 mm) and 40 mesh (0.43 mm). It should be remembered that, due to the irregular shape of the particles, certain elongated particles with average particle dimensions larger than the 16-mesh sieve (1.4 mm) could unavoidably pass through the sieve.

As shown in Figure 3A–C, the CBS-400 granules exhibit an irregular, rough and porous morphology with numerous macro pores. Figure 4A shows that the granules have sizes predominantly between 300 µm and 1700 µm. The measurements indicate that less than 5% of the particles are smaller than 300 µm, and about 90% of the particles are between 300 to 1300 µm, close to our design. Raw data of particle sizes are provided in Table S3.

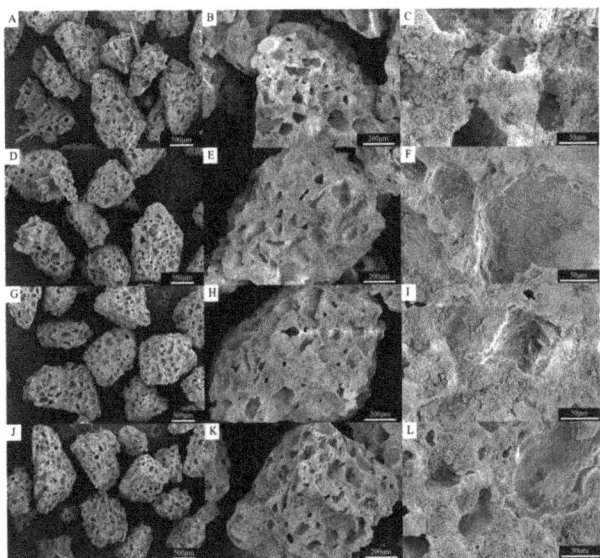

Figure 3. Surface morphology, shape and integrity of CBS-400 granules without immersion (**A**–**C**) and after immersion in Hank's solution for 1 (**D**–**F**), 3 (**G**–**I**) and 7 (**J**–**L**) days.

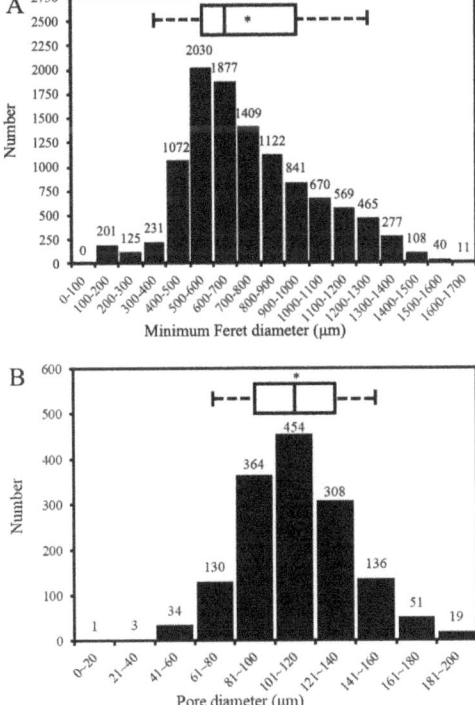

Figure 4. CBS-400 particle size distribution (**A**) and pore size distribution (**B**). Asterisk indicates the average number, and the two boundaries of the box plot above the bar graph define the 5th (left) and 95th (right) percentiles.

Figure 3D–L represents the morphology of CBS-400 granules after immersion in Hank's solution for 1, 3 and 7 days. After immersion for 7 days, the overall morphology, shape and integrity of the granules remain similar to that of the non-immersed samples, without showing an apparent collapse or disintegration under SEM.

3.3. Porosity Volume Fraction and Pore Size

The measured average porosity value of the CBS-400 granules, 77.8% v/v, was found to be higher than the theoretical volume fraction of the pore-former (~73% v/v), indicating that the pore-forming particles had been substantially rinsed out during processing. As shown in Figure 4B, the pore sizes of CBS-400 are majorly distributed between about 60 μm and 180 μm, with an average pore size of 112 μm. The distribution data indicate that 98–99% of the measured pores of the granules are between 40 μm and 180 μm; and about 90% between 60 μm and 160 μm. It should be noted that the SEM-measured pore sizes would always be somewhat smaller than the "actual" pore sizes, due to the fact that the largest dimensions of the pores being examined under SEM are probably "hidden" underneath the viewing surface. Raw data of pore sizes and porosity values are provided in Tables S4 and S5, respectively.

3.4. pH Value

The pH values of the daily refreshed Hank's solution, wherein CBS-400 was immersed for 1, 2, 3, 4, 5, 6, 7 and 14 days, are given in Figure 5. As seen in the figure, the average pH value of the 1-day solution drops to 5.28 (mildly acidic). With increasing immersion time, the pH value continues to increase significantly with immersion time. The pH value increases to 6.55 after 7 days, and 7.08 after 14 days. Raw data of pH values are provided in Table S6.

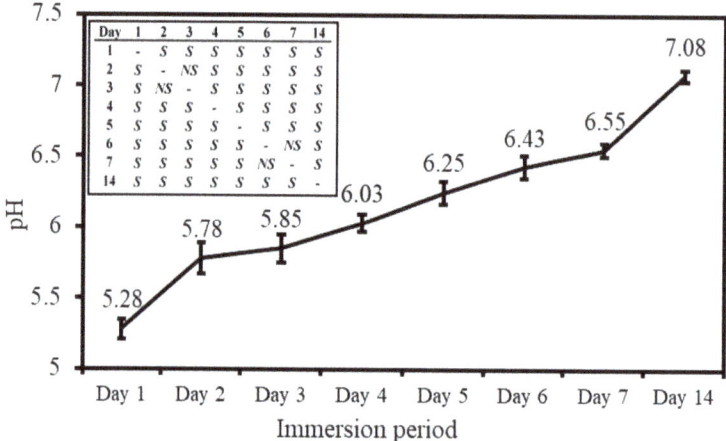

Figure 5. pH values in daily refreshed Hank's solution, wherein CBS-400 is immersed. The significance of differences between two pH values at different immersion days are shown in the upper left table; S and NS represent $p < 0.05$ and $p > 0.05$, respectively.

3.5. Solubility Tests in Buffered Citric Acid and TRIS-HCl Solutions

The concentrations and calculated masses of Ca, P and S released from CBS-400 in buffered citric acid solution and TRIS-HCl solution are shown in Figure 6. In the extreme solution test, the average concentrations of Ca, P, and S in buffered citric acid solution are 1208.1, 1285.4 and 570.9 ppm, respectively, and their respective calculated masses of Ca, P, and S elements released from CBS-400 are 24.2, 25.7 and 10.2 mg.

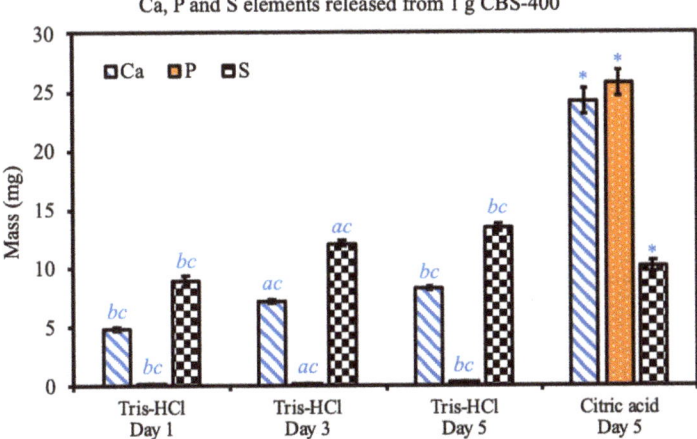

Figure 6. Ca, P and S elements released from 1 g CBS-400 in 20 mL TRIS-HCl at 37 °C for 1, 3, and 5 days; and in 20 mL buffered citric acid solution at 37 °C for 5 days. Symbols a, b and c indicate that the mean value has a significant difference ($p < 0.05$) compared to Day 1, Day 3 and Day 5 groups, respectively. Asterisks indicate that there are significant differences between TRIS-HCl and citric acid groups at day 5.

In the simulation solution test, after a 1-day immersion, the concentrations of Ca, P, and S in TRIS-HCl are 243.8, 4.6 and 443.6 ppm, respectively, and their respective calculated masses of Ca, P and S elements are 4.9, 0.09 and 8.9 mg. After a 3-day immersion, the concentrations of Ca, P and S in TRIS-HCl are 360.2, 9.4 and 603.3 ppm, respectively, and their respective calculated masses of Ca, P, and S elements are 7.2, 0.19 and 12.1 mg. After a 5-day immersion, the concentrations of Ca, P and S in TRIS-HCl are 413.3, 14.5 and 670.2 ppm, respectively, and their respective calculated masses of Ca, P and S elements are 8.3, 0.29 and 13.4 mg. At day 5, compared to the TRIS-HCl group, the citric acid group shows significant differences in Ca, P and S elements released from CBS-400. Within the TRIS-HCl group, the released Ca, P and S elements significantly increase with immersion time. Raw data of particle sizes are provided in Table S7.

3.6. Biocompatibility

3.6.1. Cytotoxicity

As shown in Figure 7, the NIH/3T3 cells treated with the extract of CBS-400 for 24 h depict adherent and extended morphology (Figure 7C), similar to that treated with medium (Figure 7B) or Al_2O_3 extract (Figure 7E), confirming that CBS-400 extract did not inhibit cell growth. On the other hand, when treated with 0.3% phenol, the cell morphology was seen to change to round-shaped (Figure 7D), indicating that the positive control had strong cytotoxicity. The average optical density (O.D.) values of the blank control, CBS-400 extract, negative control and positive control are 1.10, 0.90, 1.04 and 0.49, respectively, as shown in Figure 7F. The cell viability value of CBS-400 group is 84% (>70%), indicating that CBS-400 does not have cytotoxic potential. The cell viability values of the negative control (Al_2O_3) and positive control (0.3% phenol) are 94% and 44% of the blank control, respectively, indicating that the test was valid. Raw data of optical density values are provided in Table S8.

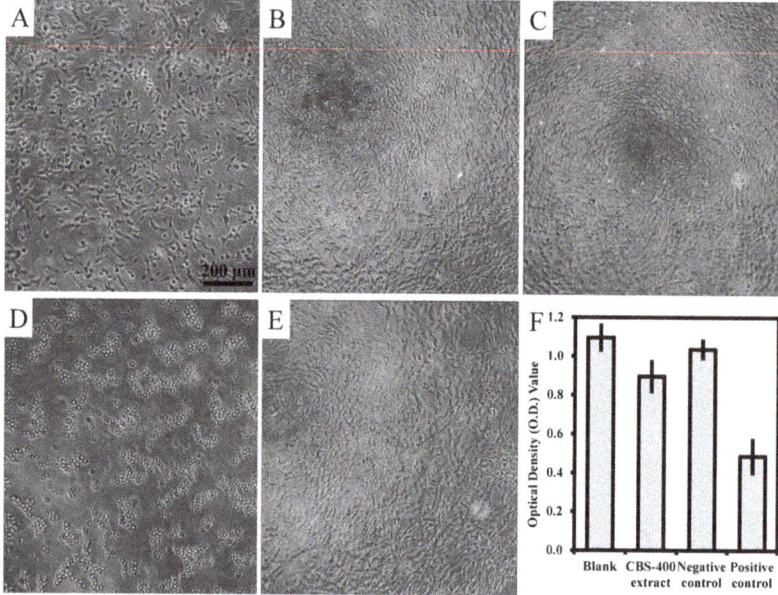

Figure 7. Morphologies of NIH/3T3 cells before treatment (**A**), treated with culture medium as blank (**B**), treated with CBS-400 extract (**C**), treated with 0.3% phenol as positive control (**D**), treated with Al_2O_3 extract as negative control (**E**), and their O.D. values (**F**).

3.6.2. Intracutaneous Reactivity Test

The observations of skin irritation at 0, 24, 48 and 72 h post intracutaneous injection showed no skin redness or swelling on either test sample-injected sites or vehicle-injected sites in any observation period, indicating that, at any observation time, the average reaction to CBS-400 extracts was the same as that of the blank control. As shown in Table 1, the final scores of both polar and non-polar CBS-400 extracts are zero (<1.0), indicating that CBS-400 does not give rise to any sign of skin irritation. The positive scores of the positive controls (3.9 for polar and 3.2 for non-polar) confirm the validity of the test (Table 2). Datasets of all scores according to Draize scale are provided in Tables S9 and S10.

Table 1. Scores of CBS-400 extract and blank control sites for intracutaneous (intradermal) reactions (Draize scale).

		Score of Polar Solvent Extraction Group					
		Animal No.1		Animal No.2		Animal No.3	
		Extract	Vehicle	Extract	Vehicle	Extract	Vehicle
Im.	Erythema	0 [1]	0	0	0	0	0
	Oedema	0	0	0	0	0	0
24 h	Erythema	0	0	0	0	0	0
	Oedema	0	0	0	0	0	0
48 h	Erythema	0	0	0	0	0	0
	Oedema	0	0	0	0	0	0
72 h	Erythema	0	0	0	0	0	0
	Oedema	0	0	0	0	0	0
Avg. of 3 time points (erythema + oedema)		0	0	0	0	0	0
Avg. of all extract sites (A) = 0; Avg. of all vehicle sites (B) = 0							
Final score of irritation of CBS-400 extract (A−B)						0	

Table 1. *Cont.*

		Score of Non-Polar Solvent Extraction Group					
		Animal No.1		Animal No.2		Animal No.3	
		Extract	Vehicle	Extract	Vehicle	Extract	Vehicle
Im.	Erythema	0	0	0	0	0	0
	Oedema	0	0	0	0	0	0
24 h	Erythema	0	0	0	0	0	0
	Oedema	0	0	0	0	0	0
48 h	Erythema	0	0	0	0	0	0
	Oedema	0	0	0	0	0	0
72 h	Erythema	0	0	0	0	0	0
	Oedema	0	0	0	0	0	0
Avg. of 3 time points (erythema + oedema)		0	0	0	0	0	0
Avg. of all extract sites (A) = 0; Avg. of all vehicle sites (B) = 0							
Final score of irritation of CBS-400 extract (A−B)					0		

[1] Average score of 5 injection sites. Im.: Observe after injection immediately; Avg.: average.

Table 2. Scores of positive control and blank control sites for intracutaneous (intradermal) reactions (Draize scale).

		Score of Polar Solvent Positive Control Group					
		Animal No.1		Animal No.2		Animal No.3	
		Formalin	Vehicle	Formalin	Vehicle	Formalin	Vehicle
Im.	Erythema	0 [1]	0	0	0	0	0
	Oedema	0	0	0	0	0	0
24 h	Erythema	0.4	0	2.0	0	1.0	0
	Oedema	0.6	0	3.0	0	2.0	0
48 h	Erythema	1.2	0	2.2	0	1.4	0
	Oedema	1.6	0	3.0	0	3.0	0
72 h	Erythema	1.8	0	2.8	0	2.4	0
	Oedema	1.2	0	2.4	0	3.0	0
Avg. of 3 time points (erythema + oedema)		2.3	0	5.1	0	4.3	0
Avg. of all positive control sites (A) = 3.9; Avg. of all vehicle sites (B) = 0							
Final score of irritation of formalin (A−B)					3.9		
		Score of Non-Polar Solvent Positive Control Group					
		Animal No.1		Animal No.2		Animal No.3	
		Histamine	Vehicle	Histamine	Vehicle	Histamine	Vehicle
Im.	Erythema	0	0	0	0	0	0
	Oedema	0	0	0	0	0	0
24 h	Erythema	0.6	0	0.6	0	1.8	0
	Oedema	1.4	0	0.6	0	2.4	0
48 h	Erythema	2.0	0	1.0	0	1.4	0
	Oedema	1.4	0	1.4	0	2.6	0
72 h	Erythema	2.2	0	1.4	0	1.8	0
	Oedema	1.4	0	1.6	0	3.0	0
Avg. of 3 time points (erythema + oedema)		3.0	0	2.2	0	4.3	0
Avg. of all positive control sites (A) = 3.2; Avg. of all vehicle sites (B) = 0							
Final score of irritation of histamine (A−B)					3.2		

[1] Average score of 5 injection sites. Im.: Observe after injection immediately; Avg.: average.

3.6.3. Skin Sensitization Test

As indicated in Table 3, the difference in body weight change between the CBS-400 extract group and the blank control group is insignificant ($p > 0.05$) within the whole experiment period. No significant difference ($p > 0.05$) in lymph nodes weight is seen between the mice applied with CBS-400 extract and the mice with the extract vehicle. Neither significant difference ($p > 0.05$) in labeling indices between the mice with the CBS-400 extract and the mice with the blank control is seen. The labeling index of the lymph nodes of the mice applied with the positive control group is significantly higher ($p < 0.05$) than that of the blank control. All these results conclude that CBS-400 with SI value of 0.6 (<1.6) is not considered to be a potential sensitizer. Raw data of body weights, lymph nodes weights and labeling index (O.D. values) are provided in Tables S11–S13.

Table 3. Body weights and weights of lymph nodes, BrdU labeling index and SI values of the mice for LLNA:BrdU-ELISA test.

		CBS-400 Extract Group	Vehicle Control (AOO) Group	Positive Control (DNCB) Group
Body weight (g)	day 0	20.2 ± 0.7	20.5 ± 1.2	19.6 ± 1.3
	day 1	20.9 ± 0.5	20.6 ± 1.1	20.5 ± 1.2
	day 2	20.3 ± 1.1	20.8 ± 1.2	20.2 ± 0.9
	day 3	20.3 ± 0.8	21.0 ± 1.0	20.2 ± 1.0
	day 4	20.3 ± 0.9	20.9 ± 1.2	19.5 ± 0.9
	day 5	20.0 ± 0.8	20.7 ± 1.0	19.8 ± 0.4
Lymph nodes weight (mg)		5.1 ± 1.3	4.7 ± 0.7	22.0 ± 2.5 *
Labeling index		0.100 ± 0.056	0.167 ± 0.026	0.431 ± 0.050 *
Stimulation Index		0.60	1	2.58

*The value has significant difference compared to the vehicle control group ($p < 0.05$).

3.7. Animal Implantation Study and Resorption of CBS-400 Implant

Typical H&E-stained histological micrographs of CBS-400 implantation group at 4W, 8W and 12W post-operation are presented in Figure 8. At 4W, new bone is observed to extensively grow into the macro-pores of the implanted granules, while the struts of the porous granules are fragmented into pieces. At this stage, abundant new bone formation-involved osteoblasts and few bone-resorption-involved osteoclasts are seen throughout the sample. It is also observed that osteoclasts can penetrate into the intra-strut micro-pores of the implanted porous granules (Figure 8B). At 8W, extensive bone remodeling is ongoing, evidenced by the abundant presence of both osteoblasts and osteoclasts adjacent to the new bone and residual implant. At this stage, the new bone is remodeled to a trabecula-liked structure with more lamella structures, cement lines and lining cells. At 12W, the morphology of the new bone has become more like native bone, wherein fewer osteoblasts and osteoclasts are observed.

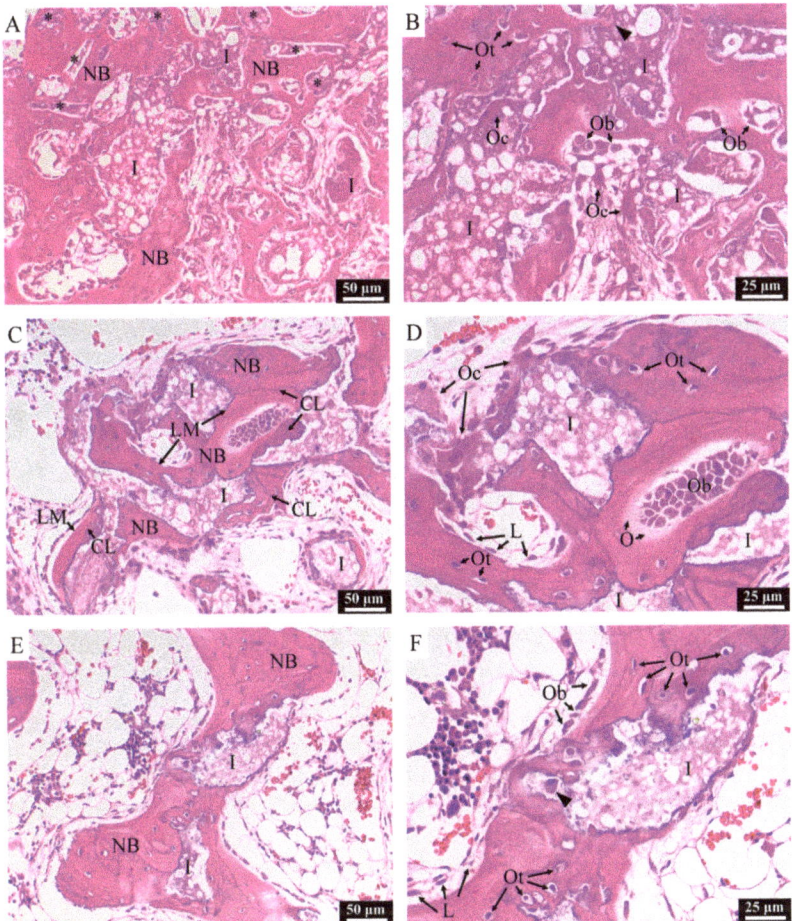

Figure 8. H&E-stained histological images of NZW rabbit femur condyle of CBS-400-implanted group at 4W (**A**,**B**), 8W (**C**,**D**) and 12W (**E**,**F**) post-operation. I: residual implant; NB: new bone; LM: lamellar matrix; CL: cement line; Ob: osteoblast; Oc: osteoclast; Ot: osteocyte; O: osteoid; L: lining cell. Asterisks indicate implant residues embedded in surrounding new bone. Arrow heads indicate new bone grown in implant micro-pores.

Typical TB-stained histological micrographs of the implantation group at 3D, 4W, 8W and 12W post-operation are presented in Figure 9. At 3D, the morphology of the porous implant is well maintained within the defect, and no bone regeneration is evidenced, due to its short implantation time. At 4W, in general, excellent bonding between the host bone and the implant is seen substantially without interposition of fibrous tissue. At this stage, numerous residual CBS-400 granules are observed to embed in the surrounding new bone which grow on the surfaces of the granules, as well as penetrate into the defect site, connecting to each other to form a dense new bone network with substantially random orientations. This new bone network becomes more porous at 8W, accompanied with a thickening of the trabecular bone and a decrease in the number and size of implant residues. With the bone remodeling process continuing at 12W, bone and implant residues both decrease, while many CBS-400 implant granules become too small to distinguish from the surrounding bone.

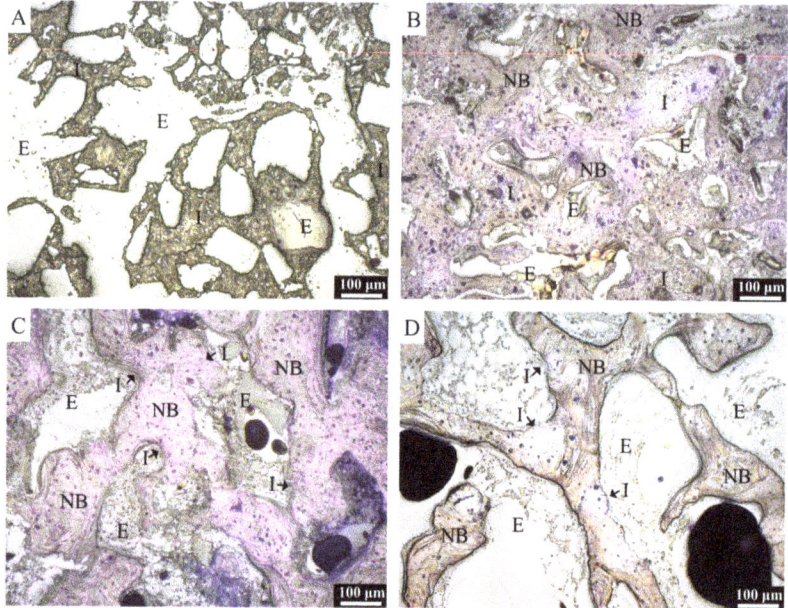

Figure 9. TB-stained histological images of NZW rabbit femur condyle in CBS-400-implanted group at 3D (**A**), 4W (**B**), 8W (**C**) and 12W (**D**) post-operation. I: residual implant; NB: new bone; E: epoxy.

According to Materials and Methods 2.7, a 5 mm dia. circle (TZ) and its radially-reduced 4 mm dia. circle (IZ) were drawn on TB-stained superimposed composite images, and the measurements of residual implant and new bone within these two circles were compared. As shown in Figure 10, the number and size of implant residues both decrease with increasing implantation time. Comparison between TZ and IZ data indicates that the number and size of implant residues are not markedly different between these two zones. Statistical analyses of the residual implant and new bone obtained from TZ and IZ are presented in Figure 11. In the TZ zone, the average area ratio of residual CBS-400 decreases from 25.4% at 3D to 11.1% at 4W, 5.9% at 8W, and 3.8% at 12W post-operation. In other words, about 85% of the implant has readily been resorbed after 12 weeks. In the IZ zone, the average area ratio of residual CBS-400 decreases from 32.3% at 3D to 15.5% at 4W, 8.3% at 8W, and 5.1% at 12W, meaning that about 84% of the implant has been resorbed after 12 weeks. On the other hand, the average area ratio of new bone in the TZ zone decreases from 43.3% at 4W to 39.4% at 8W and 30.0% at 12W. The average area ratio of new bone in IZ zone decreases from 42.4% at 4W to 37.3% at 8W, and 30.4% at 12W. The new bone area at 3D was not measured, due to its substantial absence at this early stage. Datasets of residual implant ratio, new bone formation ratio, native bone area ratio at different time points within TZ and IZ are provided in Table S14.

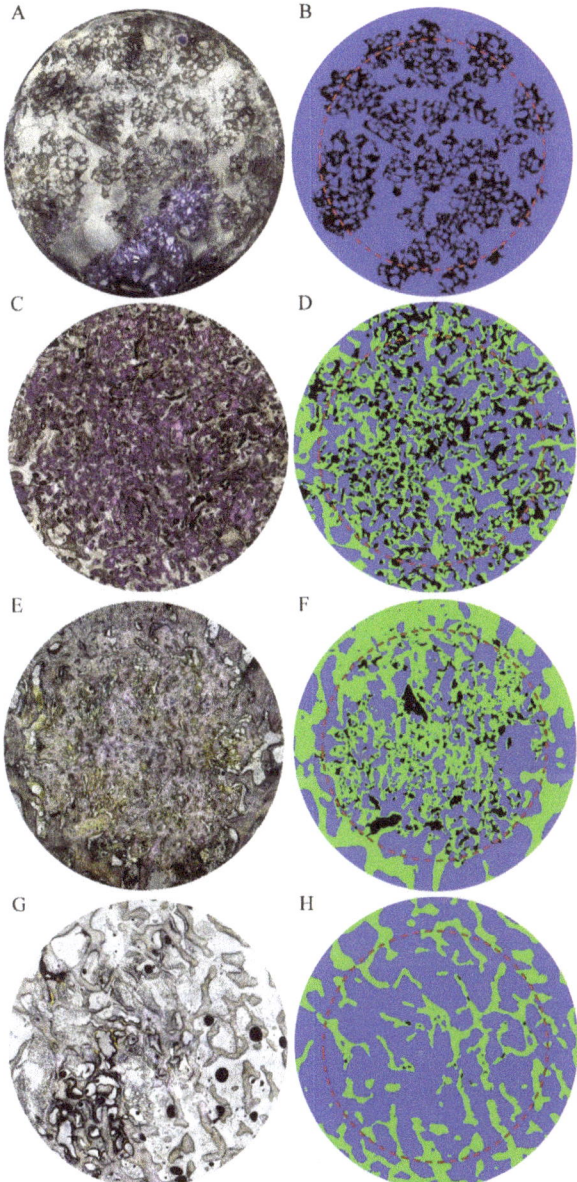

Figure 10. Overall, 5 mm dia. TB-stained superimposed composite images of NZW rabbit femur condyle in CBS-400-implanted group at 3D (**A**), 4W (**C**), 8W (**E**) and 12W (**G**) post-operation and their corresponding manually drawn images (**B**,**D**,**F** and **H**), illustrating residual CBS-400 (black), new bone (green) and marrow space (blue). Moreover, 4 mm dia. inner zones are outlined by red dash circles.

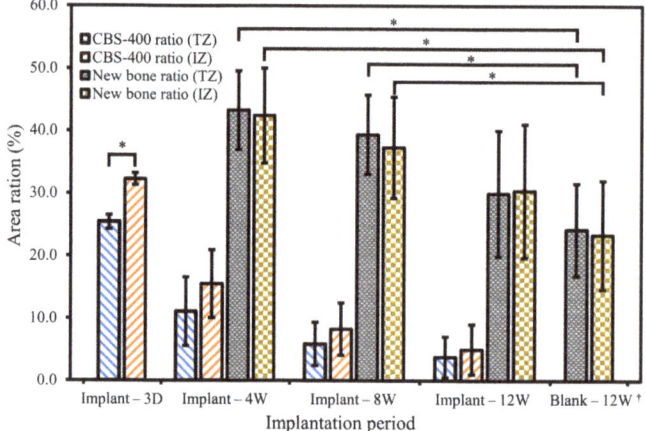

Figure 11. Area ratios of residual CBS-400 implant and new bone at 3D, 4W, 8W and 12W post-operation. The area ratios of native bone in blank group at 12W are also included. Data from TZ and IZ zones are both presented for comparison. * Significant ($p < 0.05$). † Native bone area ratio instead of new bone area ratio.

Linear regression of the residual implant ratio and quadratic regression of the new bone formation ratio versus logarithm of implantation period were performed, and the results are shown in Figure 12. Both regression analyses exhibit $p < 0.001$ for f-test, and all the regression coefficients and constants exhibit $p < 0.001$ for t-test. The correlation coefficients (R^2) are 0.766 and 0.834, for bone formation ratio and residual implant ratio, respectively. The powers (1–β error probability) of both regressions are 99%. The correlation coefficients (about 0.8) seem to be high enough [22] to suggest that these analyses are convincing, and that the percentages of residual CBS-400 and new bone are quite predictable. According to the model, the expected implantation period for CBS-400 to be totally resorbed would be about 23W, and the bone area ratio would recover to the level of the blank at about 16W post-operation.

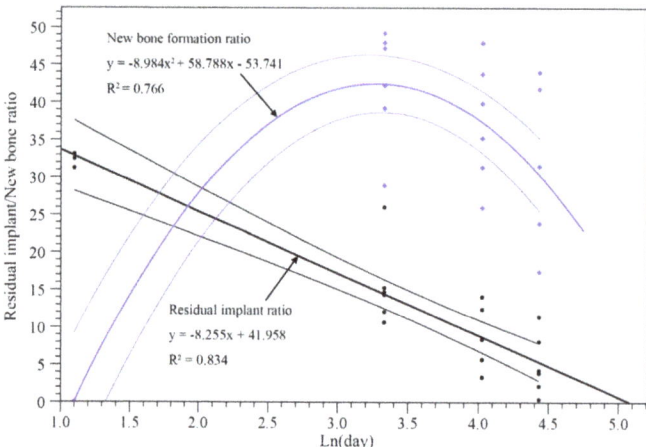

Figure 12. Residual implant and new bone ratios. Linear regression (black lines) and quadratic regression (blue lines) are applied to new bone ratio and residual implant ratio, respectively. The regression lines and their 95% confidence intervals are drawn.

4. Discussion

All the major phases of the CBS-400 granules, including low-crystalline HA, CSD, DCPA and TTCP, are well-recognized as highly biocompatible and resorbable materials. Compared to the stoichiometric HA, the structure of the low-crystalline HA in CBS-400 is more toward the natural bone tissue [23,24] and could resorb faster in vivo [24,25]. TTCP and DCPA would eventually dissolve and form HA in a neutral pH environment. Since low-crystalline HA could have a wide Ca/P range from 1.2 to 2.2 [25,26], calculations of the element contents of CBS-400 by XRD-demonstrated phase contents based on a stoichiometric HA (Ca/P = 1.67) could lead to inaccurate results. Instead, the more time-consuming SEM/EDS technique was used for the determination of Ca/P and Ca/P/S ratios of the present CBS-400.

The CBS-400 granules were designed to have particle sizes ranging from about 400 µm to 1200 µm exhibit an irregular, rough and porous morphology, with numerous macro-pores and micro-pores. According to Hirschhorn et al. [27], the preferred minimum particle size would be about 400 µm for new bone formation. The selection of 1200 µm as an upper limit is due to the practical consideration that the particles may be delivered into bone cavities using a minimally invasive delivery tool, wherein particles that are too large would be difficult to deliver. After immersion in Hank's solution for 7 days, the overall morphology, shape and integrity of the granules remain similar to that of the non-immersed samples, which is a good indication that the highly porous granules would probably survive the early implantation stage without a premature disintegration of the structure.

The pore sizes of CBS-400 are majorly distributed between about 60 µm and 180 µm. According to the animal studies of Hulbert et al. [28], Klawitter et al. [29] and Galois and Mainard [30], the minimum pore size for facilitating bone ingrowth would be about 75–100 µm. The far majority of the pores of CBS-400 have sizes larger than this, indicating that the porous morphology of the present granules would be able to facilitate new bone ingrowth. Furthermore, according to Bohner and Baumgart [31], the pore size between about 100 µm and 200 µm would be optimal to result in a proper specific surface area (SSA), which may help new bone ingrowth by increasing protein (such as vitronectin and fibronectin) adhesion to facilitate osteoblastic cell adhesion and spreading [32].

The micro-pores within the struts of CBS-400 might also promote new bone generation. The study of Coathup et al. [33] indicated that bone formation was evident within the strut, and greater bone formation was seen in scaffolds with increased strut porosity. In the present study, new bone formation was found within the struts of the implanted granules, indicating that the strut porosity could partially contribute to the observed fast resorption and new bone formation in CBS-400.

The early drop in pH of the Hank's solution is considered to be a result of the dissolution of calcium sulfate in the presence of calcium phosphate to form octacalcium phosphate (OCP) and HA [34]. A number of studies have discussed the pH effects on bone healing/regeneration, but their results are not consistent. Walsh et al. [35] suggested that the local acidity and subsequent demineralization of adjacent bone and release of matrix-bound BMPs lead to a stimulatory effect on bone regeneration. Arnett [36] reported that acidosis exerts a reciprocal inhibitory effect on the mineralization of bone matrix by cultured osteoblasts. Shen et al. [37] found that, in a slightly basic microenvironment, the proliferation and alkaline phosphatase (ALP) activity of osteoblasts could be enhanced.

The large difference in 5-day P release between the extreme solution test and the simulation solution test (25 mg vs. 0.29 mg) could be explained by the high sensitivity of P release to the pH value of the solution. The pH values of the solutions used for the extreme solution and simulation solution tests are 3.0 and 7.4, respectively. According to Chow [38], the dissolution rates of calcium phosphates can be hundreds of times larger when the solution changes from neutral to acidic. The dissolution rates of calcium sulfates, on the other hand, are rather insensitive to the pH value of the solution [39].

The dissolution/solubility data may be helpful in assessing the safety of CBS-400 under clinical conditions. First of all, calcium, phosphorus and sulfur are all essential elements which are required at levels greater than 100 mg/day by adults [40]. Calcium ions are a major component for the FDA-approved infusions such as calcium gluconate and calcium chloride, to treat acute symptomatic hypocalcemia. In an early study, [41] used calcium infusions in the diagnosis of metabolic bone disease.

An infusion of calcium gluconate (15 mg of calcium per kg of body weight) was given in 500 mL of physiological saline solution to 98 patients for 4 h. The results indicated that none of the patients showed any signs of calcium intoxication, and the serum calcium level had almost invariably returned to its original value after 20 h. The dosage used in this study was much higher than the Ca released from a 40 mL (10 g) grafted CBS-400 (the majority of clinical cases require less than 10 mL) for an average 70 kg person a day. In a relatively recent study of [42], intravenous calcium gluconate was given for 3 consecutive days to female patients undergoing assisted reproduction technique (ART) cycles for the prevention of ovarian hyperstimulation syndrome (OHSS). Such side effects as allergic reactions, anaphylaxis and symptoms of hypercalcemia were not observed in the study group. Again, the dosages of calcium used in these studies were twice higher than the mass of Ca released from 40 mL (10 g) grafted CBS-400 for an average person a day.

Most of the phosphorus in blood exists as phosphates or esters. It is known that phosphate is required for the generation of bony tissue and functions in the metabolism of glucose and lipids. The normal phosphorus levels in the blood of adults are 2.7–4.5 mg/dL [43], or 135–225 mg if the total volume of blood is 5 L. The released mass of phosphorus from CBS-400 for one person per day would be too small (<2 mg) to cause any significant fluctuation of normal phosphorus level. Even if all the P released from CBS-400 is taken by blood into the cardiovascular system, it is still far lower than the normal phosphorus level in the blood.

Sulfate ions exist in the magnesium sulfate infusion for replacement therapy in magnesium deficiency, especially in acute hypomagnesemia accompanied by signs of tetany similar to those observed in hypocalcemia. Van Norden et al. [44] used $MgSO_4$ infusion to treat aneurysmal subarachnoid hemorrhage (SAH). With an intravenous dosage of 64 mmol magnesium sulfate a day for 14 days, no severe side effects were observed in the vast majority of patients with SAH. Magnesium sulfate injection is also used for the prevention and control of seizures in pre-eclampsia and eclampsia. The mass of sulfur element released from 40 mL (10 g) CBS-400 for an average person per day is about 90 mg, or 270 mg of sulfate ion, a level much lower than the dosages used for these indications.

Although the released calcium and sulfate ions could lower the surrounding pH value when immersed in aqueous solution, CBS-400 demonstrates good biocompatibility features, confirmed by the experimental results of cytotoxicity, intracutaneous reactivity and skin sensitization tests. The adherent and extended morphology of NIH/3T3 cells along with good cell viability found in the cytotoxicity test indicates that CBS-400 does not have a cytotoxic potential. In addition, CBS-400 does not induce skin irritation or skin sensitization that can stimulate the inflammation system and cause lymphocyte proliferation.

In the international standard ISO 10993-6, rabbit, femur bone and cylindrical-shaped defect are recommended as one of the appropriate species, implant site and defect shape respectively for the bone implantation test. Furthermore, according to the FDA 510(k) Special Guidance for Resorbable Calcium Salt Bone Void Filler Device, a critical size defect should be used. According to the studies of Prieto et al. [45] and Chen et al. [46], using the rabbit femur condyle model, the critical bone defect size would be about 5 mm in diameter × 8 mm in depth. The bone defect size used in the present study is 5 mm in diameter × 10 mm in depth. The large empty space appearing in the negative control (Figure S1 in Supplementary Materials) clearly indicates that the surgically created defect is large enough so that it cannot be repaired by normal bone healing mechanisms, consistent with the findings of Prieto et al. [45] and Chen et al. [46].

Except at the very early stage (3D), the residual implant and new bone ratios obtained from the TZ and IZ are not statistically different, indicating that implant resorption and new bone formation proceed quite uniformly throughout the entire implantation site, due to the highly porous feature of the implant. The similarity between the TZ and IZ data also indicates that the drawing of the original 5 mm circles was performed quite reliably. The somewhat lower residual implant ratio in TZ than IZ at 3D is not unexpected, due to the unavoidable presence of a thin gap between the implant and the void wall.

According to Bohner et al. [47], the resorption of CSD is mainly by physicochemical processes, specifically by dissolution. On the other hand, the resorption processes of calcium phosphates are apparently more complicated, which involve dissolution, phase transformation, osteoclast-mediated resorption and fragmentation, with particles being phagocytized by macrophages or engulfed by giant cells [48]. Once osteoclasts attach an implant surface, an intimate structure between osteoclasts and bone is formed, and a 'sealing zone' is developed. Within the sealing zone, demineralization of the implant material involving acidification of the isolated extracellular microenvironment takes place. Through a series of ion transport events, osteoclasts secrete HCl to form a low pH (~4.5), resorptive microenvironment to dissolve calcium phosphates. According to Xia and Triffitt [49], macrophages respond to small particles (typically <10 μm) by internalization, via phagocytosis and intracellular digestion. For larger particles, the macrophages fuse together, forming giant cells engulfing and digesting the particles, or through release of enzymes and/or lowering in pH for bulk digestion.

It is known that the resorption rate of HA is highly dependent on the crystallinity of HA. Klein et al. [50] reported that their highly crystalline HA did not show any degradation even at 9 months post-operation in a rabbit tibia model. However, a poor or nano-crystalline HA was found to be largely resorbed within 12 weeks post-operation in a beagle alveolar bone model [51] or NZW rabbit radial defect model [52].

CSD generally has a much higher degradation rate than HA. Liu et al. [53] investigated CSD containing 10% nano-crystalline HA in a NZW rabbit femur condyle model, and found that over 50% of the implant was resorbed after 4 weeks and almost totally resorbed within 8 weeks post-operation, yet with poor bone regeneration. In a relatively small NZW rabbit femur defect model (Ø 3.1 × 4 mm), Sheikh et al. [54] found that 40–50% DCPA resorbed with considerable hard callus formation within 4 weeks post-operation, and above 90% DCPA resorbed with apparent trabeculae formation after 12 weeks. Among the very few studies on the resorption rate of monolithic TTCP, it was observed that about 40% and 60% monolithic TTCP were resorbed in NZW rabbit condyle, at 4 and 24 weeks post-operation, respectively [55].

As mentioned earlier, the major phases of CBS-400 are low-crystalline HA and CSD with minor phases of DCPA and TTCP. According to the literature, a high resorption rate would be expected in CBS-400 during the early healing stage, whereas CSD in the material would quickly dissolve, followed by a relatively low resorption stage as calcium phosphates dominate. To a large extent, the observed resorption behavior of CBS-400 with the present animal model is consistent with the findings of the literature.

Little et al. [56] proposed that the bone healing process may be briefly divided into anabolic (new bone formation) and catabolic (bone resorption) responses, wherein the anabolic response dominates at the early stage, whereas the catabolic response dominates at the later stage. The combination of the two effects would logically lead to something like a Gaussian distribution in bone volume-bone healing time profile. In the present study, the maximum bone volume (anabolic response) is observed in the 4W samples, much faster than other studies using similar animal models [57,58].

Studies [59–61] using different calcium phosphate-based bone substitutes (beta-TCP, biphasic calcium phosphate (60% HA/40% TCP), deproteinized bovine bone, etc.) implanted in NZW rabbit femur condyle indicate that, at 6–8 weeks post implantation, the new bone ratios are 20–30%, while the residual implant ratios are 45–60%. Compared to these studies, the residual CBS-400 implant ratios (11–15% at 4W and 6–8% at 8W) are much lower, while the new bone ratios (42–43% at 4 weeks and 37–39% at 8 weeks) are much higher. The fast bone regeneration can help stabilize implant fixation, which is one key requirement to achieving fast and effective osteointegration [62]. The proper intergranular and intragranular spaces built in the CBS-400 implant are critical to the fast osteogenesis and resorption processes [63].

The regression line of residual implant versus natural log of implantation period indicates a decreasing resorption rate in vivo. This decrease may be attributed to the less surface area of CBS-400 granules exposed to cells with implantation time. The high osteoconductivity and fast bone formation observed in CBS-400 at 4W post-operation could be partly attributed to the numerous macro-pores, as well as the interconnectivity of these pores. In addition, the osteoclasts entering the struts through micro-pores could accelerate the cell-mediated resorption, thereby increasing interconnectivity for bone regeneration at an early healing stage.

From a biochemical point of view, calcium sulfate and HA, the two major components of CBS-400, play a major role in bone healing. According to Ricci et al. [64], calcium sulfate implanted in a bone defect can rapidly dissolve in body fluid and release calcium ions, which react with phosphate ions to form a layer of calcium phosphate, forming an osteoblast-friendly environment. According to Walsh et al. [35], the acidity caused by the dissolution of calcium sulfate and the precipitation of calcium phosphate may demineralize adjacent bone and release matrix bound BMPs, resulting in a stimulatory effect on bone regeneration. More recently, Aquino-Martínez et al. [65] reported that calcium sulfate could promote in vitro mesenchymal stem cell (MSC) migration and bone regeneration in vivo, by attracting the host's osteoprogenitors into the implanted cell-free scaffold. Although calcium sulfate may dissolve too fast to allow bone to deposit on it, causing a non-continuous new bone formation in vivo [66], the porous CBS-400 maintains its structural integrity by the HA-dominated "scaffold" when its calcium sulfate is dissolved.

HA is considered to have the ability to directly bond to bone, possibly by natural bone cementing mechanisms [67]. This intimate contact has been observed throughout the entire implantation period in the present study. Bagambisa et al. [68] pointed out that, when the surface of HA is exposed to an aqueous environment, a process of elution and concomitant reprecipitation brings forth the formation of a transformational layer composed of spherocrystallites of colloidal dimensions. This elution/re-precipitation layer resulting from the extensive degradation/recrystallization events leads to a wide bonding layer for direct bone tissue deposition. Furthermore, the study of Chen et al. [69] indicated that calcium phosphate could attract and promote the differentiation of MSCs toward vascular endothelial cells to help the crucial revascularization process, and toward osteoblasts to regenerate new bone [70]. It seems that an appropriate combination of calcium sulfate and HA, such as CBS-400, may inherit the advantageous features of both components in implant resorption and bone regeneration processes. Furthermore, the soluble phases and highly porous morphology of CBS-400 could accelerate the process of biological apatite deposition and enhance bone-forming activity by providing more sites for cellular interaction. Other factors, such as pH and extracellular calcium ion concentration, could possibly also influence the cell-mediated resorption behavior of CBS-400 [71].

As a final remark, the H&E and TB-stained histological observations of the implanted CBS-400 and its surrounding bone morphology at 3D, 4W, 8W, and 12W post-implantation indicate that the implanted CBS-400 granules are intimately bonded to the surrounding new bone at all times. The measurements of residual implant material and the newly formed cancellous bone indicate that CBS-400 is not merely a degradable bone substitute. The resorption of CBS-400 appears simultaneously replaced by the formation of new cancellous bone. During the implantation time from 3D to 12W, the average residual CBS-400 ratio decreases with time from 25.4% down to 5.1%, meaning that about 85% of the implant has readily been resorbed after 12W. Within the same time frame, the newly formed cancellous bone ratio quickly increases to 42.4% at 4W (indicating a speedy new bone formation process at the early stage of implantation), followed by a bone remodeling process toward normal cancellous bone, wherein the new cancellous bone ratio gradually tapers down to 30.4% at 12W (the cancellous bone ratio in the normal condyle is 23.3%). These data show that the bone remodeling process toward normal cancellous bone is very close to completion at 12W. The detailed mechanisms behind the speedy new bone formation process at the crucial, early stage of implantation are apparently worth further investigation, including such topics as the early inflammation process, cells recruited for new tissue regeneration, bone mineralization process and neovascularization.

5. Conclusions

1. XRD patterns show that CBS-400 is comprised majorly of HA and CSD, with relatively small amounts of TTCP and DCPA. The SEM/EDS-determined Ca/P/S and Ca/P atomic ratios of the material are 54.61/39.21/6.18 and 1.39, respectively.
2. Structural integrity test results show that, after immersion for 7 days, the overall morphology, shape and integrity of the Hank's solution-immersed CBS-400 granules remain similar to that of non-immersed samples, without showing apparent collapse or disintegration under SEM.
3. The average pH value of the Hank's solution wherein CBS-400 is immersed for 1 day drops to 5.28. With immersion time, the pH value continues to increase to 6.55 after 7 days, and 7.08 after 14 days.
4. In the extreme solution test, the average concentrations of Ca, P, and S in buffered citric acid solution are respectively 1208.1, 1285.4 and 570.9 ppm. In the simulation solution test, the concentrations of Ca, P, and S in TRIS-HCl are respectively 243.8, 4.6 and 443.6 ppm after immersion for 1 day; and respectively, 413.3, 14.5 and 670.2 ppm after 5 days.
5. Cytotoxicity, intracutaneous reactivity and skin sensitization tests demonstrate the good biocompatibility features of CBS-400.
6. The rabbit implantation results indicate that implanted CBS-400 granules are intimately bonded to the surrounding new bone at all times. The measurements of residual implant material and newly-formed cancellous bone reveal that the resorption of the implant is simultaneously replaced by the formation of new cancellous bone. During the implantation time from 3D to 12W, the average residual CBS-400 ratio decreases with time from 25.4% to 5.1%, meaning that about 85% of the implant has been resorbed after 12W. Within the same time frame, the newly formed cancellous bone ratio quickly increases to 42.4% at 4W, followed by a bone remodeling process toward normal cancellous bone, wherein the new cancellous bone ratio gradually tapers down to 30.4% after 12W.

Supplementary Materials: The following are available online at http://www.mdpi.com/1996-1944/13/16/3458/s1, Figure S1: TB-stained superimposed composite image of NZW rabbit femur condyle in negative control at 12 weeks post-operation. Table S1: Semi-quantitatively-determined phase contents (wt %) of CBS-400, Table S2: Normalized compositions (at%) of CBS-400 determined by SEM/EDS, Table S3: Minimum Feret diameter (μm) of CBS-400 (unit: number of particles), Table S4: Pore size measurement data (number of pores in different size ranges) of CBS-400. Table S5: Porosity values (vol%) of CBS-400 granular samples. Table S6: pH values of the daily-refreshed Hanks' solution wherein CBS-400 were immersed for 1, 2, 3, 4, 5, 6, 7 and 14 days. Table S7: Concentrations of Ca, P, and S in the Tris-HCl and buffered citric acid solutions wherein CBS-400 were immersed, Table S8: Optical density of the medium, CBS-400, Al_2O_3 and phenol groups in cytotoxicity test, Table S9: Scores of CBS-400 extract and blank control sites for intracutaneous (intradermal) reactions (Draize scale), Table S10: Scores of positive control and blank control sites for intracutaneous (intradermal) reactions (Draize scale), Table S11: Body weights (g) of the mice applied with CBS-400 extract, vehicle and positive control, Table S12: Weights (mg) of the lymph nodes of the mice applied with CBS-400 extract, vehicle and positive control, Table S13: Absorbance (O.D. value) of the lymph node suspension of the mice applied with CBS-400 extract, vehicle and positive control at 450 and 630 nm, Table S14: Area ratios of residual CBS-400 implant and new bone at 3D, 4W, 8W and 12W post-operation for implantation group and area ratio of native bone at 12W for blank.

Author Contributions: Conceptualization, J.-H.C.L. and C.-P.J.; methodology, J.-H.C.L., C.-P.J. and J.-W.L.; validation, C.-P.J. and B.-C.Y.; formal analysis, B.-C.Y.; investigation, B.-C.Y. and J.-W.L.; resources, J.-H.C.L. and C.-P.J.; data curation, B.-C.Y.; writing—original draft preparation, B.-C.Y.; writing—review and editing, C.-P.J. and B.-C.Y.; visualization, B.-C.Y.; supervision, J.-H.C.L. and C.-P.J.; project administration, J.-H.C.L.; funding acquisition, J.-H.C.L., C.-P.J. and B.-C.Y. All authors have read and agreed to the published version of the manuscript.

Funding: This research was funded by Southern Taiwan Science Park Bureau Smart Biotech Medical Cluster, grant number CY-05-08-38-107.

Conflicts of Interest: J.-H.C.L. and C.-P.J. initiated the government-funded research which was later transferred to JMD. NCKU is holding the related patent rights. B.-C.Y., a graduate student, J.-H.C.L. and C.-P.J. continue to assist JMD in research.

References

1. Weiser, T.G.; Haynes, A.B.; Molina, G.; Lipsitz, S.R.; Esquivel, M.M.; Uribe-Leitz, T.; Fu, R.; Azad, T.; Chao, T.E.; Berry, W.R.; et al. Size and distribution of the global volume of surgery in 2012. *Bull. World Health Organ.* **2016**, *94*, 201F–209F. [CrossRef] [PubMed]
2. Hall, M.J.; Schwartzman, A.; Zhang, J.; Liu, X. Ambulatory surgery data from hospitals and ambulatory surgery centers: United States, 2010. *Natl. Health Stat. Rep.* **2017**, *102*, 1–15.
3. Deev, R.V.; Drobyshev, A.Y.; Bozo, I.Y.; Isaev, A.A. Ordinary and activated bone grafts: Applied classification and the main features. *BioMed Res. Int.* **2015**, *2015*, 365050. [CrossRef] [PubMed]
4. Straumann, Annual Report 2018, Pushing Boundaries. Available online: https://www.straumann.com/content/dam/media-center/group/en/documents/annual-report/2018/2018_Straumann_Annual_report.pdf (accessed on 28 March 2019).
5. Cha, H.S.; Kim, J.W.; Hwang, J.H.; Ahn, K.M. Frequency of bone graft in implant surgery. *Maxillofac. Plast. Reconstr. Surg.* **2016**, *38*, 19. [CrossRef]
6. Egermann, M.; Schneider, E.; Evans, C.H.; Baltzer, A.W. The potential of gene therapy for fracture healing in osteoporosis. *Osteoporos. Int.* **2005**, *16*, S120–S128. [CrossRef]
7. Almaiman, M.; Al-Bargi, H.H.; Manson, P. Complication of anterior iliac bone graft harvesting in 372 adult patients from may 2006 to may 2011 and a literature review. *Craniomaxillofac. Trauma Reconstr.* **2013**, *6*, 257–266. [CrossRef]
8. Truumees, E.; Herkowitz, H. Alternatives to autologous bone harvest in spine surgery. *Univ. Pa. Orthop. J.* **1999**, *12*, 77–88.
9. Hinsenkamp, M.; Muylle, L.; Eastlund, T.; Fehily, D.; Noel, L.; Strong, D.M. Adverse reactions and events related to musculoskeletal allografts: Reviewed by the World Health Organisation Project NOTIFY. *Int. Orthop.* **2012**, *36*, 633–641. [CrossRef]
10. Kim, Y.; Nowzari, H.; Rich, S.K. Risk of prion disease transmission through bovine-derived bone substitutes: A systematic review. *Clin. Implant. Dent. Relat. Res.* **2013**, *15*, 645–653. [CrossRef]
11. Schwartz, Z.; Weesner, T.; van Dijk, S.; Cochran, D.L.; Mellonig, J.T.; Lohmann, C.H.; Carnes, D.L.; Goldstein, M.; Dean, D.D.; Boyan, B.D. Ability of deproteinized cancellous bovine bone to induce new bone formation. *J. Periodontol.* **2000**, *71*, 1258–1269. [CrossRef]
12. Murugan, R.; Rao, K.P.; Kumar, T.S.S. Heat-deproteinated xenogeneic bone from slaughterhouse waste: Physico-chemical properties. *Bull. Mater. Sci.* **2003**, *26*, 523–528. [CrossRef]
13. Kim, Y.; Rodriguez, A.E.; Nowzari, H. The risk of prion infection through bovine grafting materials. *Clin. Implant. Dent. Relat. Res.* **2016**, *18*, 1095–1102. [CrossRef] [PubMed]
14. Santin, M. 14—Bone tissue engineering. In *Bone Repair Biomaterials*; Planell, J.A., Best, S.M., Lacroix, D., Merolli, A., Eds.; Woodhead Publishing: Cambridge, UK, 2009; pp. 378–422. [CrossRef]
15. Blokhuis, T.J. 4—Bioresorbable bone graft substitutes. In *Bone Substitute Biomaterials*; Mallick, K., Ed.; Woodhead Publishing: Cambridge, UK, 2014; pp. 80–92. [CrossRef]
16. Moore, W.R.; Graves, S.E.; Bain, G.I. Synthetic bone graft substitutes. *ANZ J. Surg.* **2001**, *71*, 354–361. [CrossRef] [PubMed]
17. Heinemann, F.; Mundt, T.; Biffar, R.; Gedrange, T.; Goetz, W. A 3-year clinical and radiographic study of implants placed simultaneously with maxillary sinus floor augmentations using a new nanocrystalline hydroxyapatite. *J. Physiol. Pharmacol.* **2009**, *60*, 91–97. [PubMed]
18. Beaman, F.D.; Bancroft, L.W.; Peterson, J.J.; Kransdorf, M.J.; Menke, D.M.; DeOrio, J.K. Imaging characteristics of bone graft materials. *Radiographics* **2006**, *26*, 373–388. [CrossRef] [PubMed]
19. Xia, M.; Huang, R.; Witt, K.L.; Southall, N.; Fostel, J.; Cho, M.-H.; Jadhav, A.; Smith, C.S.; Inglese, J.; Portier, C.J.; et al. Compound cytotoxicity profiling using quantitative high-throughput screening. *Environ. Health Perspect.* **2008**, *116*, 284–291. [CrossRef]
20. Takamatsu, N. The new colorimetric assay (WST-1) for cellular growth with normal aging and Alzheimer's disease. *Nihon Ronen Igakkai Zasshi* **1998**, *35*, 535–542. [CrossRef]
21. Takeyoshi, M.; Noda, S.; Yamasaki, K.; Kimber, I. Advantage of using CBA/N strain mice in a non-radioisotopic modification of the local lymph node assay. *J. Appl. Toxicol.* **2006**, *26*, 5–9. [CrossRef]
22. Mukaka, M.M. Statistics corner: A guide to appropriate use of correlation coefficient in medical research. *Malawi Med. J.* **2012**, *24*, 69–71.

23. Adam, M.; Ganz, C.; Xu, W.; Sarajian, H.-R.; Götz, W.; Gerber, T. In vivo and in vitro investigations of a nanostructured coating material—A preclinical study. *Int. J. Nanomed.* **2014**, *9*, 975–984. [CrossRef]
24. Bayani, M.; Torabi, S.; Shahnaz, A.; Pourali, M. Main properties of nanocrystalline hydroxyapatite as a bone graft material in treatment of periodontal defects. A review of literature. *Biotechnol. Biotechnol. Equip.* **2017**, *31*, 1–6. [CrossRef]
25. Levingstone, T.J.; Herbaj, S.; Dunne, N.J. Calcium phosphate nanoparticles for therapeutic applications in bone regeneration. *Nanomaterials* **2019**, *9*. [CrossRef] [PubMed]
26. Drouet, C. Apatite Formation: Why it may not work as planned, and how to conclusively identify apatite compounds. *BioMed. Res. Int.* **2013**, *2013*, 490946. [CrossRef] [PubMed]
27. Hirschhorn, J.S.; McBeath, A.A.; Dustoor, M.R. Porous titanium surgical implant materials. *J. Biomed. Mater. Res.* **1971**, *5*, 49–67. [CrossRef]
28. Hulbert, S.F.; Young, F.A.; Mathews, R.S.; Klawitter, J.J.; Talbert, C.D.; Stelling, F.H. Potential of ceramic materials as permanently implantable skeletal prostheses. *J. Biomed. Mater. Res.* **1970**, *4*, 433–456. [CrossRef]
29. Klawitter, J.J.; Hulbert, S.F. Application of porous ceramics for the attachment of load bearing internal orthopedic applications. *J. Biomed. Mater. Res.* **1971**, *5*, 161–229. [CrossRef]
30. Galois, L.; Mainard, D. Bone ingrowth into two porous ceramics with different pore sizes: An experimental study. *Acta Orthop. Belg.* **2004**, *70*, 598–603.
31. Bohner, M.; Baumgart, F. Theoretical model to determine the effects of geometrical factors on the resorption of calcium phosphate bone substitutes. *Biomaterials* **2004**, *25*, 3569–3582. [CrossRef]
32. Li, X.; van Blitterswijk, C.A.; Feng, Q.; Cui, F.; Watari, F. The effect of calcium phosphate microstructure on bone-related cells in vitro. *Biomaterials* **2008**, *29*, 3306–3316. [CrossRef]
33. Coathup, M.J.; Hing, K.A.; Samizadeh, S.; Chan, O.; Fang, Y.S.; Campion, C.; Buckland, T.; Blunn, G.W. Effect of increased strut porosity of calcium phosphate bone graft substitute biomaterials on osteoinduction. *J. Biomed. Mater. Res. A* **2012**, *100*, 1550–1555. [CrossRef]
34. Eidelman, N.; Chow, L.C.; Brown, W.E. Calcium phosphate phase transformations in serum. *Calcif. Tissue Int.* **1987**, *41*, 18–26. [CrossRef] [PubMed]
35. Walsh, W.R.; Morberg, P.; Yu, Y.; Yang, J.L.; Haggard, W.; Sheath, P.C.; Svehla, M.; Bruce, W.J. Response of a calcium sulfate bone graft substitute in a confined cancellous defect. *Clin. Orthop. Relat. Res.* **2003**, *406*, 228–236. [CrossRef]
36. Arnett, T.R. Extracellular pH regulates bone cell function. *J. Nutr.* **2008**, *138*, 415S–418S. [CrossRef] [PubMed]
37. Shen, Y.; Liu, W.; Lin, K.; Pan, H.; Darvell, B.W.; Peng, S.; Wen, C.; Deng, L.; Lu, W.W.; Chang, J. Interfacial pH: A critical factor for osteoporotic bone regeneration. *Langmuir* **2011**, *27*, 2701–2708. [CrossRef] [PubMed]
38. Chow, L.C. Solubility of calcium phosphates. *Monogr. Oral. Sci.* **2001**, *18*, 94–111. [CrossRef]
39. Shukla, J.; Mohandas, V.P.; Kumar, A. Effect of pH on the solubility of $CaSO_4 \cdot 2H_2O$ in aqueous NaCl solutions and physicochemical solution properties at 35 °C. *J. Chem. Eng. Data* **2008**, *53*, 2797–2800. [CrossRef]
40. Stipanuk, M.H.; Caudill, M.A. *Biochemical, Physiological, and Molecular Aspects of Human Nutrition*, 3rd ed.; Elsevier: St. Louis, MO, USA, 2013; pp. 719–720.
41. Finlay, J.M.; Nordin, B.E.; Fraser, R. A calcium-infusion test. II. Four-hr. skeletal retention data for recognition of osteoporosis. *Lancet* **1956**, *270*, 826–830. [CrossRef]
42. Gurgan, T.; Demirol, A.; Guven, S.; Benkhalifa, M.; Girgin, B.; Li, T.C. Intravenous calcium infusion as a novel preventive therapy of ovarian hyperstimulation syndrome for patients with polycystic ovarian syndrome. *Fertil. Steril.* **2011**, *96*, 53–57. [CrossRef]
43. Fischbach, F.T.; Dunning, M.B. *A Manual of Laboratory and Diagnostic Tests*, 9th ed.; Wolters Kluwer Health: Philadelphia, PA, USA, 2015; pp. 914–983.
44. van Norden, A.G.; van den Bergh, W.M.; Rinkel, G.J. Dose evaluation for long-term magnesium treatment in aneurysmal subarachnoid haemorrhage. *J. Clin. Pharm. Ther.* **2005**, *30*, 439–442. [CrossRef]
45. Prieto, E.M.; Talley, A.D.; Gould, N.R.; Zienkiewicz, K.J.; Drapeau, S.J.; Kalpakci, K.N.; Guelcher, S.A. Effects of particle size and porosity on in vivo remodeling of settable allograft bone/polymer composites. *J. Biomed. Mater. Res. A* **2015**, *103*, 1641–1651. [CrossRef]
46. Chen, Y.J.; Pao, J.L.; Chen, C.S.; Chen, Y.C.; Chang, C.C.; Hung, F.M.; Chang, C.H. Evaluation of New Biphasic Calcium Phosphate Bone Substitute: Rabbit Femur Defect Model and Preliminary Clinical Results. *J. Med. Biol. Eng.* **2017**, *37*, 85–93. [CrossRef]

47. Bohner, M.; Galea, L.; Doebelin, N. Calcium phosphate bone graft substitutes: Failures and hopes. *J. Eur. Ceram. Soc.* **2012**, *32*, 2663–2671. [CrossRef]
48. Sheikh, Z.; Abdallah, M.N.; Hanafi, A.A.; Misbahuddin, S.; Rashid, H.; Glogauer, M. Mechanisms of in Vivo Degradation and Resorption of Calcium Phosphate Based Biomaterials. *Materials* **2015**, *8*, 7913–7925. [CrossRef]
49. Xia, Z.; Triffitt, J.T. A review on macrophage responses to biomaterials. *Biomed. Mater.* **2006**, *1*, R1–R9. [CrossRef] [PubMed]
50. Klein, C.P.; Driessen, A.A.; de Groot, K.; van den Hooff, A. Biodegradation behavior of various calcium phosphate materials in bone tissue. *J. Biomed. Mater. Res.* **1983**, *17*, 769–784. [CrossRef] [PubMed]
51. Huang, M.-S.; Wu, H.-D.; Teng, N.-C.; Peng, B.-Y.; Wu, J.-Y.; Chang, W.-J.; Yang, J.-C.; Chen, C.-C.; Lee, S.-Y. In vivo evaluation of poorly crystalline hydroxyapatite-based biphasic calcium phosphate bone substitutes for treating dental bony defects. *J. Dent. Sci.* **2010**, *5*, 100–108. [CrossRef]
52. Zhu, W.; Xiao, J.; Wang, D.; Liu, J.; Xiong, J.; Liu, L.; Zhang, X.; Zeng, Y. Experimental study of nano-HA artificial bone with different pore sizes for repairing the radial defect. *Int. Orthop.* **2009**, *33*, 567–571. [CrossRef] [PubMed]
53. Liu, J.; Mao, K.; Liu, Z.; Wang, X.; Cui, F.; Guo, W.; Mao, K.; Yang, S. Injectable biocomposites for bone healing in rabbit femoral condyle defects. *PLoS ONE* **2013**, *8*, e75668. [CrossRef]
54. Sheikh, Z.; Zhang, Y.L.; Tamimi, F.; Barralet, J. Effect of processing conditions of dicalcium phosphate cements on graft resorption and bone formation. *Acta Biomater.* **2017**, *53*, 526–535. [CrossRef]
55. Tsai, C.H.; Lin, R.M.; Ju, C.P.; Chern Lin, J.H. Bioresorption behavior of tetracalcium phosphate-derived calcium phosphate cement implanted in femur of rabbits. *Biomaterials* **2008**, *29*, 984–993. [CrossRef] [PubMed]
56. Little, D.G.; Ramachandran, M.; Schindeler, A. The anabolic and catabolic responses in bone repair. *J. Bone Joint Surg. Br.* **2007**, *89*, 425–433. [CrossRef] [PubMed]
57. Duan, R.; Barbieri, D.; de Groot, F.; de Bruijn, J.D.; Yuan, H. Modulating bone regeneration in rabbit condyle defects with three surface-structured tricalcium phosphate ceramics. *ACS Biomater. Sci. Eng.* **2018**, *4*, 3347–3355. [CrossRef] [PubMed]
58. He, Y.; Li, Q.; Ma, C.; Xie, D.; Li, L.; Zhao, Y.; Shan, D.; Chomos, S.K.; Dong, C.; Tierney, J.W.; et al. Development of osteopromotive poly (octamethylene citrate glycerophosphate) for enhanced bone regeneration. *Acta Biomater.* **2019**, *93*, 180–191. [CrossRef] [PubMed]
59. Gauthier, O.; Goyenvalle, E.; Bouler, J.M.; Guicheux, J.; Pilet, P.; Weiss, P.; Daculsi, G. Macroporous biphasic calcium phosphate ceramics versus injectable bone substitute: A comparative study 3 and 8 weeks after implantation in rabbit bone. *J. Mater. Sci.: Mater. Med.* **2001**, *12*, 385–390. [CrossRef]
60. Le Guehennec, L.; Goyenvalle, E.; Aguado, E.; Pilet, P.; Bagot D'Arc, M.; Bilban, M.; Spaethe, R.; Daculsi, G. MBCP biphasic calcium phosphate granules and tissucol fibrin sealant in rabbit femoral defects: The effect of fibrin on bone ingrowth. *J. Mater. Sci. Mater. Med.* **2005**, *16*, 29–35. [CrossRef]
61. Chakar, C.; Naaman, N.; Soffer, E.; Cohen, N.; El Osta, N.; Petite, H.; Anagnostou, F. Bone Formation with Deproteinized Bovine Bone Mineral or Biphasic Calcium Phosphate in the Presence of Autologous Platelet Lysate: Comparative Investigation in Rabbit. *Int. J. Biomater.* **2014**, *2014*, 10. [CrossRef]
62. Sprio, S.; Sandri, M.; Panseri, S.; Iafisco, M.; Ruffini, A.; Minardi, S.; Tampieri, A. 1—Bone substitutes based on biomineralization. In *Bone Substitute Biomaterials*; Mallick, K., Ed.; Woodhead Publishing: Cambridge, UK, 2014; pp. 3–29. [CrossRef]
63. Eggli, P.S.; Müller, W.; Schenk, R.K. Porous hydroxyapatite and tricalcium phosphate cylinders with two different pore size ranges implanted in the cancellous bone of rabbits. A comparative histomorphometric and histologic study of bony ingrowth and implant substitution. *Clin. Orthop. Relat. Res.* **1988**, *232*, 127–138. [CrossRef]
64. Ricci, J.L.; Weiner, M.J.; Iorio, D.D.; Mamidwar, S.; Alexander, H. Evaluation of timed release calcium sulfate (CS-TR) bone graft substitutes. *Microsc. Microanal.* **2005**, *11*, 1256–1257. [CrossRef]
65. Aquino-Martínez, R.; Angelo, A.P.; Pujol, F.V. Calcium-containing scaffolds induce bone regeneration by regulating mesenchymal stem cell differentiation and migration. *Stem. Cell Res. Ther.* **2017**, *8*, 265. [CrossRef]
66. Orsini, G.; Ricci, J.; Scarano, A.; Pecora, G.; Petrone, G.; Iezzi, G.; Piattelli, A. Bone-defect healing with calcium-sulfate particles and cement: An experimental study in rabbit. *J. Biomed. Mater. Res. B. Appl. Biomater.* **2004**, *68*, 199–208. [CrossRef]

67. Jarcho, M. Calcium phosphate ceramics as hard tissue prosthetics. *Clin. Orthop. Relat. Res.* **1981**, *157*, 259–278. [CrossRef]
68. Bagambisa, F.B.; Joos, U.; Schilli, W. Mechanisms and structure of the bond between bone and hydroxyapatite ceramics. *J. Biomed. Mater. Res.* **1993**, *27*, 1047–1055. [CrossRef] [PubMed]
69. Chen, Y.; Wang, J.; Zhu, X.; Chen, X.; Yang, X.; Zhang, K.; Fan, Y.; Zhang, X. The directional migration and differentiation of mesenchymal stem cells toward vascular endothelial cells stimulated by biphasic calcium phosphate ceramic. *Regen. Biomater.* **2018**, *5*, 129–139. [CrossRef] [PubMed]
70. Song, G.; Habibovic, P.; Bao, C.; Hu, J.; van Blitterswijk, C.A.; Yuan, H.; Chen, W.; Xu, H.H. The homing of bone marrow MSCs to non-osseous sites for ectopic bone formation induced by osteoinductive calcium phosphate. *Biomaterials* **2013**, *34*, 2167–2176. [CrossRef] [PubMed]
71. Arnett, T. Regulation of bone cell function by acid-base balance. *Proc. Nutr. Soc.* **2003**, *62*, 511–520. [CrossRef]

© 2020 by the authors. Licensee MDPI, Basel, Switzerland. This article is an open access article distributed under the terms and conditions of the Creative Commons Attribution (CC BY) license (http://creativecommons.org/licenses/by/4.0/).

Article

Mechanical Characterization of Human Trabecular and Formed Granulate Bone Cylinders Processed by High Hydrostatic Pressure

Janine Waletzko-Hellwig [1,*], Michael Saemann [2], Marko Schulze [3], Bernhard Frerich [1], Rainer Bader [2] and Michael Dau [1]

[1] Department of Oral, Maxillofacial and Plastic Surgery, Rostock University Medical Center, 18057 Rostock, Germany; bernhard.frerich@med.uni-rostock.de (B.F.); michael.dau@med.uni-rostock.de (M.D.)

[2] Biomechanics and Implant Technology Research Laboratory, Department of Orthopaedics, Rostock University Medical Center, 18057 Rostock, Germany; michael.saemann@med.uni-rostock.de (M.S.); rainer.bader@med.uni-rostock.de (R.B.)

[3] Department of Anatomy Rostock University Medical Center, 18057 Rostock, Germany; marko.schulze@uni-bielefeld.de

* Correspondence: janine.waletzko-hellwig@med.uni-rostock.de; Tel.: +49-381-494-9336; Fax: +49-381-494-6698

Abstract: One main disadvantage of commercially available allogenic bone substitute materials is the altered mechanical behavior due to applied material processing, including sterilization methods like thermal processing or gamma irradiation. The use of high hydrostatic pressure (HHP) might be a gentle alternative to avoid mechanical alteration. Therefore, we compressed ground trabecular human bone to granules and, afterwards, treated them with 250 and 300 MPa for 20 and 30 min respectively. We characterized the formed bone granule cylinders (BGC) with respect to their biomechanical properties by evaluating stiffness and stress at 15% strain. Furthermore, the stiffness and yield strength of HHP-treated and native human trabecular bone cylinders (TBC) as control were evaluated. The mechanical properties of native vs. HHP-treated TBCs as well as HHP-treated vs. untreated BGCs did not differ, independent of the applied HHP magnitude and duration. Our study suggests HHP treatment as a suitable alternative to current processing techniques for allogenic bone substitutes since no negative effects on mechanical properties occurred.

Keywords: high hydrostatic pressure; mechanical characterization; uniaxial compression test; bone substitutes; allograft; bone regeneration

1. Introduction

The reconstruction of severe bone defects, which originate, e.g., from infections, pathologic fractures, tumors or trauma, still remains a clinical challenge [1]. Although there are a number of different possibilities for reconstructing bone, including xenografts like demineralized bone matrices, autologous bone is still considered to be the gold standard [2–4]. For reconstruction surgery, autologous bone can be obtained from various donor sites such as the iliac crest, and it is specified as osteogenic, osteoinductive, osteoconductive and biocompatible, with low immunological potential and adequate mechanical strength [5]. Most other bone substitutes cannot comply with all of these requirements. Nevertheless, harvesting autologous bone is naturally limited, the occurrence of donor-side morbidities is not unusual and it requires an advanced surgical procedure [2]. The most frequently chosen alternatives to autografts are allografts [5]. Allogenic bone grafts have osteoconductive properties and avoid donor-side morbidity in the recipient. Additionally, customized types of allografts like blocks, stripes or granules are possible [2]. However, postoperative infections due to residual microbiota, proteins, etc. following allograft transplantation of human origin are a risk [6]. To prevent this, different sterilization methods including thermal processing, gamma radiation or physical and chemical decellularization have been established in recent years. Due to the removal of cellular components, any osteogenic properties are lost. However, this circumstance can be overcome with the revitalization of the graft using the recipient's own stem cells [6,7]. Unfortunately, the mechanical strength of the allografts usually suffers when using common decellularization and sterilization methods [6].

A reasonable alternative to current decellularization methods could be treatment with high hydrostatic pressure (HHP). HHP is commonly associated with processing of food and beverages [8]. This process inactivates microbes by membrane modifications, deactivation of key enzymes and inhibition of relevant metabolic processes like protein biosynthesis [9]. In comparison to conventional thermal food processing, HHP has the advantage that flavors and vitamins are unaffected by pressures up to 800 MPa [9,10]. In recent years, HHP has gained attention in pharmaceutical research. Rigaldie et al. have shown that HHP can be used to sterilize sensitive drugs like insulin with no effect on molecular integrity [11]. Furthermore, it was shown that HHP had a devitalizing effect on different mammalian cell lines, ex situ and in situ [12,13]. In the latter, it was already shown that this form of devitalization had no negative influence on the mechanical behavior of e.g., blood vessels [13].

The aim of our present study was to evaluate the mechanical properties of human trabecular bone cylinders (TBC) and bone granules pressed to cylinders (bone granule cylinders, BGC), both treated with HHP. While TBCs with an interconnected, trabecular structure can be used for larger bone defects, the use of BGCs as filling material for non-load-bearing bone defects is conceivable. A previous study at the cellular level showed that osteoblasts, as part of trabecular bone, follow either apoptotic or necrotic means of cell death, depending on the applied HHP magnitude. A pressure range of 100–150 MPa for 10 min did not have a negative influence on the metabolic activity and cell death could not be detected. Applied pressures of 250 MPa and more led to a significant reduction in metabolic activity compared to the control group. However, it was found that a pressure of 250–300 MPa tended to lead to apoptosis, while a pressure of 450–500 MPa had a necrotic effect on the osteoblasts [14]. The level and duration of HHP applied to tissues should be selected carefully, as necrosis can be a crucial factor in clinical transplantation due to the conceivably strong immunological response of the recipient [12]. Relying on the previous cell-based study, pressures of 250 and 300 MPa were used in the present experiments, as it was assumed that the biological effects would be similar. However, due to the changes in sample geometry compared to the cell pellets, the treatment periods for TBCs and BGCs were increased from 10 min to 20 and 30 min, respectively. The mechanical properties were analyzed by performing uniaxial compression tests and comparing stiffness and strength.

2. Materials and Methods

2.1. Sample Preparation, HHP Treatment and Creation Granules-Based Bone Cylinders

Trabecular bone specimens were taken post-mortem from human femur condyles and the femoral heads of body donors (Institute of Anatomy, Rostock University Medical Center; ethics approval A 2016-0083). Both were harvested within 72 h post-mortem in order to prevent the samples from being affected by decomposition processes. Afterward, all samples were rinsed once with sterile phosphate-buffered saline (PBS) (Sigma Aldrich, Munich, Germany), supplemented with 1% penicillin/streptomycin (Sigma Aldrich, Munich, Germany). Femur condyles and femoral heads were stored at $-20\,°C$ and covered with cling film until further preparation for HHP treatment and mechanical testing.

Before preparation of the TBCs for the compression tests, femoral condyles were slowly thawed at 4 °C. The defrosted condyles were partitioned into different sections (Figure 1). Within each predefined section, cylinders with a diameter of 6 mm were obtained from the proximal side using a trepan drill (Ustomed Instrumente, Tuttlingen, Germany). This was performed at room temperature under constant cooling with physiological saline solution (B. Braun, Melsungen, Germany) to prevent damage from heat. The plane ends of the cylinders were rectified with the help of a scalpel to achieve parallel end faces perpendicular to the drilling axis and to shorten cylinders to a length of approximately 10 mm. Care was taken to ensure that specimens consisted of only trabecular bone, and that the ends of the cylinders were parallel to each other with a deviation of less than 5°. Specimens that did not satisfy these criteria were discarded.

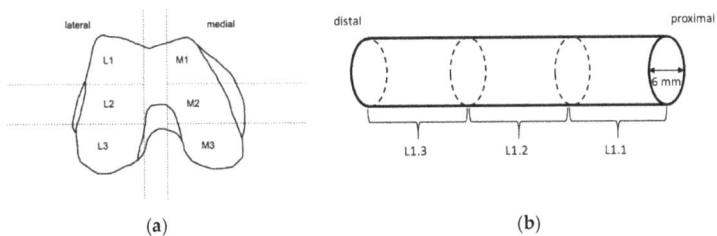

Figure 1. Sample preparation: (**a**) Knee condyles of human femurs were partitioned into the shown sections. Within shown sections, long cylinders were drilled along the femoral axis from proximal to distal using a trepan drill. (**b**) Long cylinders were sectioned into smaller ones with a length of 10 mm and a diameter of 6 mm using a scalpel, and care was taken that the ends were parallel to each other with a deviation of less than 5°.

For the compression test, trabecular cylinders from the identical harvesting location underwent different HHP treatments (control: $n = 20$; group A: 250 MPa, 20 min, $n = 18$; group B: 250 MPa, 30 min, $n = 19$; group C: 300 MPa, 20 min, $n = 16$; group D: 300 MPa, 30 min, $n = 14$) and were tested afterwards and compared one by one. Therefore, if a cylinder from the lateral region of a left femur condyle was taken for HHP treatment, the corresponding cylinder from the lateral region of a right femur was used as a control. The different group sizes arose due to the rejection of samples that did not satisfy the above-mentioned criteria and the alternating approach described before.

To investigate the influence of HHP on the mechanical properties of the bone granule cylinders (BGCs), the femoral heads were sawed into bone blocks with a size ranging between 0.05 and 0.1 cm^3. Afterward, the bone blocks were processed by a bone mill (Ustomed Instrumente, Tuttlingen, Germany) to granules with a size between 1 and 2 mm.

For HHP treatment, the granules were transferred into 2 mL cryogenic tubes filled with sterile PBS. Different treatment protocols (control: $n = 7$; group A: 250 MPa, 20 min, $n = 9$; group B: 250 MPa, 30 min, $n = 10$; group C: 300 MPa, 20 min, $n = 8$; group D: 300 MPa, 30 min, $n = 6$) were applied at a constant temperature of 30 °C. The untreated specimens of the control group were stored for the same time in PBS at 30 °C.

Before performing the uniaxial compression test, the bone granules were pressed to cylinders with a diameter of 6 mm and a length of about 10 to 12 mm. To generate cylinders of similar density, between 0.75 and 1 g of granules per cylinder were put into a hollow cylinder (Figure 2) and compressed with a uniaxial testing machine (ZwickRoell, Ulm, Germany) using a predefined compression regime (Figure 3). A compression speed of 0.5 mm/s was applied, and the compression stopped after reaching an end load of 1000 N for 5 min. Afterwards, the cylinders formed from the granules were taken out of the hollow cylinders and stored at room temperature until performing the unconfined uniaxial compression test.

2.2. Unconfined Uniaxial Compression Test

The unconfined uniaxial compression tests were conducted at room temperature using a uniaxial testing machine (Z050, ZwickRoell, Ulm, Germany) and a 2.5 kN load cell (Zwick-Roell, Ulm, Germany). A preload of 0.1 N was applied at a test speed of 0.05 mm/s, which was chosen based on previous studies, and which represents a physiological range [15,16]. The test runs were terminated at an engineering strain of 80%. The test setup is shown in

Figure 4. TBCs and BGCs that lost their axial alignment during the uniaxial compression test were discarded.

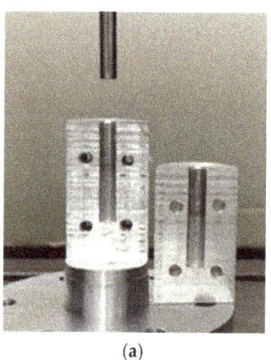
(a)

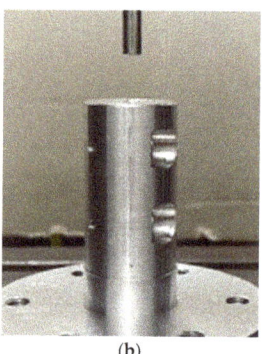
(b)

Figure 2. (a) Two segments of the hollow cylindrical body with a diameter of 6 mm. (b) Both segments screwed together to form a hollow cylinder. Bone granules were placed into the hollow cylinder and pressed together with a testing machine.

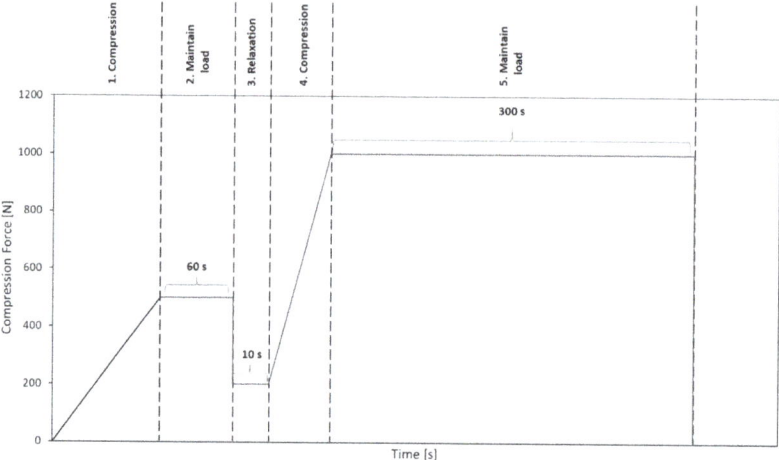

Figure 3. Predefined compression regime to compact bone granules to cylinders.

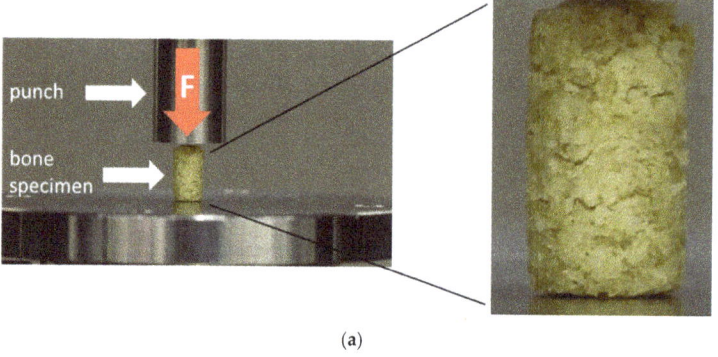
(a)

Figure 4. *Cont.*

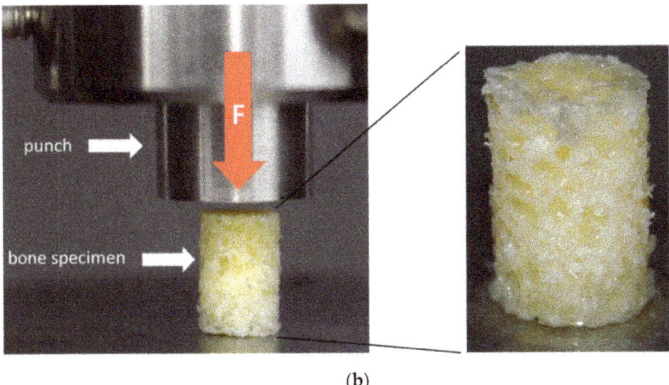

(b)

Figure 4. Test setup for the uniaxial unconfined compression test for the bone granule cylinders (BGCs) (**a**) and trabecular bone cylinders (TBCs) (**b**) with a representation of the respective test specimens.

2.3. Evaluation of the Results and Statistics

For the TBCs, the stiffness and the yield strength (first stress maximum after linear behavior) were compared. For BGCs, the stiffness and the stress at 15% strain were compared because the stress–strain curves of the bone granule specimens did not exhibit local maxima due to the lack of an interconnected trabecular structure.

For all human bone specimens, the generated engineering stress–strain curves were analyzed using a self-developed MATLAB script (v. R2018a, MathWorks, Natick, MA, USA). The linear-elastic region of the stress-strain curves was automatically identified and used for calculation of the stiffness via regression. For TBCs, the yield strength was identified as the first local dominant maximum after linear behavior. For BGCs, the engineering stress at 15% strain was calculated as a comparable alternative to the yield strength. Additionally, all curves of each group were averaged using Origin (v. 2018b, OriginLab, Northampton, MA, USA).

Statistical analyses were done by one-way ANOVA tests using GraphPad Prism Version 7 (GraphPad Software, San Diego, CA, USA), and results are presented as box-and-whisker plots. p-values ≤ 0.05 were seen as significant.

3. Results

3.1. Effects of HHP Treatment on the Mechanical Properties of Trabecular Bone Cylinders (TBCs)

To evaluate the effects of HHP treatment on the mechanical properties of TBCs, samples were treated with HHPs of 250 and 300 MPa for 20 and 30 min each. The parameters of stiffness and yield strength were used for mechanical characterization. Results from samples that were tilted and/or slipped during testing or showed macroscopic defects were excluded. All results are shown as box plots in Figure 5 and summarized in Table 1. Additionally, a summary of the averaged stress-strain curves for all tested groups is shown in Figure 6.

Analyses showed no significant differences between the untreated and HHP-treated specimens (TBCs and BGCs), neither for stiffness nor yield strength. Comparing the different HHP magnitudes and durations applied to the specimens, no significant differences were determined within the treated groups. Considering the curves averaged within each group in Figure 6, it is shown that the courses of the stress–strain curves are similar. No effects of the HHP treatments can be observed in the stress–strain curves.

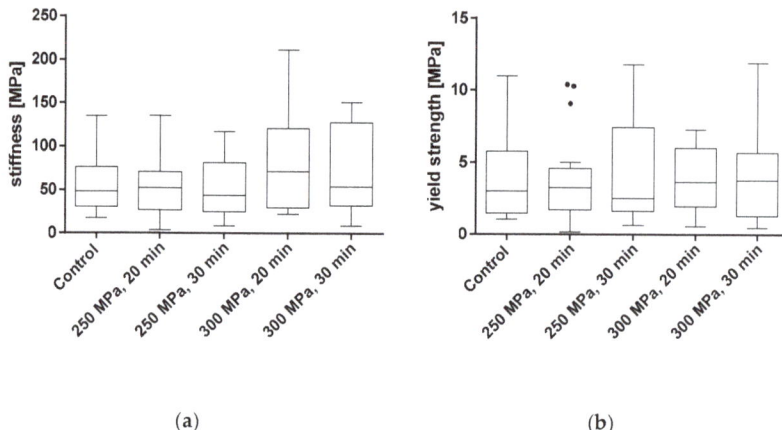

(a) (b)

Figure 5. Analysis of stiffness (**a**) and yield strength (**b**) of trabecular bone cylinders (TBC) treated with and without high hydrostatic pressure (HHP). Mechanical properties were tested using a uniaxial compression test. Data are shown as box plots with median and interquartile ranges from 25 to 75%. Statistical analyses were performed using a one-way ANOVA. Sample size: control group (n = 20); 250 MPa, 20 min (n = 18); 250 MPa, 30 min (n = 19); 300 MPa, 20 min (n = 16); 300 MPa, 30 min (n = 14).

Table 1. Overview of the results, including sample size n, the mean and the standard deviation for trabecular bone cylinders after the uniaxial compression test for stiffness and yield strength.

Treatment	n	Stiffness [MPa]		Yield Strength [MPa]	
		Mean	Standard Deviation	Mean	Standard Deviation
control	20	58.432	±35.916	3.767	±2.676
250 MPa, 20 min	18	57.364	±35.697	3.881	±3.080
250 MPa, 30 min	19	54.691	±34.732	4.461	±3.557
300 MPa, 20 min	16	80.366	±58.505	3.775	±2.272
300 MPa, 30 min	14	75.071	±49.520	4.238	±3.184

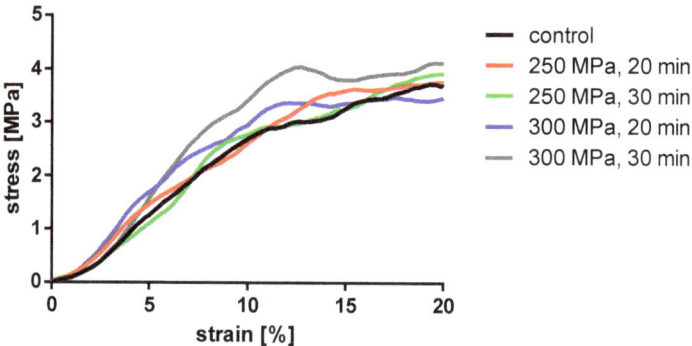

Figure 6. Averaged stress-strain curves of the compressed TBCs.

3.2. Compression of Granules to Cylindrical Samples

Pressed granulate bone cylinders were prepared using the described setup and technique and resulted in a length between 8 and 12 mm. An example of the compressed granules can be found in Figure 7.

Figure 7. Bone granules with an average size of 1 to 2 mm (**a**). These were compressed to cylindrical samples using a hollow cylinder (**b**).

3.3. Effect of HHP Treatment on the Mechanical Properties of Granules Bone Cylinders

To assess the influence of HHP on the mechanical properties of the BGCs, stiffness and stress at 15% strain were chosen as comparative parameters. The results are shown in Figure 8 and in Table 2. The averaged stress–strain curves for granulated bone cylinders are shown in Figure 9.

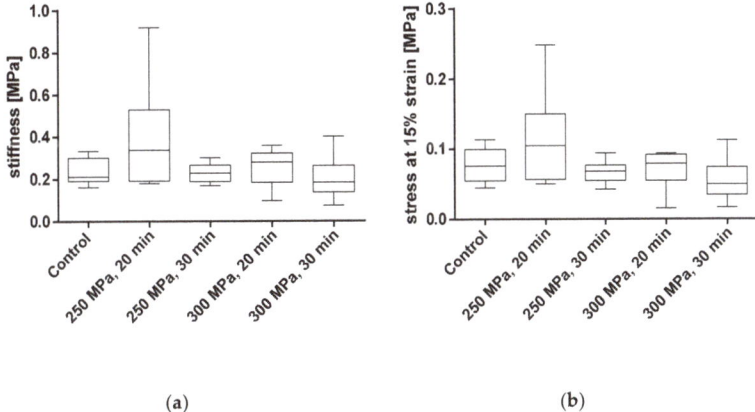

Figure 8. Analysis of stiffness (**a**) and stress at 15% strain (**b**) of pressed bone granules treated with and without HHP. Mechanical properties were tested using a uniaxial compression test. Data are shown as box plots with median and interquartile ranges from 25 to 75%. Statistical analyses were performed using a one-way ANOVA. Sample sizes: control group ($n = 7$); 250 MPa, 20 min ($n = 10$); 250 MPa, 30 min ($n = 10$); 300 MPa, 20 min ($n = 8$); 300 MPa, 30 min ($n = 6$).

Table 2. Overview of the results, including the sample size n, the mean and standard deviation for BGCs after the uniaxial compression test for the stiffness and stress at 15% strain.

Treatment	Group Size	Stiffness		Stress at 15% Strain	
		Mean	Standard Deviation	Mean	Standard Deviation
control	7	0.239	±0.062	0.078	±0.025
250 MPa, 20 min	9	0.381	±0.246	0.110	±0.066
250 MPa, 30 min	10	0.227	±0.044	0.067	±0.015
300 MPa, 20 min	8	0.253	±0.087	0.070	±0.026
300 MPa, 30 min	6	0.203	±0.108	0.055	±0.032

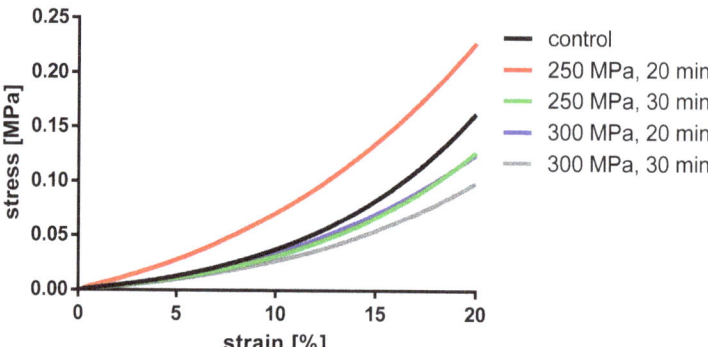

Figure 9. Averaged engineering stress-strain curves of pressed BGCs.

For BGCs, neither the stiffness nor the strain at 15% stress showed any significant differences between the groups. Figure 9 shows, as for the TBCs, the averaged stress-strain curves for granulate bone cylinders. Here, too, a similar course for all groups can be determined.

Comparing the stiffness of TBCs and BGCs, the latter comprises only a fraction of the native cylinders due to its missing intertrabecular structure.

4. Discussion

The reconstruction of bone defects is still challenging. In particular, cases that are correlated with severe bone loss or cases of patients with disorders in healing processes are clinically demanding [5]. Autologous bone is still considered to be the gold standard for bone defect reconstruction despite the known donor site morbidity and the limited amount of harvestable autologous bone [17]. A major drawback of the alternative, allogenic bone, which is less limited in quantity, is the alteration of mechanical properties due to current devitalization and sterilization methods, including thermal processing and gamma irradiation [18–20].

HHP as a gentle devitalization and sterilization method has been used in the food industry for several years, but it has also gained attention in the medical and pharmaceutical sectors [9]. Depending on the applied pressure, various studies have shown that mammalian cells can be devitalized while preserving an intact tissue matrix, which has already been shown for blood vessels and uterine tissues [12,13,21]. HHP has the potential to serve as a novel way to process allogenic bone substitute materials.

Within this study, it was shown that HHP had no effect on the macroscopic mechanical properties of human trabecular bone, as had already been shown for other tissues [13,21,22]. Specifically, HHP-treated TBCs showed no significant differences in either stiffness or yield strength. It was noticeable that all groups showed high variance in their mechanical properties, which is typical for biological samples, as gender and physical conditions of tissue donors can influence the results. Nevertheless, when looking at the averaged stress–strain curves of the individual groups with very similar curve progressions, it was shown that the eventual effects of HHP were small compared to the naturally occurring effects. The range for the compressive strength of trabecular bone is specified as 2 to 48 MPa according to the literature [7]. The measured strength of TBCs in this study was at the lower end of this range, at around 4 MPa for all groups. For this reason, the effect of HHP on trabecular bone specimens from other regions that have typically denser bone than the femoral condyles should be analyzed in further studies.

For pressed BGCs, no significant differences between the groups were found with stiffness and stress at 15% strain, i.e., no changes in the mechanical properties when comparing untreated and HHP-treated groups could be observed, but they were clearly below the values of the TBCs. The stress-strain curves of the BGCs also differ substantially from that of the TBCs. These curves exhibit a continuous, monotonic progression without local maxima due to the lack of an interconnected trabecular structure. Furthermore, this results in the significantly lower strength of BGCs when compared to native trabecular specimens. In addition, with these BGCs as well as the TBCs, it is noticeable that the values for both stiffness and stress at 15% strain vary widely. Many providers of bone substitute materials advertise both granules and bone blocks, which have been processed thermally or with gamma radiation [23,24]. According to various studies, these sterilization methods greatly reduce the mechanical properties of the allografts [19]. Commonly used irradiation doses between 20 and 30 kGy do not reduce the stiffness of bone, but they significantly reduce the ultimate stress and, to a smaller extent, the bending strength [18,25]. Thermal sterilization up to 60 °C has no effect on the mechanical properties, but higher temperatures (up to 100 °C) reduce the mechanical strength significantly [19].

There are several indications as to why HHP does not seem to affect the mechanical properties of treated biological tissues, which could explain the results observed in this study. An important role in the toughness of bone is played by collagen type I, which makes up the main part of extracellular matrix proteins that could be found in trabecular bone [26]. Pivotal to the mechanical properties of bone tissue is the formation of calcium-apatite crystals in the collagen fibrils interface [27,28]. This can also be seen in the correlation between calcification and bone stiffness [29]. Diehl et al. evaluated the effect of HHP on the biological properties of extracellular matrix (ECM) proteins and showed that HHP treatment had no influence on collagen type I and other common ECM proteins, such as fibronectin and vitronectin, in regards to their biological behavior in when compared to untreated ECM proteins [30].

These findings are supported by various studies at the molecular level [31,32]. Proteins are a complex organization of subunits with primary, secondary, tertiary and quaternary structures. The successive hierarchy describes increasing complexity of organization, which is affected by HHP treatment in different ways [31,32]. The primary structure (polypeptide chain) consists of covalent bonds, which are not affected by HHP. The secondary structure of proteins, created by the formation of hydrogen bonds between the polypeptides, is irreversibly degraded at pressures higher than 700 MPa. Tertiary structures, made up of hydrophobic interactions and ionic bonds, are broken up at HHPs higher than 200 MPa, and quaternary structures, with non-covalent bonds including Van der Waal's forces, are dissolved at HHPs between 100 and 150 MPa [31,32]. In contrast to the complete protein destruction that occurs at very high or low temperatures or during gamma irradiation, these structural changes caused by HHP are reversible at pressures between 100 and 300 MPa. This means proteins can rearrange after HHP treatment [18]. Based on this observation, some literature also describes a new association of previously incorrectly folded protein structures after HHP treatment [33–35].

In the present study, HHP between 250 and 300 MPa was applied, which induced reversible changes in protein structure. This could be the reason for the maintained mechanical properties of the specimens. However, our study is limited by several factors. As mentioned above, only TBCs from femoral condyles were studied. The same applies to BGCs, which were solely extracted from femoral heads. Furthermore, only one mechanical test was performed, and the specimen size did not reflect the influence of HHP on an entire femur. Additionally, only trabecular bone was analyzed in the study at hand, and the effect of HHP on cortical bone tissue should be analyzed in further studies. Another limitation is due to the use of biological samples; different tissue donors vary in physiological characteristics such as age, physical activity or pre-existing diseases. The structure and morphology of bone varies as well, which is reflected in the mechanical properties of whole bones and also bone tissue. A further limitation is contamination with bacteria and germs. The sterilization effect of HHD on these organisms was not the subject of this study.

The results presented fit well with the investigations described above. Further mechanical tests, such as a three-point-bending test, could be performed in addition to the uniaxial compression test already shown here. Samples composed of both cortical and trabecular bone tissue or whole bones could also be analyzed with regard to the effects of HHP on their mechanical properties. This could give a good overall view of the influence of HHP on bone at the macroscopic level. Simultaneously, the effects of HHP at the microscopic and molecular levels should not be neglected. Here, a structural analysis of proteins after HHP treatment and an analysis of the inorganic bone components via electron microscopy is conceivable. If HHP is discussed as an alternative to the previous methods of processing allografts, the inherent mechanical properties are of great importance. With regard to clinical applications, the devitalizing efficiency and immunological safety should also be studied in the future. In further studies, it will still be necessary to assess the effects of HHP from a microbiological and virological point of view. Here, the study of bacterial and virological load before and after HHP treatment could be conceivable following previous works [36,37].

In the case of the BGCs, the compression method could also be optimized depending on the targeted clinical application. Loosely packed granules may well have an advantage for the ingrowth of cells, and more densely packed granules could possibly be used as a load-bearing structure. In addition, compression parameters or shapes (e.g., blocks) deviating from those shown here should be investigated.

As shown in the study, the clinical use of HHP-treated TBCs and BGCs is conceivable. Although the formed granulated bone cylinders are more fragile than native trabecular bone, granulate cylinders might be used as shaping filler material for non-load-bearing bone defects, acting as osteoinductive and osteogenic scaffolds.

In conclusion, this study showed that HHP treatment has a pivotal advantage over conventional processing methods of bone substitute materials by maintaining the mechanical properties in combination with effective cell devitalization.

5. Patents

A patent application with the number DE 10 2020 131 181.8 was submitted to the German Patent and Trademark Office.

Author Contributions: Conceptualization, J.W.-H. and M.S. (Michael Saemann); methodology, M.S. (Michael Saemann), M.S. (Marko Schulze); software, M.S. (Michael Saemann); formal analysis, J.W.-H.; investigation, J.W.-H. and M.S. (Michael Saemann); resources, R.B.; data curation, J.W.-H. and M.S. (Miachel Saemann); writing—original draft preparation, J.W.-H.; writing—review and editing, M.S. (Michael Saemann), M.S. (Marko Schulze), M.D., R.B., B.F.; visualization, J.W.-H., M.S. (Michael Saemann), M.D.; supervision, M.D., R.B.; project administration, R.B., M.D.; funding acquisition, R.B., M.D., B.F. All authors have read and agreed to the published version of the manuscript.

Funding: This joint research project HOGEMA is supported by the European Social Fund (ESF), reference: ESF/14-BM-A55-0012/18, and the Ministry of Education, Science and Culture of Mecklenburg-Vorpommern, Germany.

Institutional Review Board Statement: The study was conducted according to the guidelines of the Declaration of Rostock University and approved by the Institutional Ethics Committee Rostock University, Germany. Prior to preparation of human femoral condyles and femoral heads, ethical approval (A 2016-0083) including IRB information was obtained.

Informed Consent Statement: Not applicable.

Data Availability Statement: The data presented in this study are available on request from the corresponding author. The data are not publicly available due to the nondisclosure agreement with the project sponsor.

Acknowledgments: We thank Mario Jackszis for his excellent technical support during preparation and mechanical testing of bone specimens. We acknowledge financial support by Deutsche Forschungsgemeinschaft and Universität Rostock/Universitätsmedizin Rostock within the funding programme Open Access Publishing.

Conflicts of Interest: The authors declare no conflict of interest. The funders had no role in the design of the study; in the collection, analyses, or interpretation of data; in the writing of the manuscript, or in the decision to publish the results.

References

1. Fernandez de Grado, G.; Keller, L.; Idoux-Gillet, Y.; Wagner, Q.; Musset, A.M.; Benkirane-Jessel, N.; Bornert, F.; Offner, D. Bone substitutes: A review of their characteristics, clinical use, and perspectives for large bone defects management. *J. Tissue Eng.* **2018**, *9*. [CrossRef] [PubMed]
2. Giannoudis, P.V.; Dinopoulos, H.; Tsiridis, E. Bone substitutes: An update. *Injury* **2005**, *36*, 20–27. [CrossRef]
3. Stumbras, A.; Krukis, M.M.; Januzis, G.; Juodzbalys, G. Regenerative bone potential after sinus floor elevation using various bone graft materials: A systematic review. *Quintessence Int.* **2019**, *50*, 548–558. [CrossRef] [PubMed]
4. Janicki, P.; Schmidmaier, G. What should be the characteristics of the ideal bone graft substitute? Combining scaffolds with growth factors and/or stem cells. *Injury* **2011**, *42*, S77–S81. [CrossRef] [PubMed]
5. Giannoudis, P.V.; Chris Arts, J.J.; Schmidmaier, G.; Larsson, S. What should be the characteristics of the ideal bone graft substitute? *Injury* **2011**, *42*, S1. [CrossRef] [PubMed]
6. Mikhael, M.M.; Huddleston, P.M.; Zobitz, M.E.; Chen, Q.; Zhao, K.D.; An, K.N. Mechanical strength of bone allografts subjected to chemical sterilization and other terminal processing methods. *J. Biomech.* **2008**, *41*, 2816–2820. [CrossRef]
7. Hannink, G.; Arts, J.J.C. Bioresorbability, porosity and mechanical strength of bone substitutes: What is optimal for bone regeneration? *Injury* **2011**, *42*, S22–S25. [CrossRef]
8. Rendueles, E.; Omer, M.K.; Alvseike, O.; Alonso-Calleja, C.; Capita, R.; Prieto, M. Microbiological food safety assessment of high hydrostatic pressure processing: A review. *LWT Food Sci. Technol.* **2011**, *44*, 1251–1260. [CrossRef]
9. Masson, P.; Tonello, C.; Balny, C. High-pressure biotechnology in medicine and pharmaceutical science. *J. Biomed. Biotechnol.* **2001**, *2001*, 85–88. [CrossRef]
10. Yamamoto, K. Food processing by high hydrostatic pressure. *Biosci. Biotechnol. Biochem.* **2017**, *81*, 672–679. [CrossRef]
11. Rigaldie, Y.; Largeteau, A.; Lemagnen, G.; Ibalot, F.; Pardon, P.; Demazeau, G.; Grislain, L. Effects of High Hydrostatic Pressure on Several Sensitive Therapeutic Molecules and a Soft Nanodispersed Drug Delivery System. *Pharm. Res.* **2003**, *20*, 2036–2040. [CrossRef]
12. Rivalain, N.; Roquain, J.; Demazeau, G. Development of high hydrostatic pressure in biosciences: Pressure effect on biological structures and potential applications in Biotechnologies. *Biotechnol. Adv.* **2010**, *28*, 659–672. [CrossRef] [PubMed]
13. Funamoto, S.; Nam, K.; Kimura, T.; Murakoshi, A.; Hashimoto, Y.; Niwaya, K.; Kitamura, S.; Fujisato, T.; Kishida, A. The use of high-hydrostatic pressure treatment to decellularize blood vessels. *Biomaterials* **2010**, *31*, 3590–3595. [CrossRef] [PubMed]
14. Waletzko, J.; Dau, M.; Seyfarth, A.; Springer, A.; Frank, M.; Bader, R.; Jonitz-Heincke, A. Devitalizing effect of high hydrostatic pressure on human cells—influence on cell death in osteoblasts and chondrocytes. *Int. J. Mol. Sci.* **2020**, *21*, 3836. [CrossRef]
15. Linde, F.; Nørgaard, P.; Hvid, I.; Odgaard, A.; Søballe, K. Mechanical properties of trabecular bone. Dependency on strain rate. *J. Biomech.* **1991**, *24*, 803–809. [CrossRef]
16. Linde, F.; Hvid, I. Stiffness behaviour of trabecular bone specimens. *J. Biomech.* **1987**, *20*, 83–89. [CrossRef]
17. Daculsi, G.; Fellah, B.H.; Miramond, T.; Durand, M. Osteoconduction, Osteogenicity, Osteoinduction, what are the fundamental properties for a smart bone substitutes. *Irbm* **2013**, *34*, 346–348. [CrossRef]
18. Nguyen, H.; Morgan, D.A.F.; Forwood, M.R. Sterilization of allograft bone: Effects of gamma irradiation on allograft biology and biomechanics. *Cell Tissue Bank.* **2007**, *8*, 93–105. [CrossRef] [PubMed]
19. Shin, S.; Yano, H.; Fukunaga, T.; Ikebe, S.; Shimizu, K.; Kaku, N.; Nagatomi, H.; Masumi, S. Biomechanical properties of heat-treated bone grafts. *Arch. Orthop. Trauma Surg.* **2005**, *125*, 1–5. [CrossRef]
20. Vastel, L.; Meunier, A.; Siney, H.; Sedel, L.; Courpied, J.P. Effect of different sterilization processing methods on the mechanical properties of human cancellous bone allografts. *Biomaterials* **2004**, *25*, 2105–2110. [CrossRef]
21. Santoso, E.G.; Yoshida, K.; Hirota, Y.; Aizawa, M.; Yoshino, O.; Kishida, A.; Osuga, Y.; Saito, S.; Ushida, T.; Furukawa, K.S. Application of detergents or high hydrostatic pressure as decellularization processes in uterine tissues and their subsequent effects on in vivo uterine regeneration in murine models. *PLoS ONE* **2014**, *9*, e0103201. [CrossRef]
22. Steinhauser, E.; Diehl, P.; Hadaller, M.; Schauwecker, J.; Busch, R.; Gradinger, R.; Mittelmeier, W. Biomechanical investigation of the effect of high hydrostatic pressure treatment on the mechanical properties of human bone. *J. Biomed. Mater. Res. Part B Appl. Biomater.* **2006**, *76*, 130–135. [CrossRef] [PubMed]
23. Geistlich. Available online: https://www.geistlich.de/de/dental/knochenersatz/bio-oss/vorteile-fuer-den-anwender/ (accessed on 5 January 2021).
24. Botiss-Dental. Available online: https://botiss-dental.com/de/products/cerabone-de/ (accessed on 5 January 2021).
25. Cornu, O.; Boquet, J.; Nonclercq, O.; Docquier, P.L.; Van Tomme, J.; Delloye, C.; Banse, X. Synergetic effect of freeze-drying and gamma irradiation on the mechanical properties of human cancellous bone. *Cell Tissue Bank.* **2011**, *12*, 281–288. [CrossRef] [PubMed]
26. Wang, X.; Bank, R.A.; TeKoppele, J.M.; Mauli Agrawal, C. The role of collagen in determining bone mechanical properties. *J. Orthop. Res.* **2001**, *19*, 1021–1026. [CrossRef]
27. Viguet-Carrin, S.; Garnero, P.; Delmas, P.D. The role of collagen in bone strength. *Osteoporos. Int.* **2006**, *17*, 319–336. [CrossRef] [PubMed]
28. Le, B.Q.; Nurcombe, V.; Cool, S.M.K.; van Blitterswijk, C.A.; de Boer, J.; LaPointe, V.L.S. The Components of bone and what they can teach us about regeneration. *Materials* **2017**, *11*, 14. [CrossRef]

29. Matinfar, M.; Mesgar, A.S.; Mohammadi, Z. Evaluation of physicochemical, mechanical and biological properties of chitosan/carboxymethyl cellulose reinforced with multiphasic calcium phosphate whisker-like fibers for bone tissue engineering. *Mater. Sci. Eng. C* **2019**, *100*, 341–353. [CrossRef] [PubMed]
30. Diehl, P.; Schmitt, M.; Schauwecker, J.; Eichelberg, K.; Gollwitzer, H.; Gradinger, R.; Goebel, M.; Preissner, K.T.; Mittelmeier, W.; Magdolen, U. Effect of high hydrostatic pressure on biological properties of extracellular bone matrix proteins. *Int. J. Mol. Med.* **2005**, *16*, 285–289. [CrossRef] [PubMed]
31. Breda, A.; Valadares, N.F.; de Souza, O.N.; Garratt, R.C. *Bioinformatics in Tropical Disease Research: A Practical and Case-Study Approach*; Gruber, A., Durham, A.M., Huynh, C., Eds.; 2008; pp. 1–34, Chapter A06. Protein Structure, Modelling and Applications 1. Why Is It Important to Study Proteins? 2. Explosion of Biological Sequence and Structure Data. Available online: https://www.ncbi.nlm.nih.gov/books/NBK6824/ (accessed on 2 February 2021).
32. Rodiles-López, J.O.; Arroyo-Maya, I.J.; Jaramillo-Flores, M.E.; Gutiérrez-López, G.F.; Hernández-Arana, A.; Barbosa-Cánovas, G.V.; Niranjan, K.; Hernández-Sánchez, H. Effects of high hydrostatic pressure on the structure of bovine α-lactalbumin. *J. Dairy Sci.* **2010**, *93*, 1420–1428. [CrossRef] [PubMed]
33. Gorovits, B.M.; Horowitz, P.M. High hydrostatic pressure can reverse aggregation of protein folding intermediates and facilitate acquisition of native structure. *Biochemistry* **1998**, *37*, 6132–6135. [CrossRef]
34. Burton, B.; Gaspar, A.; Josey, D.; Tupy, J.; Grynpas, M.D.; Willett, T.L. Bone embrittlement and collagen modifications due to high-dose gamma-irradiation sterilization. *Bone* **2014**, *61*, 71–81. [CrossRef] [PubMed]
35. Gross, M.; Jaenicke, R. Proteins under pressure: The influence of high hydrostatic pressure on structure, function and assembly of proteins and protein complexes. *Eur. J. Biochem.* **1994**, *221*, 617–630. [CrossRef] [PubMed]
36. Kingsley, D.H.; Hoover, D.G.; Papafragkou, E.; Richards, G.P. Inactivation of hepatitis A virus and a calicivirus by high hydrostatic pressure. *J. Food Prot.* **2002**, *65*, 1605–1609. [CrossRef] [PubMed]
37. Abe, F. Exploration of the effects of high hydrostatic pressure on microbial growth, physiology and survival: Perspectives from piezophysiology. *Biosci. Biotechnol. Biochem.* **2007**, *71*, 2347–2357. [CrossRef] [PubMed]

Article

The Influence of Hyaluronic Acid Biofunctionalization of a Bovine Bone Substitute on Osteoblast Activity In Vitro

Solomiya Kyyak [1], Andreas Pabst [2], Diana Heimes [1] and Peer W. Kämmerer [1,*]

[1] Department of Oral- and Maxillofacial Surgery, University Medical Center Mainz, 55131 Mainz, Germany; solomiya.kyyak@unimedizin-mainz.de (S.K.); diana.heimes@unimedizin-mainz.de (D.H.)
[2] Department of Oral- and Maxillofacial Surgery, Federal Armed Forces Hospital, 56072 Koblenz, Germany; andipabst@me.com
* Correspondence: peer.kaemmerer@unimedizin-mainz.de; Tel.: +49-6131-17-5458

Abstract: Bovine bone substitute materials (BSMs) are used for oral bone regeneration. The objective was to analyze the influence of BSM biofunctionalization via hyaluronic acid (HA) on human osteoblasts (HOBs). BSMs with ± HA were incubated with HOBs including HOBs alone as a negative control. On days 3, 7 and 10, cell viability, migration and proliferation were analyzed by fluorescence staining, scratch wound assay and MTT assay. On days 3, 7 and 10, an increased cell viability was demonstrated for BSM+ compared with BSM− and the control (each $p \leq 0.05$). The cell migration was enhanced for BSM+ compared with BSM− and the control after day 3 and day 7 (each $p \leq 0.05$). At day 10, an accelerated wound closure was found for the control compared with BSM+/− (each $p < 0.05$). The highest proliferation rate was observed for BSM+ on day 3 ($p \leq 0.05$) followed by BSM− and the control (each $p \leq 0.05$). At day 7, a non-significantly increased proliferation was shown for BSM+ while the control was higher than BSM− (each $p < 0.05$). The least proliferation activity was observed for BSM− ($p < 0.05$) at day 10. HA biofunctionalization of the BSMs caused an increased HOB activity and might represent a promising alternative to BSM− in oral bone regeneration.

Keywords: bone substitute; bovine; xenograft; oral regeneration; biofunctionalization; hyaluronic acid; osteoblasts

1. Introduction

Presently, the demand for soft tissue and hard tissue regeneration is frequently increasing where bone is one of the most transplanted tissues because of a multitude of congenital or acquired diseases [1]. Nevertheless, the field of bone transplantation and regeneration faces limitations regarding infections, immunological reactions, failed osteointegration and graft resorption [2]. To avoid graft harvesting and to support a better and faster regeneration, numerous materials are combined to find suitable alternatives to autogenous bone grafts [3]. Bone substitute materials (BSMs) of a xenogeneic, an allogeneic and an alloplastic origin are well established and widely used as suitable alternatives in numerous fields of medicine [4–7]. In the range of craniomaxillofacial regeneration, BSMs can cover a wide variety of clinical indications such as alveolar ridge preservation and augmentation, sinus floor elevation and the bony reconstruction of congenital or acquired maxillofacial malformations and defects [8–11].

Xenogeneic BSMs of bovine origin are long-term established and widely spread [12]. The hydroxyapatite-based substance [13] is known for its biocompatibility, sufficient osteoconduction and low up to no resorption [14,15] and its similarity to human bone due to its microstructure [16,17] and crystalline phase [18]. In contrast to autogenous grafts, BSMs do not contain organic components such as osteogenic cells or growth factors such as BMP-2 (bone morphogenic protein-2) and a VEGF (vascular endothelial growth factor) and they also may not contain collagen structures and fibers, enabling an osteoconductive and inductive regenerative potential in autogenous grafts. Thus, different BSM preparation methods and processes could affect the regeneration and surface characteristics of xenogeneic BSMs [19–22]. Accordingly, BSM sintering under a temperature >1000 °C seems to remove all organic compounds, thereby excluding an immune reaction and disease transmission and increasing crystallinity and volume stability [13,21,23–25]. Furthermore, it has been observed that even after a high temperature treatment, xenogeneic BSMs preserve their surface characteristics and a good biological performance [20–22,26]. Additionally, the carbonate content of high temperature treated hydroxyapatite stimulates human osteoblast (HOB) attachment and proliferation [27]. Nevertheless, it appears that BSMs may not be able to perform with an equal regenerative potency compared with autogenous grafts caused by the acellular and inorganic matrix. To overcome this limitation, BSM biofunctionalization has become more and more popular and has been tested in different ways. Recent studies analyzed combinations of BSMs with growth factors (e.g., BMP-2, VEGF) and PRF (platelet-rich fibrin). The findings of these studies illustrated that such biofunctionalized BSMs have the potency to accelerate and increase bone formation and vascularization as characteristic hallmarks of fast and sufficient bone regeneration [28–33]. As BSM modification with growth factors is technically challenging and restricted by legal requirements in most countries, further substances might be of interest for BSM biofunctionalization.

Hyaluronic acid (HA) is one of the largest components of the extracellular matrix. It is a long polysaccharide composed of macromolecules of many repetitive units of glucuronic acid and N-acetyl-glucosamine, remaining the same within all species [34–36]. It is stated that HA may regulate cell proliferation, differentiation, adhesion and gene expression [37]. These characteristics have aroused interest in HA in cutaneous research, cartilage grafting [38,39] and even bone reconstruction [34,40,41]. Thus, Kawano et al. reported that HA enhanced BMP-2 osteogenic bioactivity [35]. It has been discussed that HA retards bone resorption and osteoclast genesis through its receptor, CD44 [42]. HA may demonstrate lubricity under peculiar circumstances [43] and has been studied to have a bacteriostatic effect [44]. Sasaki et al. suggested that high molecular HA serves as a retainer for osteoinductive growth factors, thus stimulating osteogenic cell differentiation [45]. In addition, HA may positively influence angiogenesis and (neo-) vascularization because of its possible effects on endothelial cells, thus in turn indirectly stimulating new bone formation [45,46].

Different variations of HA molecules and their possible influence on tissue formation have been discussed. Guo et al. suggested that the molecular weight of HA strongly influences pro- and/or anti-inflammatory reactions of various tissues as far as peculiar angiogenic processes [47,48]. For example, Pilloni et al. observed that HA of a high molecular weight is dose-independent and not able to present any significant effects on bone formation [49]. However, further studies showed opposite findings [50]. This led to a significant interest in HA as an additive to different polymers and BSMs in bone engineering and regeneration.

Thus, the objective of this study was to analyze the influence of a commercially available BSM with (+) and without (−) HA biofunctionalization on viability, migration ability and the proliferation rate of human osteogenic cells. The zero hypothesis claims that this HA biofunctionalization has no influence on osteoblast activity.

2. Materials and Methods

2.1. Bovine Bone Substitutes

A commercially available xenogeneic bone substitute material (BSM−) of bovine origin (cerabone®, granularity: 1–2 mm; botiss biomaterials GmbH, Zossen, Germany) and a commercially available BSM with an HA modification (BSM+; cerabone® Plus, granularity: 0.5–1 mm; botiss biomaterials GmbH) were used.

2.2. Cell Culture

Commercially available human osteoblasts (HOBs) were applied in the present study (HOB; PromoCell, Heidelberg, Germany). A HOB medium was supplemented with Dulbecco's modified Eagle's medium (DMEM; Gibco Invitrogen, Karlsruhe, Germany), fetal calf serum (FCS; Gibco Invitrogen), streptomycin (100 mg/mL; Gibco Invitrogen), dexamethasone (100 nmol/L; Serva Bioproducts, Heidelberg, Germany) and L-glutamine (Gibco Invitrogen). The HOBs were cultured according to standard protocols in an incubator at 37 °C, 95% humidity and 5% of CO_2. Reaching a 70% confluence, the HOBs were passaged using 0.25% trypsin (Seromed Biochrom KG, Berlin, Germany) until passage five. The plates were filled with 100 mg BSM+/− together with 5×10^4 HOB per well, respectively (27 wells per group, two groups). The plates with HOBs alone served as a negative control group (overall 27 wells). A further incubation was performed under the same conditions as by cell passaging. The measures were conducted in three time points in triplicate for each group and for each time point (days 3, 7 and 10; overall 81 wells).

2.3. Cell Viability

To analyze the HOB cell viability, CellTracker staining (Life Technologies, Thermo Fisher Scientific, Darmstadt, Germany; catalog number: C34552) was performed on days 3, 7 and 10. Red dye was prepared and used according to the manufacturer's protocol. After the removal of the culture media, red dye was added into the wells. After 30 min, the red dye was removed and a serum-free medium was applied. The wells were further incubated for 30 min at 37 °C. After the removal of the serum-free medium, a fluorescence microscope (BZ-9000; Keyence, Osaka, Japan) for cell imaging was used where one image per well in ten-fold magnification was conducted. The cell quantification was managed by means of ImageJ software (ACTREC, Navi Mumbai, India) [51] by the following steps: the conversion of the images into grayscale, the correction of the background by image subtraction, automatic thresholding for cell structure extraction from the background and the final calculation of the percentage area fraction (%). The measures were carried out in triplicate for each group and for each time point by three time points (on days 3, 7 and 10; overall 9 wells per group).

2.4. Cell Migration

The cell attachment was measured by means of a scratch wound assay. A scratch wound was performed at the bottom of the wells with a sterile pipet tip (p200; Gilson, Middleton, USA) on days 3, 7 and 10 [52]. Immediately after the scratch, a fluorescence microscope (BZ-9000; Keyence, Osaka, Japan) for cell imaging was used. Twenty-four hours later, red dye staining was obtained for preparing images with the aforementioned microscope (one image for each well, 9 wells per group, ten-fold magnification). An area of migrated cells into the gap was quantified by the percentage area (%) using ImageJ software as described before [51]. The measures were carried out in triplicate for each group and for each time point (three time points).

2.5. Cell Proliferation

The proliferation activity was measured by a 3-(4,5-Dimethylthiazol-2-yl) -2,5-dipheny ltetrazolium bromide (MTT) assay on days 3, 7 and 10. An MTT solution (200 µL, 2 mg/mL) was applied to the cell culture medium in the wells followed by 4 hours of incubation at 37 °C. After the removal of the culture medium and washing up by phosphate buffered saline, a lysis buffer (Isopranol (49 mL) with 2N NCl (1 mL; 1 mL per well) was added. The measurement was performed without the BSM in separate wells using a fluorescence microplate reader with a wavelength of 570 nm (Versamax; Molecular Devices, San Jose, CA, USA). The measures were carried out in triplicate for each group on days 3, 7 and 10 (overall 9 wells for each group).

2.6. Statistics

The mean values were interpreted into a standard error of the mean (SEM) in the cases of parametric data and into median values for non-parametric data. The numbers were rounded (to two decimal places). The normal distribution was defined by a Shapiro-Wilk test. In the case of a normal distribution, to compare two subgroups a two-sided Student's *t*-test for paired samples was applied. In the case of non-normal distributions, a Mann-Whitney test was used. For a comparison of all subgroups, a Kruskal-Wallis rank sum test was performed. *p*-values ≤ 0.05 were considered to be significant. Data were illustrated with bar charts including error bars.

3. Results
3.1. Cell Viability

On day 3, the highest cell viability was observed for BSM+ when compared with BSM− ($p = 0.028$, *t*-test) and the control ($p = 0.24$, *t*-test). The cell viability of the control group was significantly higher than BSM− ($p < 0.001$, *t*-test) On day 7, the highest cell viability was seen for BSM+ compared with BSM− ($p < 0.001$, *t*-test; $p < 0.05$, KWT) and the control ($p = 0.014$, *t*-test; $p < 0.05$, KWT) followed by the control when compared with BSM− ($p = 0.006$, *t*-test; $p < 0.05$, KWT). At day 10, the cell viability of BSM+ was significantly higher when compared with the controls ($p = 0.004$, *t*-test) and BSM− ($p = 0.002$, *t*-test) (Table 1, Figures 1 and 2). Although the cell viability values for BSM+ were the highest of all groups through the whole period, the greatest tendency to increase was observed in BSM− in which the cell viability raised almost five times compared with BSM+ and the control with approximately two times (Figure 3a).

Table 1. Cell Viability. Percentage area fraction (%) of fluorescence-stained HOBs at a ten-fold magnification for BSM− (cerabone®), BSM+ (cerabone® Plus including HA) and the control (HOB alone) on days 3, 7 and 10. The mean values are for parametric data and the median values are for non-parametric data.

	Day 3		Day 7		Day 10	
	Mean Value	SEM	Mean Value	SEM	Mean Value	SEM
BSM−	4.41	±0.82	8.85	±1.14	19.86	±11.47
BSM+	15.92	±3.38	21.55	±1.32	27.84	±16.08
Control	11.22	±0.22	15.57	±0.54	20.9	±12.07

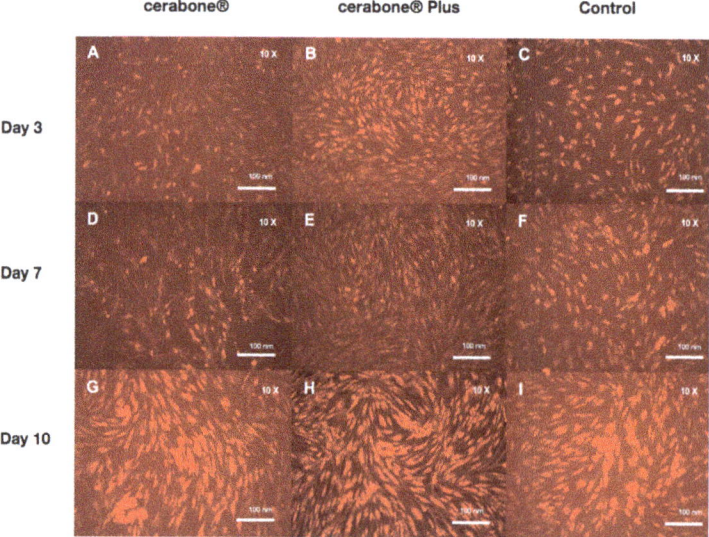

Figure 1. Fluorescence imaging (red cell tracker) in groups with BSM− (cerabone®), BSM+ (cerabone® Plus including HA) and the control (HOB alone) on days 3 (**A–C**), 7 (**D–F**) and 10 (**G–I**).

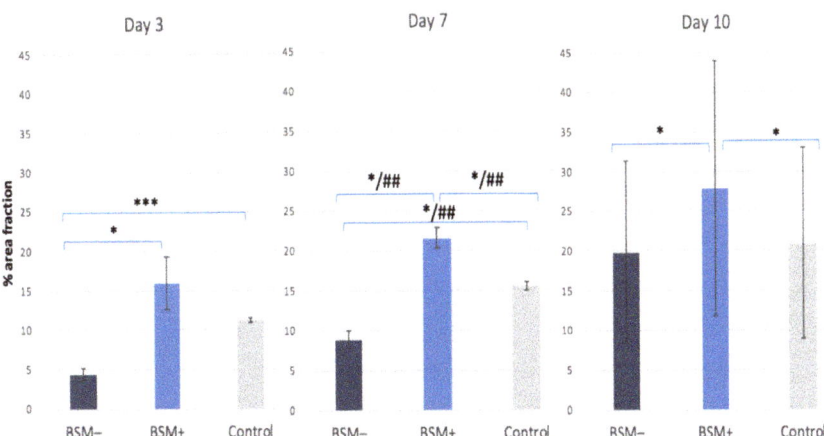

Figure 2. Cell Viability. Percentage area fraction (%) of fluorescence-stained HOBs at a ten-fold magnification for BSM− (cerabone®), BSM+ (cerabone® Plus including HA) and the control (HOB alone) on days 3, 7 and 10. * = $p \leq 0.05$, *t*-test; *** = $p \leq 0.0001$, *t*-test; ## = $p \leq 0.05$, KWT.

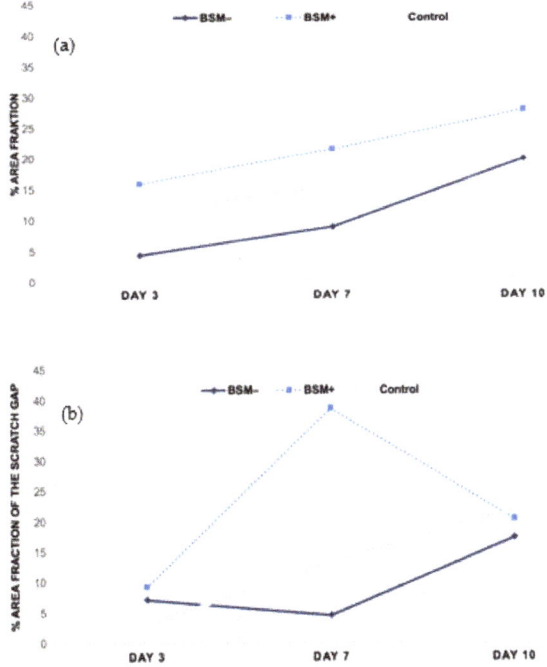

Figure 3. *Cont.*

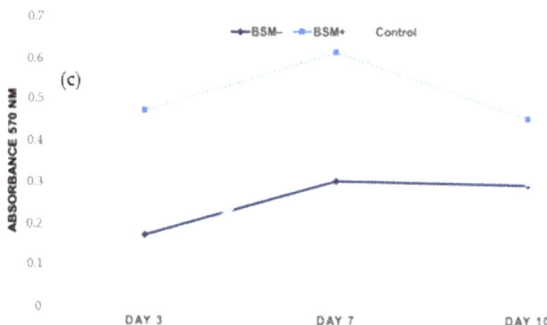

Figure 3. Tendency through the period of 3, 7 and 10 days within groups BSM− (cerabone®), BSM+ (cerabone® Plus) and the control (HOB alone). (**a**) Cell viability, (**b**) migration ability, (**c**) proliferation rate.

3.2. Cell Migration

On day 3, the highest cell migration rate was found for BSM+ followed by BSM− and the control (each $p > 0.05$, t-test). On day 7, the highest value was observed for BSM+ ($p < 0.05$, KWT). The controls showed a significantly increased proliferation rate when compared with BSM− ($p = 0.007$, t-test). On day 10, the best wound closure was observed for the control followed by BSM+ and BSM− ($p > 0.05$ each, t-test) (Table 2, Figure 4). The migration ability in BSM+ increased from day 3 to day 7 by five and a half times and then decreased almost two times until day 10, being on day 10 almost on the same level with BSM− and the control group (Figure 3b).

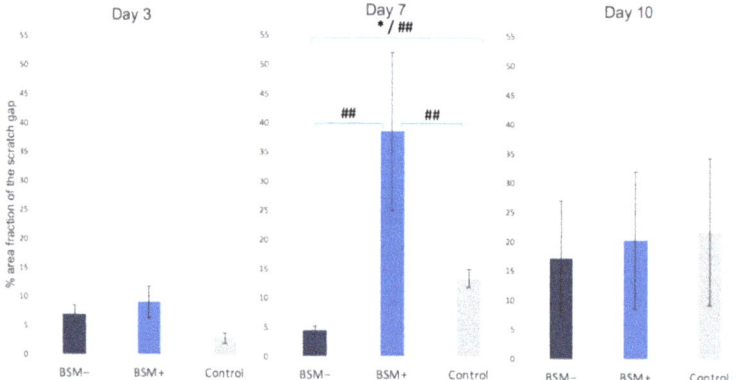

Figure 4. Migration ability: Percentage area fraction of the scratch gap (%) of fluorescence-stained HOBs at a ten-fold magnification for BSM− (cerabone®), BSM+ (cerabone® Plus including HA) and the control (HOB alone) on days 3, 7 and 10. * = $p \leq 0.05$, t-test; ## = $p \leq 0.05$, KWT.

Table 2. Migration ability: Percentage area fraction of the scratch gap (%) of fluorescence-stained HOBs at a ten-fold magnification for BSM− (cerabone®), BSM+ (cerabone® Plus including HA) and the control (HOB alone) on days 3, 7 and 10. The mean values are for parametric data and the median values are for non-parametric data.

	Day 3		Day 7		Day 10	
	Mean Value	SEM	Mean Value	SEM	Mean Value	SEM
BSM−	7.01	±1.49	4.51	±0.75	17.19	±9.93
BSM+	9.15	±2.74	38.57	±13.47	20.3	±11.72
Control	2.91	±0.92	13.46	±1.59	21.81	±12.59

3.3. Cell Proliferation

On day 3, the highest cell proliferation was observed for BSM+ in comparison with BSM− ($p = 0.011$, t-test; $p < 0.05$, KWT) and the control ($p < 0.001$, t-test; $p < 0.05$, KWT) followed by BSM− and the control ($p < 0.05$ each, KWT). On day 7, an increased proliferation rate was shown for BSM+ in comparison with BSM− ($p = 0.019$, t-test; $p < 0.05$, KWT) and the control ($p < 0.05$, KWT) while the control demonstrated increased values compared with BSM− ($p = 0.046$, t-test; $p < 0.05$, KWT). On day 10, the least proliferative activity was measured for BSM− ($p > 0.05$, MWT). Here, the highest proliferation rate was demonstrated for BSM+ ($p > 0.05$, MWT) (Table 3, Figure 5). The groups generally showed a tendency to increase up to day 7 and decrease until day 10. The highest raise rate was observed in the control on day 7. However, BSM+ stayed far on the top throughout the whole period (Figure 3c).

Table 3. Cell Proliferation: MTT assay, absorbance at 570 nm for BSM− (cerabone®), BSM+ (cerabone® Plus including HA) and the control (HOB alone) on days 3, 7 and 10. The mean values are for parametric data and the median values (*) are for non-parametric data.

	Day 3		Day 7		Day 10	
	Mean Value	SEM	Mean Value	SEM	Mean Value	SEM
BSM−	0.17	±0.06	0.3	±0.04	0.29	±0.07
BSM+	0.47	±0.03	0.61	±0.07	0.54 *	-
Control	0.06	±0.02	0.44	±0.03	0.3	0.05

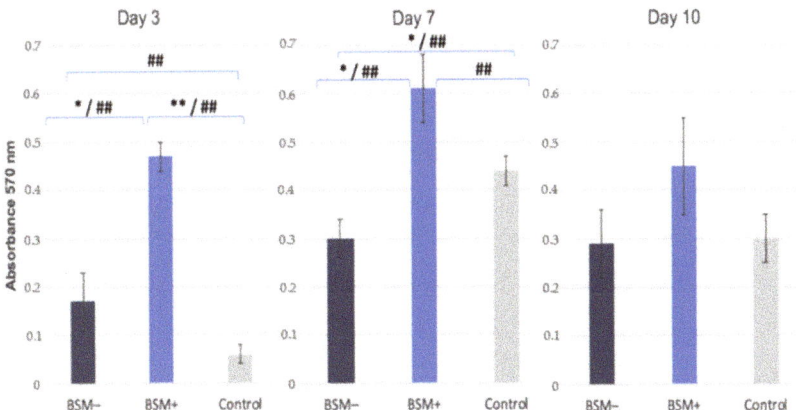

Figure 5. Cell Proliferation: MTT assay, absorbance at 570 nm for BSM− (cerabone®), BSM+ (cerabone® Plus including HA) and Control (HOB alone) on days 3, 7 and 10. * = $p \leq 0.05$, t-test; ** = $p \leq 0.001$, t-test; ## = $p \leq 0.05$, KWT.

4. Discussion

This in vitro study analyzed the effects of HA in combination with commercially available BSMs of bovine origin on human HOB cell viability, migration ability and proliferation rate. The overall findings demonstrated a significant benefit of HA biofunctionalization of BSMs on the above-mentioned HOB cell features responsible for bone regeneration. In brief, the modification of bovine BSM with HA significantly increased the biological activity of HOBs in comparison with the same BSM alone. The cell viability presented a smooth increase through the whole period where BSM+ stayed distinctly the highest of all groups. HA additivity activated the migration ability on days 3, 7 and 10. The cell proliferation in its turn was significantly affected on day 7 and presented only a slight difference among groups on day 3 and day 10.

In a previous study, we evaluated different commercially available BSMs of bovine origin in regard to their biological effect on human HOBs. Here, the high temperature (>1200 °C) sintered bovine BSM, which was included in the present study whether alone or in combination with an injectable PRF, seemed to have the best effects on HOB cell viability, metabolic activity and gene expression of alkaline phosphatase (ALP), osteonectin and BMP-2 when compared with other BSMs of bovine origin prepared at lower temperatures [29]. Hence, the aforementioned BSMs of bovine origin commercially modified with HA or pure were included in the present study. According to our results, the combination of HA manufactured by bacterial fermentation and bovine BSMs presents an increase in the biological activity of HOBs in comparison with the same BSM alone. Accordingly, the cell viability in all groups presented a smooth increase throughout the whole period where they stayed distinctly the highest in groups with HA modification. Moreover, HA biofunctionalization activated the proliferation rate of HOBs on days 3, 7 and 10. The cell proliferation in its turn was significantly affected on day 7 and presented only a slight difference among groups on days 3 and 10. Our findings, that HA positively affects HOB bioactivity, were in accordance with other in vitro and in vivo studies although, to the best of our knowledge, there are no in vitro studies dealing with information about the effects of HA in combination with BSMs of bovine origin on HOBs. Kawano et al. concluded that HA enhances the osteogenic activity of HOBs in vitro via the down-regulation of BMP-2 antagonists and the phosphorylation of extracellular signal-regulated kinase [35]. Thus, chemically cross-linked hyaluronan-based hydrogels with HA and BMP-2 demonstrated cancellous bone formation in ectopic sites after five weeks [53]. HA functionalization of a titanium surface seems to enhance HOB proliferation and alkaline phosphatase activity [54,55]. Furthermore, HA has been studied to modify the composition of the extracellular matrix, affecting its fibrillary and non-fibrillar components [56]. Sasaki et al. suggested that HA acts as a detent for growth factors even enhancing HOB activity [45]. Interestingly, HA and its side groups happen to reduce bacterial adhesion and prevent biofilm formation [57].

In spite of intensive research in this area, there are no evident studies proving a HA-specific mechanism of interactions and pathways considering osteogenesis [58]. It has been reported that HA affects wound healing by enhancing the CD44 surface marker consequently activating early inflammation and cell migration into granulation tissue [59]. Due to the similarity with the extracellular matrix, HA seems to be biocompatible inducing a low immune response. Furthermore, it accelerates cell adhesion, migration and proliferation and, as a result, to some extent new tissue formation [60]. However, HA presents a low mechanical strength and a high degradation rate, thus being limited and requiring appropriate modifications [61]. A combination of HA with gelatin and alginate into a three-dimensional composite scaffold showed to be high load bearing without fractural deformation [62]. Mathews et al. presented a scaffold with a chitosan-collagen-HA ratio of 1:1:0.1 in which lower HA concentrations and more uniform pores seemed to enhance HOB differentiation-promoting effects [63]. Furthermore, HA appears to be capable of encapsulating bioactive factors by cross-linking [64,65]. Nevertheless, the general process of HA cell bioactivation, due to its complexity, especially including osteogenesis, is still unclear [58]. Presently, the effect of HA as an enhancer of the biological properties of a synthetic scaffold, an activator of osteogenesis and as a vector for osteoinductive substances is approved [66].

It is known that HA combined with BSM+ is of a bacterial origin non-cross linked high molecular weight hyaluronic acid (h-HA) with a molecular mass of 1.9–2.1 MDa. However, it belongs to the limit of our study that the amount of it added to the BSM was not given. It has been reported that the molecular weight of HA is greatly decisive regarding the effect on the biological activity of cells and pro-inflammatory characteristics [45,67]. However, there is no consensus in the literature regarding the ideal constitution and concentration of HA for better bone regeneration [68]. Thus, Boeckel et al. observed a decrease in HOB viability under presence of HA and referred this not to chemical composition but rather to the molecular weight of HA [68]. The same findings were found in other studies [49,69,70]. Hence, it was suggested by that h-HA positively alters the cellular parameters of HOBs and influences peculiar inflammatory mediators, acting as an adjustor of HOB biological capacities [71]. Furthermore, Agarwal et al. demonstrated that h-HA in comparison with a low molecular one (l-HA) presented a significantly increased osteogenic differentiation of HOBs based on an upregulation of ALP, collagen and EM mineralization as well as the effects of l-HA, in its turn, on HOB proliferation and adhesion [50].

Another limitation of our study was sample staining using the Cell Tracker 5-chloromethylfluorescein diacetate for cell viability, which also permeates dead cell membranes. However, stained live cells are >100-fold brighter than dead cells and could be easily distinguished from the dead population [72]. It also belongs to the limits of the study that bone substitutes of two different particle sizes were compared: 0.5–1 mm versus 1–2 mm. However, the difference was not significant and may not have affected the results [73–75]. The critical difference was studied to be between the particles of less than 0.4 mm and more than 1 mm [73]. However, another study contradicted this statement, concluding that particles of 0.1–0.3 mm and 0.5–0.7 mm were not significantly different in terms of their osteogenic potential [74]. Another study suggested that the granularity was not of a significant relevance but was rather dependent on the clinical defect size. It seems that the microstructure characteristics of the material rather than its granularity plays an important role [75].

The implementation of HA in combination with bone substitute materials may be very promising to overcome any limitations in the soft and hard tissue regeneration. HA modified BSMs have the advantage of being classified as commercially available medical devices ready to use. Further in vitro and in vivo studies of HA in combination with BSMs of different origins will carve out the significance of dosage and the molecular weight of HA in bone engineering as far as there are no specific mechanisms of interactions and pathways considering HA involved in osteogenesis [58]. Clinical trials will focus on visible benefits such as the bone regeneration capacity and long-term stability in vivo.

5. Conclusions

HA biofunctionalization of BSMs enhancing the viability, migration ability and proliferation rate of human osteogenic cells on days 3, 7 and 10 might be able to accelerate and improve bone regeneration and might represent a promising alternative to native BSMs.

Author Contributions: Conceptualization, S.K. and P.W.K.; methodology, S.K. and P.W.K.; software, S.K. and P.W.K.; validation, S.K. and P.W.K.; formal analysis, S.K., D.H. and P.W.K.; investigation, S.K. and P.W.K.; resources, P.W.K.; data curation, S.K., D.H. and P.W.K.; writing—original draft preparation, S.K., A.P. and P.W.K.; writing—review and editing, S.K., A.P. and P.W.K.; visualization, S.K. and P.W.K.; supervision, P.W.K.; project administration, P.W.K.; funding acquisition, P.W.K. All authors have read and agreed to the published version of the manuscript.

Funding: Botiss biomaterials GmbH kindly provided the bone substitute material for the research.

Institutional Review Board Statement: Not applicable.

Informed Consent Statement: Not applicable.

Data Availability Statement: Data are available on request.

Conflicts of Interest: The authors A.P. and P.W.K. received speaker fees and research support from botiss biomaterials GmbH and Straumann AG for other studies. This had no influence on the current study. For this study, free samples of cerabone® and cerabone® Plus were received from botiss biomaterials GmbH.

References

1. Greenwald, A.S.; Boden, S.D.; Goldberg, V.M.; Khan, Y.; Laurencin, C.T.; Rosier, R.N. Bone-Graft Substitutes: Facts, Fictions, and Applications. *J. Bone Jt. Surg.* **2001**, *83*, 98–103. [CrossRef]
2. Agarwal, R.; García, A.J. Biomaterial strategies for engineering implants for enhanced osseointegration and bone repair. *Adv. Drug Deliv. Rev.* **2015**, *94*, 53–62. [CrossRef]
3. Turnbull, G.; Clarke, J.; Picard, F.; Riches, P.; Jia, L.; Han, F.; Li, B.; Shu, W. 3D bioactive composite scaffolds for bone tissue engineering. *Bioact. Mater.* **2018**, *3*, 278–314. [CrossRef]
4. Al-Moraissi, E.; Alkhutari, A.; Abotaleb, B.; Altairi, N.; Del Fabbro, M. Do osteoconductive bone substitutes result in similar bone regeneration for maxillary sinus augmentation when compared to osteogenic and osteoinductive bone grafts? A systematic review and frequentist network meta-analysis. *Int. J. Oral Maxillofac. Surg.* **2020**, *49*, 107–120. [CrossRef] [PubMed]
5. Buser, Z.; Brodke, D.S.; Youssef, J.A.; Meisel, H.-J.; Myhre, S.L.; Hashimoto, R.; Park, J.-B.; Yoon, S.T.; Wang, J.C. Synthetic bone graft versus autograft or allograft for spinal fusion: A systematic review. *J. Neurosurg. Spine* **2016**, *25*, 509–516. [CrossRef] [PubMed]
6. Baldwin, P.; Li, D.J.; Auston, D.A.; Mir, H.S.; Yoon, R.S.; Koval, K.J. Autograft, Allograft, and Bone Graft Substitutes: Clinical Evidence and Indications for Use in the Setting of Orthopaedic Trauma Surgery. *J. Orthop. Traum.* **2019**, *33*, 203–213. [CrossRef] [PubMed]
7. Liu, L.; Lu, S.-T.; Liu, A.-H.; Hou, W.-B.; Cao, W.-R.; Zhou, C.; Yin, Y.-X.; Yuan, K.-S.; Liu, H.-J.; Zhang, M.-G.; et al. Comparison of complications in cranioplasty with various materials: A systematic review and meta-analysis. *Br. J. Neurosurg.* **2020**, *34*, 388–396. [CrossRef]
8. Avila-Ortiz, G.; Chambrone, L.; Vignoletti, F. Effect of alveolar ridge preservation interventions following tooth extraction: A systematic review and meta-analysis. *J. Clin. Periodontol.* **2019**, *46*, 195–223. [CrossRef] [PubMed]
9. Mendoza-Azpur, G.; de la Fuente, A.; Chavez, E.; Valdivia, E.; Khouly, I. Horizontal ridge augmentation with guided bone regeneration using particulate xenogenic bone sub-stitutes with or without autogenous block grafts: A randomized controlled trial. *Clin. Implant Dent. Relat. Res.* **2019**, *21*, 521–530.
10. Fouad, W.; Osman, A.; Atef, M.; Hakam, M. Guided maxillary sinus floor elevation using deproteinized bovine bone versus graftless Schneiderian membrane elevation with simultaneous implant placement: Randomized clinical trial. *Clin. Implant. Dent. Relat. Res.* **2018**, *20*, 424–433. [CrossRef] [PubMed]
11. Naros, A.; Bayazeed, B.; Schwarz, U.; Nagursky, H.; Reinert, S.; Schmelzeisen, R.; Sauerbier, S. A prospective histomorphometric and cephalometric comparison of bovine bone substitute and autogenous bone grafting in Le Fort I osteotomies. *J. Cranio-Maxillofac. Surg.* **2019**, *47*, 233–238. [CrossRef]
12. Falacho, R.; Palma, P.; Marques, J.; Figueiredo, M.; Caramelo, F.; Dias, I.; Viegas, C.; Guerra, F. Collagenated Porcine Heterologous Bone Grafts: Histomorphometric Evaluation of Bone Formation Using Different Physical Forms in a Rabbit Cancellous Bone Model. *Molecules* **2021**, *26*, 1339. [CrossRef] [PubMed]
13. Bohner, M. Calcium orthophosphates in medicine: From ceramics to calcium phosphate cements. *Injury* **2000**, *31*, D37–D47. [CrossRef]
14. Klein, M.O.; Kämmerer, P.W.; Götz, H.; Duschner, H.; Wagner, W. Long-Term Bony Integration and Resorption Kinetics of a Xenogeneic Bone Substitute After Sinus Floor Augmentation: Histomorphometric Analyses of Human Biopsy Specimens. *Int. J. Periodontics Restor. Dent.* **2013**, *33*, e101–e110. [CrossRef] [PubMed]
15. Dau, M.; Kämmerer, P.W.; Henkel, K.-O.; Gerber, T.; Frerich, B.; Gundlach, K.K.H. Bone formation in mono cortical mandibular critical size defects after augmentation with two synthetic nanostructured and one xenogenous hydroxyapatite bone substitute-in vivo animal study. *Clin. Oral Implants Res.* **2016**, *27*, 597–603. [CrossRef] [PubMed]
16. Glowacki, J. A review of osteoinductive testing methods and sterilization processes for demineralized bone. *Cell Tissue Bank.* **2005**, *6*, 3–12. [CrossRef] [PubMed]
17. Yamada, M.; Egusa, H. Current bone substitutes for implant dentistry. *J. Prosthodont. Res.* **2018**, *62*, 152–161. [CrossRef] [PubMed]
18. Laschke, M.W.; Witt, K.; Pohlemann, T.; Menger, M.D. Injectable nanocrystalline hydroxyapatite paste for bone substitution:In vivo analysis of biocompatibility and vascularization. *J. Biomed. Mater. Res. Part B Appl. Biomater.* **2007**, *82*, 494–505. [CrossRef] [PubMed]
19. Gehrke, S.A.; Mazón, P.; Pérez-Díaz, L.; Calvo-Guirado, J.L.; Velásquez, P.; Aragoneses, J.M.; Fernández-Domínguez, M.; De Aza, P.N. Study of Two Bovine Bone Blocks (Sintered and Non-Sintered) Used for Bone Grafts: Physico-Chemical Charac-terization and In Vitro Bioactivity and Cellular Analysis. *Materials* **2019**, *12*, 452. [CrossRef]
20. De Carvalho, B.; Rompen, E.; Lecloux, G.; Schupbach, P.; Dory, E.; Art, J.-F.; Lambert, F. Effect of Sintering on In Vivo Biological Performance of Chemically Deproteinized Bovine Hydroxyapatite. *Materials* **2019**, *12*, 3946. [CrossRef]

21. Ong, J.L.; Hoppe, C.A.; Cardenas, H.L.; Cavin, R.; Carnes, D.L.; Sogal, A.; Raikar, G.N. Osteoblast precursor cell activity on HA surfaces of different treatments. *J. Biomed. Mater. Res. Off. J. Soc. Biomater. Jpn. Soc. Biomater. Aust. Soc. Biomater.* **1998**, *39*, 176–183. [CrossRef]
22. Perić Kačarević, Z.; Kavehei, F.; Houshmand, A.; Franke, J.; Smeets, R.; Rimashevskiy, D.; Wenisch, S.; Schnettler, R.; Jung, O.; Barbeck, M. Purification processes of xenogeneic bone substitutes and their impact on tissue reactions and regeneration. *Int. J. Artif. Organs* **2018**, *41*, 789–800. [CrossRef]
23. Accorsi-Mendonça, T.; Conz, M.B.; Barros, T.C.; Sena, L.Á.D.; Soares, G.D.A.; Granjeiro, J.M. Physicochemical characterization of two deproteinized bovine xenografts. *Braz. Oral Res.* **2008**, *22*, 5–10. [CrossRef]
24. Kusrini, E.; Sontang, M. Characterization of X-ray diffraction and electron spin resonance: Effects of sintering time and temperature on bovine hydroxyapatite. *Radiat. Phys. Chem.* **2012**, *81*, 118–125. [CrossRef]
25. Riachi, F.; Naaman, N.; Tabarani, C.; Aboelsaad, N.; Aboushelib, M.N.; Berberi, A.; Salameh, Z. Influence of Material Properties on Rate of Resorption of Two Bone Graft Materials after Sinus Lift Using Radiographic Assessment. *Int. J. Dent.* **2012**, *2012*, 1–7. [CrossRef]
26. Trajkovski, B.; Jaunich, M.; Müller, W.-D.; Beuer, F.; Zafiropoulos, G.-G.; Houshmand, A. Hydrophilicity, Viscoelastic, and Physicochemical Properties Variations in Dental Bone Grafting Substitutes. *Materials* **2018**, *11*, 215. [CrossRef] [PubMed]
27. Redey, S.A.; Nardin, M.; Bernache-Assolant, D.; Rey, C.; Delannoy, P.; Sedel, L.; Marie, P.J. Behavior of human osteoblastic cells on stoichiometric hydroxyapatite and type A carbonate apatite: Role of surface energy. *J. Biomed. Mater. Res.* **2000**, *50*, 353–364. [CrossRef]
28. Kyyak, S.; Blatt, S.; Pabst, A.; Thiem, D.; Al-Nawas, B.; Kämmerer, P.W. Combination of an allogenic and a xenogenic bone substitute material with injectable platelet-rich fibrin—A comparative in vitro study. *J. Biomater. Appl.* **2020**, *35*, 83–96. [CrossRef]
29. Kyyak, S.; Blatt, S.; Schiegnitz, E.; Heimes, D.; Staedt, H.; Thiem, D.G.E.; Sagheb, K.; Al-Nawas, B.; Kämmerer, P.W. Activation of Human Osteoblasts via Different Bovine Bone Substitute Materials With and Without Injectable Platelet Rich Fibrin in vitro. *Front. Bioeng. Biotechnol.* **2021**, *9*, 71. [CrossRef] [PubMed]
30. Blatt, S.; Thiem, D.G.E.; Pabst, A.; Al-Nawas, B.; Kämmerer, P.W. Does Platelet-Rich Fibrin Enhance the Early Angiogenetic Potential of Different Bone Substitute Materials? An In Vitro and In Vivo Analysis. *Biomedicines* **2021**, *9*, 61. [CrossRef]
31. Blatt, S.; Thiem, D.G.E.; Kyyak, S.; Pabst, A.; Al-Nawas, B.; Kämmerer, P.W. Possible Implications for Improved Osteogenesis? The Combination of Platelet-Rich Fibrin With Different Bone Sub-stitute Materials. *Front. Bioeng. Biotechnol.* **2021**, *9*, 640053. [CrossRef] [PubMed]
32. Teng, F.; Wei, L.; Yu, D.; Deng, L.; Zheng, Y.; Lin, H.; Liu, Y. Vertical bone augmentation with simultaneous implantation using deproteinized bovine bone block functionalized with a slow delivery of BMP-2. *Clin. Oral Implant. Res.* **2019**, *31*, 215–228. [CrossRef] [PubMed]
33. Geiger, F.; Lorenz, H.; Xu, W.; Szalay, K.; Kasten, P.; Claes, L.; Augat, P.; Richter, W. VEGF producing bone marrow stromal cells (BMSC) enhance vascularization and resorption of a natural coral bone substitute. *Bone* **2007**, *41*, 516–522. [CrossRef] [PubMed]
34. Price, R.D.; Berry, M.; Navsaria, H.A. Hyaluronic acid: The scientific and clinical evidence. *J. Plast. Reconstr. Aesthetic Surg.* **2007**, *60*, 1110–1119. [CrossRef] [PubMed]
35. Kawano, M.; Ariyoshi, W.; Iwanaga, K.; Okinaga, T.; Habu, M.; Yoshioka, I.; Tominaga, K.; Nishihara, T. Mechanism involved in enhancement of osteoblast differentiation by hyaluronic acid. *Biochem. Biophys. Res. Commun.* **2011**, *405*, 575–580. [CrossRef]
36. Knudson, C.B.; Knudson, W. Hyaluronan-binding proteins in development, tissue homeostasis, and disease. *FASEB J.* **1993**, *7*, 1233–1241. [CrossRef]
37. Lee, J.Y.; Spicer, A.P. Hyaluronan: A multifunctional, megaDalton, stealth molecule. *Curr. Opin. Cell Biol.* **2000**, *12*, 581–586. [CrossRef]
38. Aigner, J.; Tegeler, J.; Hutzler, P.; Campoccia, D.; Pavesio, A.; Hammer, C.; Kastenbauer, E.; Naumann, A. Cartilage tissue engineering with novel nonwoven structured biomaterial based on hyaluronic acid benzyl ester. *J. Biomed. Mater. Res. Off. J. Soc. Biomater. Jpn. Soc. Biomater. Aust. Soc. Biomater.* **1998**, *42*, 172–181. [CrossRef]
39. Grigolo, B.; Lisignoli, G.; Piacentini, A.; Fiorini, M.; Gobbi, P.; Mazzotti, G.; Duca, M.; Pavesio, A.; Facchini, A. Evidence for redifferentiation of human chondrocytes grown on a hyaluronan-based biomaterial (HYAFF®11): Molecular, immunohistochemical and ultrastructural analysis. *Biomaterials* **2002**, *23*, 1187–1195. [CrossRef]
40. Lisignoli, G.; Fini, M.; Giavaresi, G.; Aldini, N.N.; Toneguzzi, S.; Facchini, A. Osteogenesis of large segmental radius defects enhanced by basic fibroblast growth factor activated bone marrow stromal cells grown on non-woven hyaluronic acid-based polymer scaffold. *Biomaterials* **2002**, *23*, 1043–1051. [CrossRef]
41. Palma, P.J.; Ramos, J.; Martins, J.B.; Diogenes, A.; Figueiredo, M.H.; Ferreira, P.; Viegas, C.; Santos, J.M. Histologic Evaluation of Regenerative Endodontic Procedures with the Use of Chitosan Scaffolds in Immature Dog Teeth with Apical Periodontitis. *J. Endod.* **2017**, *43*, 1279–1287. [CrossRef]
42. Spessotto, P.; Rossi, F.M.; Degan, M.; Di Francia, R.; Perris, R.; Colombatti, A.; Gattei, V. Hyaluronan–CD44 interaction hampers migration of osteoclast-like cells by down-regulating MMP-9. *J. Cell Biol.* **2002**, *158*, 1133–1144. [CrossRef]
43. Tadmor, R.; Chen, N.; Israelachvili, J.N. Thin film rheology and lubricity of hyaluronic acid solutions at a normal physiological concentration. *J. Biomed. Mater. Res.* **2002**, *61*, 514–523. [CrossRef] [PubMed]
44. Pirnazar, P.; Wolinsky, L.; Nachnani, S.; Haake, S.; Pilloni, A.; Bernard, G.W. Bacteriostatic Effects of Hyaluronic Acid. *J. Periodontol.* **1999**, *70*, 370–374. [CrossRef]

45. Sasaki, T.; Watanabe, C. Stimulation of osteoinduction in bone wound healing by high-molecular hyaluronic acid. *Bone* **1995**, *16*, 9–15. [CrossRef]
46. West, D.C.; Kumar, S. Hyaluronan and Angiogenesis. *Novartis Found. Symp.* **2007**, *143*, 187–207. [CrossRef]
47. Guo, J.; Guo, S.; Wang, Y.; Yu, Y. Adipose-derived stem cells and hyaluronic acid based gel compatibility, studied in vitro. *Mol. Med. Rep.* **2017**, *16*, 4095–4100. [CrossRef] [PubMed]
48. Agolli, E.; Diffidenti, B.; Di Zitti, N.; Massidda, E.; Patella, F.; Santerini, C.; Beatini, A.; Bianchini, M.; Bizzarri, S.; Camilleri, V.; et al. Hybrid cooperative complexes of high and low molecular weight hyaluronans (Profhilo®): Review of the literature and presentation of the VisionHA project. *Esperienze Dermatol.* **2018**, *20*, 5–14.
49. Pilloni, A.; Bernard, G.W. The effect of hyaluronan on mouse intramembranous osteogenesis in vitro. *Cell Tissue Res.* **1998**, *294*, 323–333. [CrossRef]
50. Agarwal, S.; Duffy, B.; Curtin, J.; Jaiswal, S. Effect of High- and Low-Molecular-Weight Hyaluronic-Acid-Functionalized-AZ31 Mg and Ti Alloys on Proliferation and Differentiation of Osteoblast Cells. *ACS Biomater. Sci. Eng.* **2018**, *4*, 3874–3884. [CrossRef]
51. Fuchs, S.; Jiang, X.; Schmidt, H.; Dohle, E.; Ghanaati, S.; Orth, C.; Hofmann, A.; Motta, A.; Migliaresi, C.; Kirkpatrick, C.J. Dynamic processes involved in the pre-vascularization of silk fibroin constructs for bone regeneration using outgrowth endothelial cells. *Biomaterials* **2009**, *30*, 1329–1338. [CrossRef]
52. Liang, C.-C.; Park, A.Y.; Guan, J.-L. In vitro scratch assay: A convenient and inexpensive method for analysis of cell migration in vitro. *Nat. Protoc.* **2007**, *2*, 329–333. [CrossRef] [PubMed]
53. Stenfelt, S.; Hulsart-Billstrom, G.; Gedda, L.; Bergman, K.; Hilborn, J.; Larsson, S.; Bowden, T. Pre-incubation of chemically crosslinked hyaluronan-based hydrogels, loaded with BMP-2 and hydroxyapatite, and its effect on ectopic bone formation. *J. Mater. Sci. Mater. Electron.* **2014**, *25*, 1013–1023. [CrossRef]
54. Hwang, D.S.; Waite, J.H.; Tirrell, M. Promotion of osteoblast proliferation on complex coacervation-based hyaluronic acid–recombinant mussel adhesive protein coatings on titanium. *Biomaterials* **2010**, *31*, 1080–1084. [CrossRef] [PubMed]
55. Chua, P.-H.; Neoh, K.-G.; Kang, E.-T.; Wang, W. Surface functionalization of titanium with hyaluronic acid/chitosan polyelectrolyte multilayers and RGD for promoting osteoblast functions and inhibiting bacterial adhesion. *Biomaterials* **2008**, *29*, 1412–1421. [CrossRef]
56. Schmidt, J.R.; Vogel, S.; Moeller, S.; Kalkhof, S.; Schubert, K.; von Bergen, M.; Hempel, U. Sulfated hyaluronic acid and dexamethasone possess a synergistic potential in the differentiation of osteoblasts from human bone marrow stromal cells. *J. Cell. Biochem.* **2019**, *120*, 8706–8722. [CrossRef]
57. Palumbo, F.S.; Volpe, A.B.; Cusimano, M.G.; Pitarresi, G.; Giammona, G.; Schillaci, D. A polycarboxylic/amino functionalized hyaluronic acid derivative for the production of pH sensible hydrogels in the prevention of bacterial adhesion on biomedical surfaces. *Int. J. Pharm.* **2015**, *478*, 70–77. [CrossRef]
58. Zhai, P.; Peng, X.; Li, B.; Liu, Y.; Sun, H.; Li, X. The application of hyaluronic acid in bone regeneration. *Int. J. Biol. Macromol.* **2020**, *151*, 1224–1239. [CrossRef]
59. Kobayashi, H.; Terao, T. Hyaluronic acid-specific regulation of cytokines by human uterine fibroblasts. *Am. J. Physiol. Physiol.* **1997**, *273*, C1151–C1159. [CrossRef]
60. Zhao, W.; Jin, X.; Cong, Y.; Liu, Y.; Fu, J. Degradable natural polymer hydrogels for articular cartilage tissue engineering. *J. Chem. Technol. Biotechnol.* **2012**, *88*, 327–339. [CrossRef]
61. García-Gareta, E.; Coathup, M.J.; Blunn, G.W. Osteoinduction of bone grafting materials for bone repair and regeneration. *Bone* **2015**, *81*, 112–121. [CrossRef]
62. Singh, D.; Tripathi, A.; Zo, S.; Singh, D.; Han, S.S. Synthesis of composite gelatin-hyaluronic acid-alginate porous scaffold and evaluation for in vitro stem cell growth and in vivo tissue integration. *Colloids Surf. B Biointerfaces* **2014**, *116*, 502–509. [CrossRef]
63. Mathews, S.; Bhonde, R.; Gupta, P.K.; Totey, S. Novel biomimetic tripolymer scaffolds consisting of chitosan, collagen type 1, and hyaluronic acid for bone marrow-derived human mesenchymal stem cells-based bone tissue engineering. *J. Biomed. Mater. Res. Part B Appl. Biomater.* **2014**, *102*, 1825–1834. [CrossRef]
64. Huang, Y.; Luo, Q.; Li, X.; Zhang, F.; Zhao, S. Fabrication and in vitro evaluation of the collagen/hyaluronic acid PEM coating crosslinked with functionalized RGD peptide on titanium. *Acta Biomater.* **2012**, *8*, 866–877. [CrossRef]
65. Xu, X.; Jha, A.; Duncan, R.L.; Jia, X. Heparin-decorated, hyaluronic acid-based hydrogel particles for the controlled release of bone morphogenetic protein 2. *Acta Biomater.* **2011**, *7*, 3050–3059. [CrossRef] [PubMed]
66. Kumar, P.S.; Hashimi, S.; Saifzadeh, S.; Ivanovski, S.; Vaquette, C. Additively manufactured biphasic construct loaded with BMP-2 for vertical bone regeneration: A pilot study in rabbit. *Mater. Sci. Eng. C* **2018**, *92*, 554–564. [CrossRef] [PubMed]
67. Rayahin, J.E.; Buhrman, J.S.; Zhang, Y.; Koh, T.J.; Gemeinhart, R.A. High and Low Molecular Weight Hyaluronic Acid Differentially Influence Macrophage Activation. *ACS Biomater. Sci. Eng.* **2015**, *1*, 481–493. [CrossRef] [PubMed]
68. Boeckel, D.G.; Shinkai, R.S.A.; Grossi, M.L.; Teixeira, E. In vitro evaluation of cytotoxicity of hyaluronic acid as an extracellular matrix on OFCOL II cells by the MTT assay. *Oral Surg. Oral Med. Oral Pathol. Oral Radiol.* **2014**, *117*, e423–e428. [CrossRef]
69. Kunze, R.; Rösler, M.; Möller, S.; Schnabelrauch, M.; Riemer, T.; Hempel, U.; Dieter, P. Sulfated hyaluronan derivatives reduce the proliferation rate of primary rat calvarial osteoblasts. *Glycoconj. J.* **2010**, *27*, 151–158. [CrossRef] [PubMed]
70. Huang, L.; Cheng, Y.Y.; Koo, P.L.; Lee, K.M.; Qin, L.; Cheng, J.C.Y.; Kumta, S.M. The effect of hyaluronan on osteoblast proliferation and differentiation in rat calvarial-derived cell cultures. *J. Biomed. Mater. Res.* **2003**, *66*, 880–884. [CrossRef]

71. Lajeunesse, D.; Delalandre, A.; Martel-Pelletier, J.; Pelletier, J.-P. Hyaluronic acid reverses the abnormal synthetic activity of human osteoarthritic subchondral bone osteoblasts. *Bone* **2003**, *33*, 703–710. [CrossRef]
72. Johnson, S.; Nguyen, V.; Coder, D. Assessment of Cell Viability. *Curr. Protoc. Cytom.* **2013**, *64*, 9.2.1–9.2.26. [CrossRef] [PubMed]
73. Committee on Research, Science and Therapy of the American Academy of Periodontology. Position Paper, Tissue Banking of Bone Allografts Used in Periodontal Regeneration. *J. Periodontol.* **2001**, *72*, 834–838. [CrossRef] [PubMed]
74. Lee, S.-U.; Chung, Y.-G.; Kim, S.-J.; Oh, I.-H.; Kim, Y.-S.; Ju, S.-H. Does size difference in allogeneic cancellous bone granules loaded with differentiated autologous cultured osteoblasts affect osteogenic potential? *Cell Tissue Res.* **2013**, *355*, 337–344. [CrossRef] [PubMed]
75. Kolk, A.; Handschel, J.; Drescher, W.; Rothamel, D.; Kloss, F.; Blessmann, M.; Heiland, M.; Wolff, K.-D.; Smeets, R. Current trends and future perspectives of bone substitute materials—From space holders to innovative biomaterials. *J. Cranio-Maxillofac. Surg.* **2012**, *40*, 706–718. [CrossRef]

Article

Effects of Gamma Radiation-Induced Crosslinking of Collagen Type I Coated Dental Titanium Implants on Osseointegration and Bone Regeneration

Won-Tak Cho [1,†], So-Yeun Kim [2,†], Sung-In Jung [3], Seong-Soo Kang [4], Se-Eun Kim [4], Su-Hyun Hwang [1], Chang-Mo Jeong [1] and Jung-Bo Huh [1,*]

1. Department of Prosthodontics, Dental Research Institute, Dental and Life Sciences Institute, Education and Research Team for Life Science on Dentistry, School of Dentistry, Pusan National University, Yangsan-si 50612, Korea; joonetak@hanmail.net (W.-T.C.); hsh2942@hanmail.net (S.-H.H.); cmjeong@pusan.ac.kr (C.-M.J.)
2. Department of Prosthodontics, Biomedical Research Institute, Pusan National University Hospital, Busan 49241, Korea; soyeunkim179@gmail.com
3. Advanced Radiation Technology Institute, Korea Atomic Energy Research Institute, 29 Geumgu-gil, Jeongeup-si 56212, Korea; sijeong@kaeri.re.kr
4. Department of Veterinary Surgery and R&BD Center, College of Veterinary Medicine, Chonnam National University, Gwangju 61186, Korea; vetkang@chonnam.ac.kr (S.-S.K.); ksevet@chonnam.ac.kr (S.-E.K.)
* Correspondence: huhjb@pusan.ac.kr; Tel.: +82-10-8007-9099; Fax: +82-55-360-5134
† These authors contributed equally to this work.

Abstract: This study aimed to compare two methods of crosslinking collagen type I on implanted titanium surfaces, that is, using glutaraldehyde (GA) or gamma-rays (GRs), in a beagle dog model. For in vivo experiments, implants were allocated to three groups and applied to mandibular bone defects in beagle dogs; Group SLA; non-treated Sandblasted, large grit, acid-etched (SLA) implants, Group GA; SLA implants coated with GA crosslinked collagen type I, Group GR; SLA surface implants coated with collagen type I and crosslinked using 25 kGy of ^{60}Co gamma radiation. New bone µCT volumes were obtained, and histologic and histometric analyses were performed in regions of interest. The GR group had significantly better new bone areas (NBAs) and bone to implant contact (BIC) results than the SLA group ($p < 0.05$), but the GA and GR groups were similar in this respect. New bone volumes and inter-thread bone densities (ITBD) were non-significantly different in the three groups ($p > 0.05$). Within the limits of this study, gamma-ray collagen crosslinking on titanium implants can be considered a substitute for glutaraldehyde crosslinking.

Keywords: bone regeneration; collagen; gamma radiation; surface modification; titanium implant

Citation: Cho, W.-T.; Kim, S.-Y.; Jung, S.-I.; Kang, S.-S.; Kim, S.-E.; Hwang, S.-H.; Jeong, C.-M.; Huh, J.-B. Effects of Gamma Radiation-Induced Crosslinking of Collagen Type I Coated Dental Titanium Implants on Osseointegration and Bone Regeneration. *Materials* **2021**, *14*, 3268. https://doi.org/10.3390/ma14123268

Academic Editor: Paolo Cappare

Received: 28 April 2021
Accepted: 10 June 2021
Published: 13 June 2021

Publisher's Note: MDPI stays neutral with regard to jurisdictional claims in published maps and institutional affiliations.

Copyright: © 2021 by the authors. Licensee MDPI, Basel, Switzerland. This article is an open access article distributed under the terms and conditions of the Creative Commons Attribution (CC BY) license (https://creativecommons.org/licenses/by/4.0/).

1. Introduction

The interaction between bone and implant interfaces is the key to osseointegration, and various methods of modifying the surfaces of titanium implants have been introduced to improve this process [1–3]. Ti surface modifications influence bone regeneration and biocompatibility and facilitate successful implant fixation without soft tissue intervention [4–6]. Increasing surface roughness and coating implants with biocompatible materials or growth factors are known to increase the osseointegration of Ti implants [7]. In particular, collagen type I is used as a biocompatible polymer because it promotes osteoblast differentiation and provides a suitable environment for bone formation [8–10].

At the molecular level, collagen type I has a tangled, triple-helix structure with two α1 (I) and one α2 (I) polypeptide chains, and many years of clinical use have proven it to be a biocompatible, bioactive, bioresorbable material [11,12]. Implant surfaces coated with crosslinked collagen type I provide a favorable environment for initial osteoblast adhesion and stimulate their proliferation [9]. However, rapid absorption and decomposition by enzymes and immune reactions against animal-derived collagen cause type I collagen degradation; therefore, crosslinking is required to improve its in vivo stability [13]. Glutaraldehyde (GA) is commonly used as a crosslinker for collagen-based biomaterials, and GA cross-linking of collagen decreases its antigenicity, makes it resistant to phagocytosis, and invisible to the immune system [9,14,15]. However, like other chemical crosslinking methods, GA has been reported to produce harmful cytotoxic residues and increase proinflammatory cytokine release by macrophages [16–19]. Recently, different types of irradiation-induced crosslinking methods such as gamma-ray and ultraviolet have been used in preference to chemical crosslinkers substances to crosslink polymers like collagen [13,20,21].

Unlike ethylene oxide or GA sterilization, gamma radiation leaves no harmful residues that could potentially harm human health or the environment and is used to sterilize medical devices [6,22]. Moreover, gamma radiation-induced polymer crosslinking enables control of radiation-induced decomposition reactions, e.g., polymer chain scission, which can cause molecular weight reductions, as its effects are not dependent on material compositions [23–25]. Furthermore, when collagen is irradiated with gamma rays, peptide bonds are destroyed due to amino acid deformation, and hydrophilicity is improved by hydrogen bond formation [26]. In addition, enhancements of sandblasted, large grit, acid-etched (SLA) implant surface hydrophilicity have been reported to increase alkaline phosphatase (ALP) by more than 2-fold in cell culture experiments [27].

A previous comparative study concluded that there was no difference between the cytotoxicities of the gamma radiation crosslinked group and a GA-crosslinked group, based on absorbance data. However, gamma crosslinked collagen-coated Ti implants had significantly higher BICs than non-coated controls in a small animal model [28]. Therefore, we compared the effects of GA and gamma-ray crosslinking of collagen type I on the surfaces of SLA Ti implants in a beagle model to determine the effectiveness of gamma-induced cross-linking. The null hypothesis was that bone regeneration and osseointegration after GA or gamma crosslinking of collagen type I coated SLA implants are similar.

2. Materials and Methods

2.1. Experimental Materials

Collagen type I solution (0.5% (w/v)) was obtained by dissolving collagen (source: porcine skin, atelocollagen type I, Matrixen-PSP, Sk Bioland Co. Ltd., Cheonan, Korea) in 0.05 M acetic acid (Sigma-Aldrich, St. Louis, MO, USA) at room temperature. The Ti implant fixtures (D 4.0 mm × H 8.0 mm, SLA surface, Cowellmedi Co., Ltd., Pusan, Korea) were placed in a 0.5% (w/v) collagen type I solution. Bubbles on implant surfaces were removed by sonication (Elmasonic, S 180 H, Elma Schmidbauer, Elma, Germany) for 10 min. Then implants were placed in climate chambers (MIR-253, SANYO, Moriguchi, Japan) to dry for 1 h at 4 °C. Implants in the GA group were crosslinked by placing them in 2.5% (v/v) GA (DAEMYUNG CHEMICAL, Gyeonggi-do, Korea) for 1 h. Unreacted GA and collagen type I were then removed by washing in distilled water, dried in a vacuum oven (WOV-30, DAIHAN Scientific Co.Ltd., Gangwon-do, Korea) for 3 days [28], and sterilized with ethylene oxide (Manufacturer, City, State, Country). The implants of the GR (gamma-radiation) group were immersed in collagen solution in the same way as in the GA group, followed by ultrasonic cleaning for 10 min, and dried in a climate chamber for 1 h. The GR group implants were then irradiated with ^{60}Co gamma rays (MDS Nordion, Ottawa, ON, Canada) at 25 kGy for 1 h [28].

2.2. In Vitro Study

2.2.1. Scanning Electron Microscopy (SEM) Analysis

Surface images of implants were obtained using an SEM unit (Hitachi S3500N, Hitachi, Tokyo, Japan) at magnifications of ×40, ×5000, and ×50,000. For the SEM study, implants were splutter-coated with gold (SCD 005, BAL-TEC, Balzers, Liechtenstein). SEM images were obtained at 15 kV.

2.2.2. X-ray Photoelectron Spectroscopy (XPS)

Implant surfaces were analyzed by XPS (AXIS SUPRA, Kratos Analytical Ltd., Manchester, UK) using a monochromatic Al-Kα (1486.6 eV) X-ray source (1486.6 eV) at 15 kV and 225 W. The binding energy scale was calibrated at the C 1s level (284.5 eV). Implants in each group were subjected to a compositional survey at a pass energy of 160 eV, and core level spectra were obtained at a pass energy of 20 eV. Data analysis was performed using data reduction software (Vision 1.5, Kratos Analytical Ltd., Manchester, UK). Deconvoluted spectra were fitted using a Gaussian−Lorentzian sum function (20% Gaussian and 80% Lorentzian) using XPSPEAK Version 4.1 (Dr. Raymond Kwok, Hong Kong, China).

2.3. In Vivo Experiment

2.3.1. Experimental Animals

This study was approved by the Ethics Committee on Animal Experimentation of Chonnam National University (CNU IACUC-YB-2018-94). Six beagles (males, three years old, 12 kg) were used in the study.

2.3.2. Surgical Procedure

Beagles were anesthetized with a medetomidine (Tomidin®, Provet, Istanbul, Turkey) 10 μg/kg and tiletamine-zolazepam (Zoletil 50®, Virbac Laboratories, Carros, France) at 5 mg before the procedure and followed by isoflurane inhalation anesthesia (Sevoflurane®, Hana Pharm Co., Seoul, Korea). Anesthesia was maintained using tramadol (Maritrol®, Cheil Pharmaceutical, Uiwang, Korea) 2 mg/kg and carprofen (Rimadyl® inj, Zoetis, Parsippany, NJ, USA) 2.2 mg/kg IV. In addition, infiltration anesthesia at surgical sites was performed using 0.4 mL bupivacaine (Bupivacaine HCl 0.5% Inj., Myungmoon Pharm Co., Seoul, Korea). To prevent infection, 20 mg/kg of cefazolin sodium (Cefazolin®, Chongkundang Pharm Co., Seoul, Korea) was injected subcutaneously.

Mandibular premolars (P1–P4) and M1 molar were extracted after full mouth scaling. Implants were placed after extraction sites had healed for 8 weeks [28,29]. General anesthesia and local infiltration anesthesia were applied as described for extractions. A mid-crestal incision was made at each premolar site, and vertical incisions were made at the mucogingival junction. After mucoperiosteal flap elevation, crestal bone was homogenized by osteoplasty using a bone file and rongeur. Buccal cuboid defects, approximately 5 mm in height from crestal bone, 5 mm deep from the surface of the buccal bone, and 8 mm in width mesiodistally, were created using a straight fissure carbide bur under saline irrigation (JW Pharmaceutical Co. Ltd., Gyeonggi-do, Korea) (Figure 1A). Animals were allocated randomly to the three study groups, which were as follows:

- Group SLA (n = 12): Non-treated SLA implants.
- Group GA (n = 12): SLA implants coated with GA crosslinked collagen type I.
- Group GR (n = 12): SLA implants coated with 25 kGy ^{60}Co gamma radiation crosslinked collagen type I.

Then, 36 implants (Cowell Medi Co., Ltd., Busan, Korea), 4 mm in diameter and 8 mm high, were implanted in the mandibular defects of 6 animals to expose three threads (Figure 1B). Peri-implant defect sites were grafted with porcine xenografts (Bone-XP, MedPark, Busan, Korea) (Figure 1C), and bone regeneration was guided using resorbable collagen membrane (Bone-D, MedPark, Busan, Korea) (Figure 1D). Surgical sites were sutured with 4-0 Vicryl (Mersilk, Ethicon Co., Livingston, UK). Post-operative care consisted of oral amoxicillin-clavulanate (Amocla®, Kuhnil Pharm Co., Seoul, Korea) 12.5 mg/kg, firocoxib (Previcox, Merial, France) 5 mg/kg, and famotidine (Famotidine®, Nelson, Seoul, Korea) at 0.5 mg/kg for 2 weeks.

Eight weeks after implant placements, animals were sacrificed by potassium chloride intravenous injection (JW Pharmaceutical Co. Ltd., Gyeonggi-do, Korea) under general anesthesia, and mandibular bones were harvested and fixed in neutral buffered formalin (Duksan Pure Chemical. Co. Ltd, Gyeonggi-do, Korea) for 2 weeks.

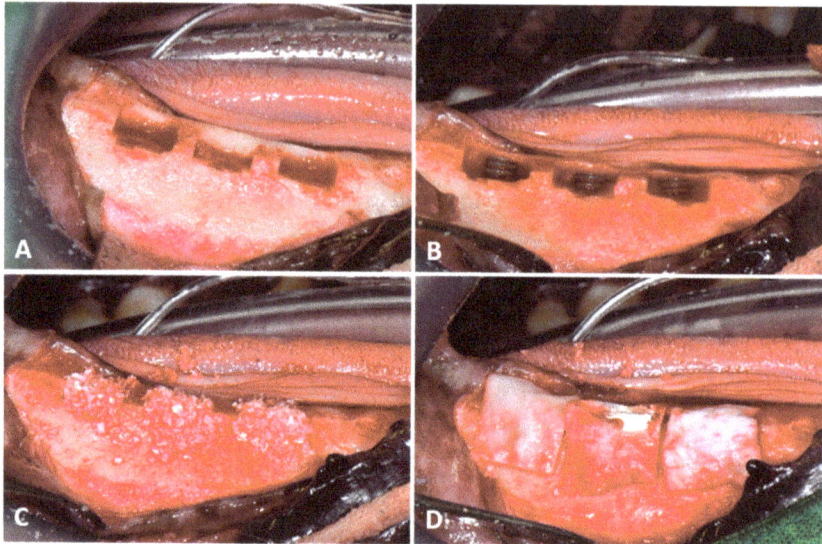

Figure 1. Surgical procedures used to place surface-treated implants in beagle mandibles. (**A**) Creation of buccal cubic defects, (**B**) Implant placement, (**C**) Distribution of bone graft material, (**D**) Collagen membrane placement.

2.3.3. Micro-Computed Tomography (µCT) Analysis

Mandibles were wrapped with Parafilm M® (Heathrow Scientific, Vernon Hills, IL, USA) and scanned by µCT (Skyscan-1173, ver. 1.6, Bruker-CT Co., Kontich, Belgium) at 130 kV and an intensity of 60 µA to obtain the µCT images of regions of interest (ROIs). We used a pixel resolution of 24.15 µm to determine new bone volumes (NBVs) in defect areas around implants. µCT image reconstructions were performed using Nrecon reconstruction software ver. 1.7.0.4 (Bruker-CT Co., Kotich, Belgium). The study used 1 mm diameter ROIs around implants (Figure 2).

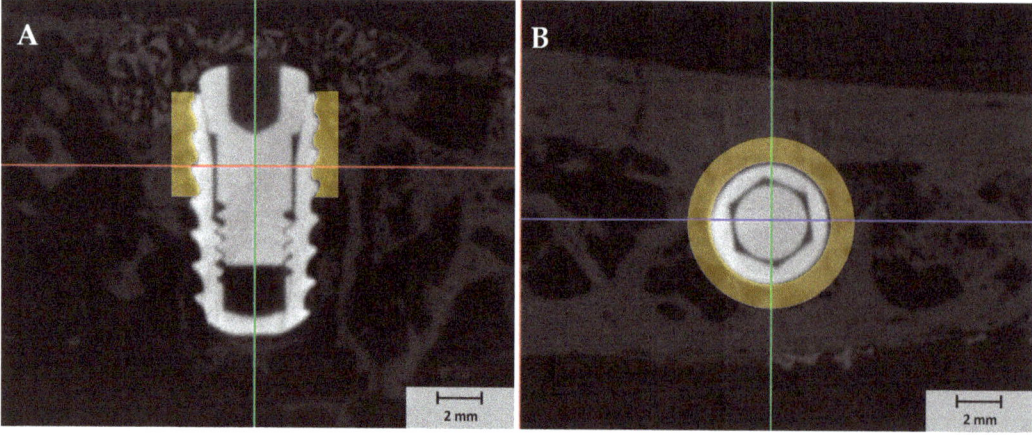

Figure 2. The µCT images of regions of interest (ROIs) which included 1 mm around each implant. (**A**) buccal view, (**B**) occlusal view.

2.3.4. Histologic Analysis

After μCT analysis, mandibular bone specimens were dehydrated in an ethanol series (Duksan Pure Chemical. Co. Ltd, Gyeonggi-do, Korea) 70, 80, 90, and 100%, infiltrated with resin (Technovit 7200, Heraeus KULZER, Hanau, Germany) for a week, fixed to an embedding frame, and embedded using a UV curing system (KULZER EXAKT 520, Heraeus Kulzer, Norderstedt, Germany). Polymerized specimens were sectioned at 400 μm at implant centers using a diamond cutter (KULZER EXAKT 300 CP Band System, Exakt Apparatebau, Norderstedt, Germany). Then, they were polished to a thickness of 30 μm using an EXAKT grinding machine (KULZER EXAKT 400CS, Exakt Apparatebau, Norderstedt, Germany), mounted on slides, and stained with hematoxylin and eosin (H&E). Images of stained specimens were obtained using a light microscope (Olympus BX, Olympus, Tokyo, Japan). BIC and ITBD values and new bone areas (NBAs) were measured using an image analysis program (ver. 7.5, i-solution, IMT i-solution. Inc., Vancouver, BC, Canada) by a trained investigator (Figure 3). ROIs were set at exposed three upper threads and 1 mm around fixtures, as shown in Figure 2.

$$\text{NBAs (\%)} = \text{New bone area (mm}^2\text{)} / \text{Total ROI area (mm}^2\text{)} \times 100 \tag{1}$$

$$\text{BIC (\%)} = \text{Length of the new bone to implant contact (mm}^2\text{)} / \text{Total ROI length of implant (mm}^2\text{)} \times 100 \tag{2}$$

$$\text{ITBDs (\%)} = \text{New bone area of inter thread (mm}^2\text{)} / \text{Total area of inter thread (mm}^2\text{)} \times 100 \tag{3}$$

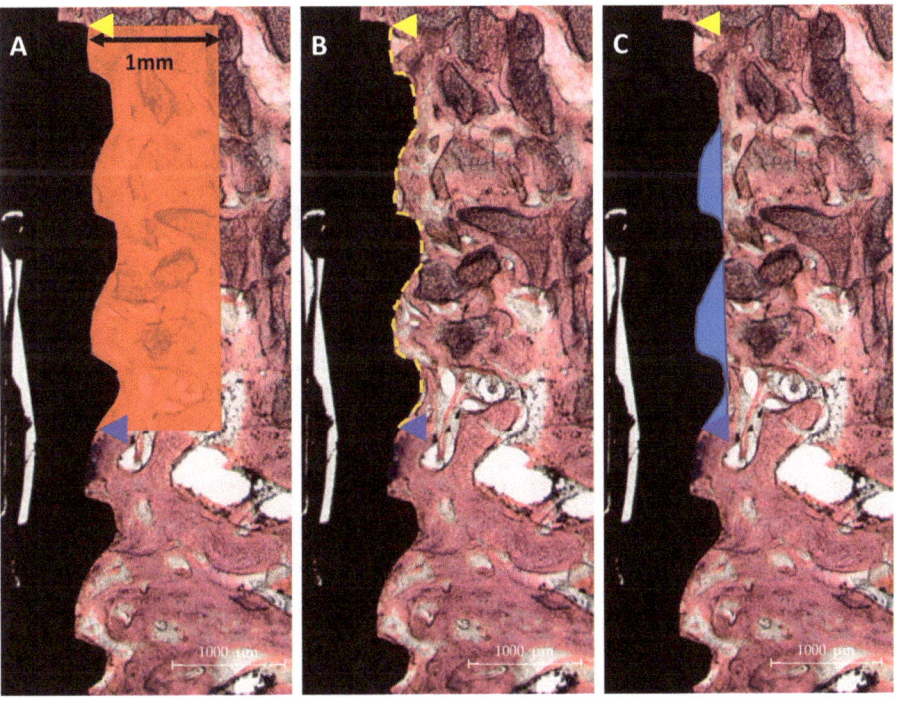

Figure 3. Histometric measurements in regions of interest (ROIs). ROIs were fixed from implant platforms to the third thread and at 1 mm around implants in occlusal view. (**A**) NBA: New bone area, (**B**) BIC: Bone-to-implant contact, (**C**) ITBD: Inter-thread bone density.

2.3.5. Statistical Analysis

Results are presented as means ± standard deviations (SDs), and the analysis was performed using SPSS Ver. 25 (SPSS Inc., Chicago, IL, USA). Since NBAs, ITBDs, NBV, and BIC values were not normally distributed by the normality test, the Kruskal-Wallis one-way analysis was used to determine the significances of intergroup differences. The Mann-Whitney U test was applied as a post hoc test. Statistical significance was accepted for p values < 0.05.

3. Results
3.1. In-Vitro Study
3.1.1. Collagen Crosslinked Ti Implant Surface Morphologies

When collagen was crosslinked using GA or 25 kGy Gamma rays on SLA implant surfaces, surface morphologies were similar due to their rough SLA surfaces (Figure 4).

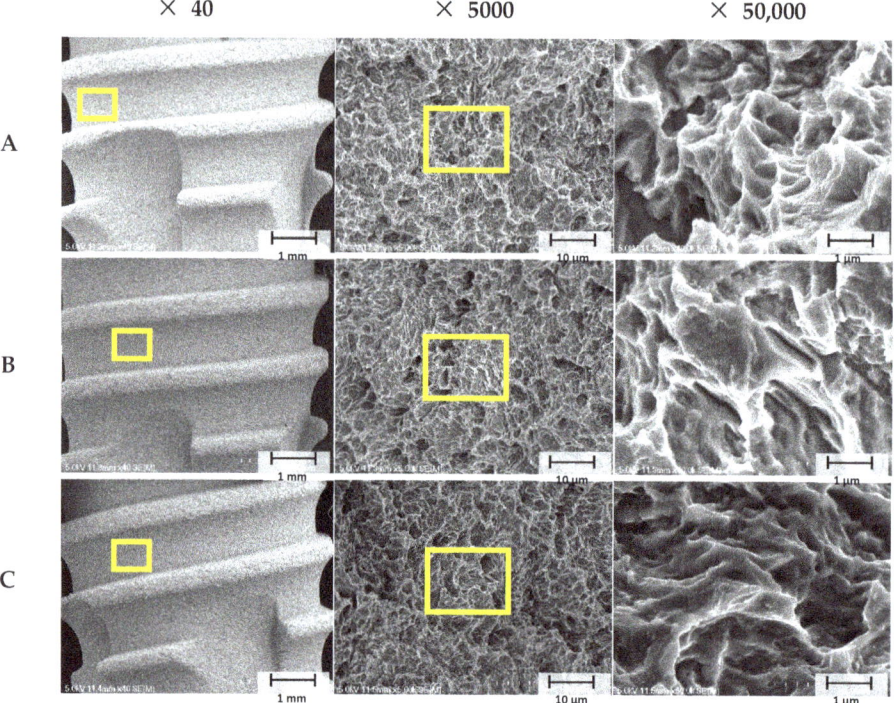

Figure 4. The scanning electron microscopy (SEM) images. (**A**) The SLA (Sandblasted, large grit, acid-etched implant surface) group, (**B**) The GA (glutaraldehyde) group, and (**C**) the GR (gamma-radiation) group. [Original magnifications: ×40, ×5000, and ×50,000].

3.1.2. XPS Findings

Surface elemental compositions were determined by XPS (Figure 5).

The SLA group had the lowest nitrogen content (0.33%), followed by the GA group (6.22%) and the GR group (17.64%). Since the major component of collagen is gelatin (a protein), a large amount of nitrogen indicates good crosslinking [30] (Table 1).

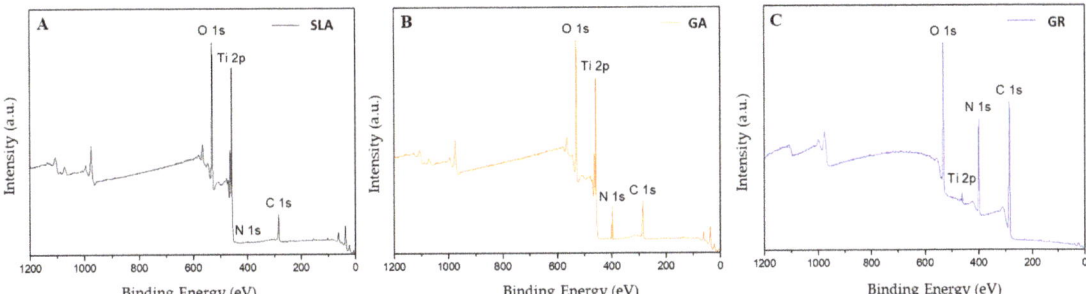

Figure 5. Surface XPS spectra of the three study groups. (**A**) The SLA group, (**B**) the GA group, and (**C**) the GR group.

Table 1. Atomic concentrations (at. %) on implant surfaces as determined by XPS.

Elements	Group		
	SLA	GA	GR
C	20.5 ± 0.33	29.87 ± 0.25	64.77 ± 0.42
O	58.22 ± 0.79	46.64 ± 0.49	16.67 ± 0.09
Ti	20.96 ± 0.36	17.27 ± 0.27	0.93 ± 0.03
N	0.33 ± 0.17	6.22 ± 0.13	17.64 ± 0.30

3.2. In Vivo Study

3.2.1. Clinical Findings

All beagles survived the surgical procedures without complications, such as inflammation or infection. Mandibular jaw segments were harvested after sacrifice.

3.2.2. Micro-Computed Tomography (μCT) Findings

In regions of interest, NBV was 64.78 ± 3.24% in the GR group, 61.42 ± 7.07% in the GA group, and 56.06 ± 7.31% in the SLA group. Thus, although NBV was relatively high in the GR group, differences were not significant (Figure 6).

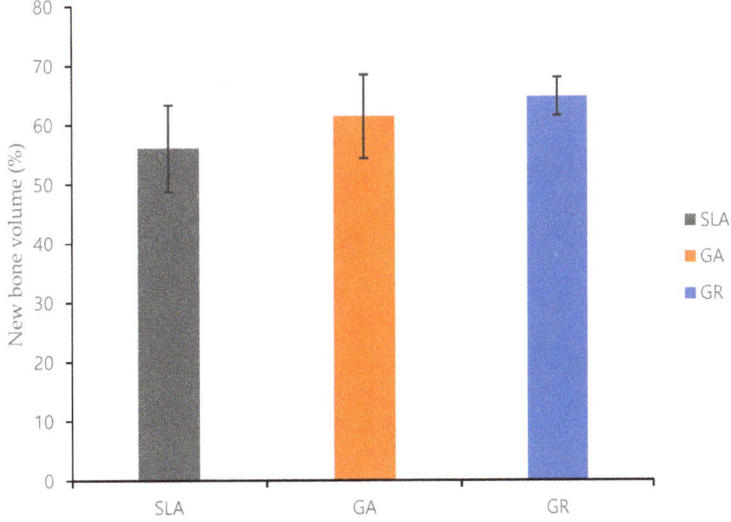

Figure 6. The volumetric ROI analysis of new bone.

3.2.3. Histological Findings

The histological results of the SLA, GA, and GR groups are shown in Figure 7. No abnormal inflammatory cells or singularities were found in any group. However, new bone formation was observed between the third and second threads in the SLA group but distributed evenly in all the GA and GR groups. The crosslinked groups exhibited more new bone formation than the SLA group, but the new bone formation was similar in the GA and GR groups.

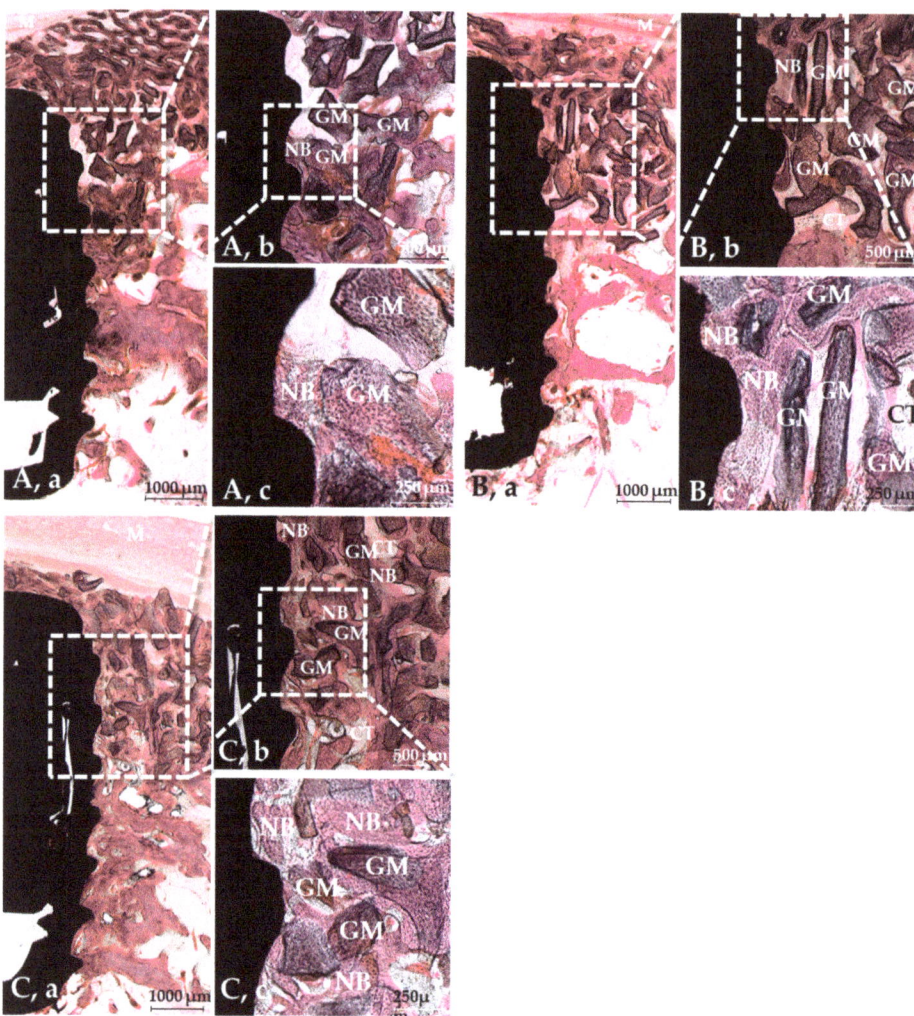

Figure 7. H&E stained sections at 8 weeks post-implantation. (**A**) The SLA group, (**B**) the GA group, (**C**) the GR group, (**a**) ×12.5, (**b**) ×40, (**c**) ×100. Note: NB = New bone, GM = Bone graft material, CT = Connective tissue, M = Membrane.

3.2.4. Histometric Findings

Histometric results are summarized in Table 2 and Figure 8. NBA values of the SLA, GA, and GR groups were $38.27 \pm 9.34\%$, $52.37 \pm 7.93\%$, and $43.77 \pm 8.81\%$, respectively, and were significantly higher in the GR group than in the SLA group ($p < 0.05$).

Table 2. Mean values of new bone areas (NBAs), inter-thread bone densities (ITBDs), and bone to implant contacts (BICs) as determined by histometric analysis.

Measurement	Group	Mean ± SD	p-Value
NBA (%)	SLA	38.27 ± 9.34	0.033 *
	GA	43.77 ± 8.81	
	GR	52.37 ± 7.93	
ITBD (%)	SLA	49.52 ± 5.11	0.053
	GA	58.10 ± 12.33	
	GR	64.10 ± 5.65	
BIC (%)	SLA	47.3 ± 6.58	0.046 *
	GA	54.61 ± 9.4	
	GR	60.19 ± 11.23	

* Indicates statistical significance ($p < 0.05$).

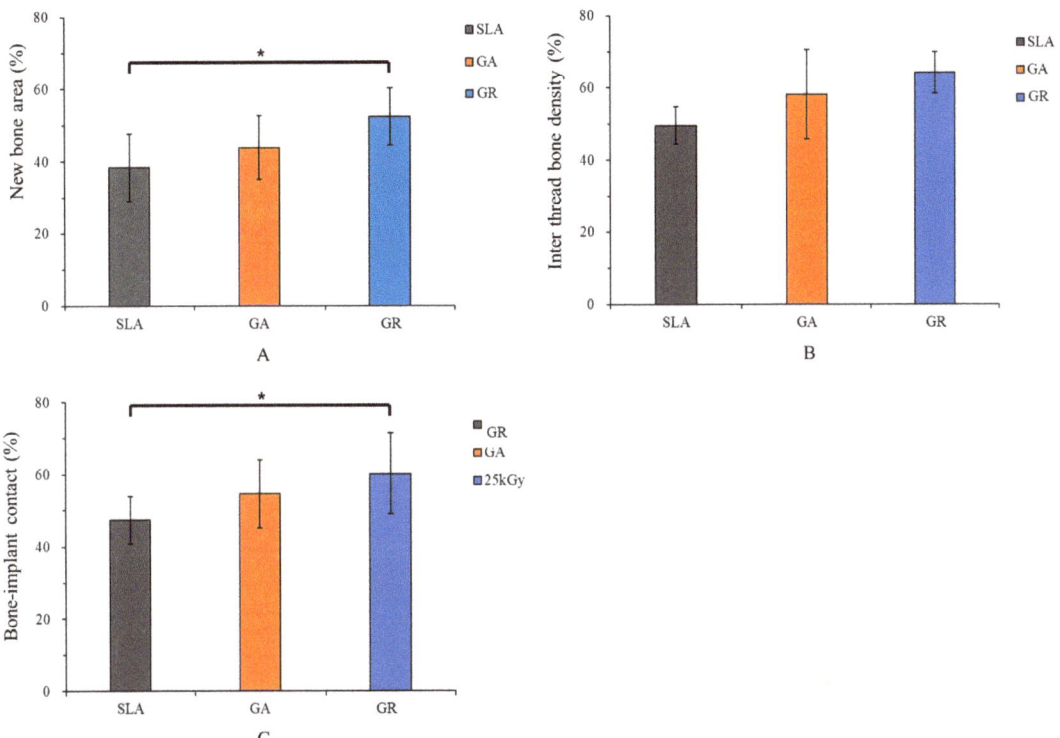

Figure 8. Histometric analysis within regions of interest (ROI). (**A**) New bone area (%), (**B**) Inter thread bone density (%), (**C**) Bone-implant contact (%). * Indicates statistical significance ($p < 0.05$).

ITBD results of the SLA, GA, and GR groups were 49.52 ± 5.11%, 58.10 ± 12.33%, and 64.10 ± 5.65%, respectively, and no significant intergroup difference was found ($p > 0.05$).

Corresponding BIC results were 47.3 ± 6.58%, 54.61 ± 9.4%, and 60.19 ± 11.23%, and BIC was significantly greater in the GR group than in the SLA group ($p < 0.05$). On the other hand, the results of the GR group were similar to the values of the GA group in NBA, ITBD, and BIC ($p > 0.05$).

4. Discussion

Commercially available dental implants are generally considered to have high biocompatibility and surfaces suitable for bone regeneration [4,31], and this is supported by the results of prospective and retrospective clinical studies, which reported implant 10-year survival rates exceeding 90% [32–35]. Nevertheless, dental implant failure due to osseointegration failure often occurs unexpectedly and remains an important clinical problem [36,37]. Therefore, studies on implant surface modification methods have also been conducted to improve osseointegration using surface treatments and collagen as bioactive material [38,39]. However, extracted collagen's mechanical properties and stabilities are inferior; thus, its potential is limited [40,41]. GA has been used as a collagen crosslinking agent for several decades, but some GA probably remains in situ after crosslinking. Protocols for removing unreacted GA have been proposed to solve this problem, but unfortunately, these methods have also been reported to have cytotoxic side effects [42,43]. On the other hand, gamma-ray-based crosslinking does not leave harmful residues and has recently been used to crosslink polymers, including collagen [6]. Therefore, this study was conducted to evaluate and compare the merits of crosslinking collagen type I on the surfaces of SLA implants with gamma-ray radiation or GA in a large animal model.

Collagen type I is a useful biopolymer and widely used clinically due to its low immunogenicity, biocompatibility, and biomedical potential [42]. In addition, collagen is known to promote osteoblast adhesion when coated on implant surfaces [9]. Previous in vivo studies have confirmed that collagen treatment promotes bone regeneration following implantation of crosslinked collagen-coated Ti implants and that collagen treatment enhances bone to implant adhesion to bone and accelerates bone formation [44,45]. Likewise, in the present study, NBAs and BIC values were higher in the GR group than in the SLA group, similar to the GA and SLA groups, which suggests 25 kGy gamma-ray exposure provides better crosslinking than GA. Furthermore, XPS analysis showed surface nitrogen levels (17.64%) were higher in the GR group than in the GA group (6.22%). However, the GR group did not significantly differ compared to the GA group ($p > 0.05$).

After machining Ti, its surface is contaminated by adsorbed organic entities such as atmospheric hydrocarbons, water, or cleaning fluids [46,47]. Previous studies that analyzed the chemical compositions of different implant surfaces by XPS have reported carbon deposition percentages ranging from 17.9 to 76.5% [48]. Therefore, gamma irradiation at 25 to 35 kGy has been recommended for the rapid disinfection and sterilization of medical devices. Ueno et al. [49] found that deposited hydrocarbons can be removed by high-energy UV or gamma radiation and that the removal of hydrocarbons improves Ti biocompatibility and induces osseointegration. Our XPS results returned surface carbon figures in the GR, GA, and SLA groups of 0.93, 6.22, and 20.96%, respectively, suggesting that surface carbon was removed by gamma irradiation [30]. This observed reduction in surface carbon levels by gamma irradiation is consistent with the results of previous studies [6].

Accordingly, the present study suggests that gamma irradiation-induced collagen crosslinking enhances Ti implant biocompatibility and bone adhesion in beagle mandible models. Collagen cross-linked implants using gamma irradiation may improve the osseointegration in adverse circumstances requiring transcrestal sinus lift procedures [50]. Besides, in patients with a history of systemic disease, increased implant-bone osseointegration may be an important factor for long-term implant survival [51]. Meanwhile, Misch [52] recommended that the occlusal implant area be made small. Since the increased osseointegration increases the mechanical strength of the bone tissue, the occlusion of the implant prosthesis can be properly distributed [53].

Furthermore, if a substance that induces a stem cell response, such as rhBMP-2, is attached to the collagen-crosslinked implant with gamma rays, better osteoinductivity can be expected. However, the study was limited by the model used, the number of beagles involved, and its short duration. Furthermore, there was no difference in the histological aspect compared to the GA group. In addition, it is considered necessary to compare it with other biocompatible materials other than collagen. Accordingly, we recommend additional experiments be performed to establish a scientific basis for the clinical effectiveness of crosslinking collagen on Ti implants using gamma radiation.

5. Conclusions

This study was conducted to assess the effects of gamma radiation-induced collagen crosslinking on osseointegration and bone regeneration in defect areas around SLA implants. Within the limitations of this study, gamma-ray collagen crosslinking was found to be at least as effective as GA crosslinking in terms of bone regeneration efficacy. According to our results, gamma-radiation can be used to effectively crosslink collagen on implant surfaces and not raise concerns about toxic residues. Additional animal studies are required to determine optimum gamma-radiation dose criteria and to more comprehensively evaluate the effect of irradiation on osseointegration.

Author Contributions: Conceptualization, J.-B.H.; methodology, J.-B.H., C.-M.J., S.-E.K., and S.-S.K.; formal analysis, W.-T.C. and S.-H.H.; investigation, W.-T.C. and S.-H.H.; data curation, W.-T.C., S.-H.H., S.-I.J., and S.-Y.K.; writing—original draft preparation, J.-B.H., S.-Y.K., and W.-T.C.; writing—review and editing, C.-M.J., W.-T.C., and J.-B.H.; supervision, J.-B.H. All authors have read and agreed to the published version of the manuscript.

Funding: This work was supported by the National Research Foundation of Korea (NRF) funded by the Korean government (MSIT)(Grant no. NRF-2020R1A2C1004927).

Institutional Review Board Statement: The study was approved by the Institutional Animal Care and Use Committee (IACUC) of Chonnam National University (CNU IACUC-YB-2018-94, 07-01-2019).

Informed Consent Statement: Not applicable.

Data Availability Statement: Data sharing not applicable.

Conflicts of Interest: The authors have no conflict of interest to declare.

References

1. Rocci, A.; Calcaterra, R.; Di Girolamo, M.; Rocci, M.; Rocci, C.; Baggi, L. The influence of micro and macro-geometry in term of bone-implant interface in two implant systems: An histomorphometrical study. *Oral Implantol.* **2015**, *8*, 87–95.
2. Textor, M.; Sittig, C.; Frauchiger, V.; Tosatti, S.; Brunette, D. Properties and biological significance of natural oxide films on titanium and its alloys. In *Titanium in Medicine*; Springer: Berlin/Heidelberg, Germany, 2001; pp. 171–230.
3. Morra, M. Biochemical modification of titanium surfaces: Peptides and ECM proteins. *Eur. Cells Mater.* **2006**, *12*, 1–15. [CrossRef] [PubMed]
4. Le Guéhennec, L.; Soueidan, A.; Layrolle, P.; Amouriq, Y. Surface treatments of titanium dental implants for rapid osseointegration. *Dent. Mater.* **2007**, *23*, 844–854. [CrossRef] [PubMed]
5. Müeller, W.D.; Gross, U.; Fritz, T.; Voigt, C.; Fischer, P.; Berger, G.; Rogaschewski, S.; Lange, K.P. Evaluation of the interface between bone and titanium surfaces being blasted by aluminium oxide or bioceramic particles. *Clin. Oral Implants Res.* **2003**, *14*, 349–356. [CrossRef]
6. Ueno, T.; Takeuchi, M.; Hori, N.; Iwasa, F.; Minamikawa, H.; Igarashi, Y.; Anpo, M.; Ogawa, T. Gamma ray treatment enhances bioactivity and osseointegration capability of titanium. *J. Biomed. Mater. Res. B Appl. Biomater.* **2012**, *100*, 2279–2287. [CrossRef]
7. Hwang, S.-T.; Han, I.-H.; Huh, J.-B.; Kang, J.-K.; Ryu, J.-J. Review of the developmental trend of implant surface modification using organic biomaterials. *J. Korean Acad. Prosthodont.* **2011**, *49*, 254–262. [CrossRef]
8. Reyes, C.D.; García, A.J. Engineering integrin-specific surfaces with a triple-helical collagen-mimetic peptide. *J. Biomed. Mater. Res. A* **2003**, *65*, 511–523. [CrossRef]
9. Costa, D.G.; Ferraz, E.P.; Abuna, R.P.; de Oliveira, P.T.; Morra, M.; Beloti, M.M.; Rosa, A.L. The effect of collagen coating on titanium with nanotopography on in vitro osteogenesis. *J. Biomed. Mater. Res. A* **2017**, *105*, 2783–2788. [CrossRef]
10. Nagai, M.; Hayakawa, T.; Fukatsu, A.; Yamamoto, M.; Fukumoto, M.; Nagahama, F.; Mishima, H.; Yoshinari, M.; Nemoto, K.; Kato, T. In vitro study of collagen coating of titanium implants for initial cell attachment. *Dent. Mater. J.* **2002**, *21*, 250–260. [CrossRef]
11. Gelse, K.; Pöschl, E.; Aigner, T. Collagens—Structure, function, and biosynthesis. *Adv. Drug. Deliv. Rev.* **2003**, *55*, 1531–1546. [CrossRef]
12. Brodsky, B.; Ramshaw, J.A. The collagen triple-helix structure. *Matrix Biol.* **1997**, *15*, 545–554. [CrossRef]
13. Zhang, X.; Xu, L.; Huang, X.; Wei, S.; Zhai, M. Structural study and preliminary biological evaluation on the collagen hydrogel crosslinked by γ-irradiation. *J. Biomed. Mater. Res. A* **2012**, *100*, 2960–2969. [CrossRef]

14. Damink, L.O.; Dijkstra, P.; Van Luyn, M.; Van Wachem, P.; Nieuwenhuis, P.; Feijen, J. Glutaraldehyde as a crosslinking agent for collagen-based biomaterials. *J. Mater. Sci. Mater. Med.* **1995**, *6*, 460–472. [CrossRef]
15. De Assis, A.F.; Beloti, M.M.; Crippa, G.E.; De Oliveira, P.T.; Morra, M.; Rosa, A.L. Development of the osteoblastic phenotype in human alveolar bone-derived cells grown on a collagen type I-coated titanium surface. *Clin. Oral Implants Res.* **2009**, *20*, 240–246. [CrossRef]
16. Harjula, A.; Nickels, J.; Mattila, S. Histological study of glutaraldehyde-processed vascular grafts of biological origin. *Ann. Chir. Gynaecol.* **1980**, *69*, 256–262.
17. Cooke, A.; Oliver, R.; Edward, M. An in vitro cytotoxicity study of aldehyde-treated pig dermal collagen. *Br. J. Exp. Pathol.* **1983**, *64*, 172–176.
18. Charulatha, V.; Rajaram, A. Influence of different crosslinking treatments on the physical properties of collagen membranes. *Biomaterials* **2003**, *24*, 759–767. [CrossRef]
19. Zeugolis, D.I.; Paul, G.R.; Attenburrow, G. Cross-linking of extruded collagen fibers—A biomimetic three-dimensional scaffold for tissue engineering applications. *J. Biomed. Mater. Res. A* **2009**, *89*, 895–908. [CrossRef]
20. Sionkowska, A.; Wisniewski, M.; Skopinska, J.; Poggi, G.; Marsano, E.; Maxwell, C.; Wess, T.J. Thermal and mechanical properties of UV irradiated collagen/chitosan thin films. *Polym. Degrad. Stab.* **2006**, *91*, 3026–3032. [CrossRef]
21. Cataldo, F.; Ursini, O.; Lilla, E.; Angelini, G. Radiation-induced crosslinking of collagen gelatin into a stable hydrogel. *J. Radioanal. Nucl. Chem.* **2008**, *275*, 125–131. [CrossRef]
22. Türker, N.S.; Özer, A.Y.; Kutlu, B.; Nohutcu, R.; Sungur, A.; Bilgili, H.; Ekizoglu, M.; Özalp, M. The effect of gamma radiation sterilization on dental biomaterials. *Tissue Eng. Regen. Med.* **2014**, *11*, 341–349. [CrossRef]
23. Wach, R.A.; Mitomo, H.; Nagasawa, N.; Yoshii, F. Radiation crosslinking of methylcellulose and hydroxyethylcellulose in concentrated aqueous solutions. *Nucl. Instrum. Methods B* **2003**, *211*, 533–544. [CrossRef]
24. Chen, C.-C.; Chueh, J.-Y.; Tseng, H.; Huang, H.-M.; Lee, S.-Y. Preparation and characterization of biodegradable PLA polymeric blends. *Biomaterials* **2003**, *24*, 1167–1173. [CrossRef]
25. Almeida, O.M.D.; Jorgetti, W.; Oksman, D.; Jorgetti, C.; Rocha, D.L.; Gemperli, R. Comparative study and histomorphometric analysis of bone allografts lyophilized and sterilized by autoclaving, gamma irradiation and ethylene oxide in rats. *Acta Cir. Bras.* **2013**, *28*, 66–71. [CrossRef]
26. Azorin, E.; Gonzalez-Martinez, P.; Azorin, J. Collagen I confers gamma radiation resistance. *Appl. Radiat. Isot.* **2012**, *71*, 71–74. [CrossRef] [PubMed]
27. Zhao, G.; Schwartz, Z.; Wieland, M.; Rupp, F.; Geis-Gerstorfer, J.; Cochran, D.L.; Boyan, B. High surface energy enhances cell response to titanium substrate microstructure. *J. Biomed. Mater. Res. A* **2005**, *74*, 49–58. [CrossRef]
28. Bae, E.-B.; Yoo, J.-H.; Jeong, S.-I.; Kim, M.-S.; Lim, Y.-M.; Ahn, J.-J.; Lee, J.-J.; Lee, S.-H.; Kim, H.-J.; Huh, J.-B. Effect of titanium implants coated with radiation-crosslinked collagen on stability and osseointegration in rat tibia. *Materials* **2018**, *11*, 2520. [CrossRef]
29. Anesi, A.; Di Bartolomeo, M.; Pellacani, A.; Palumbo, C.; Chiarini, L. Bone healing evaluation following different osteotomic techniques in animal models: A suitable method for clinical insights. *Appl. Sci.* **2020**, *10*, 7165. [CrossRef]
30. Chang, M.C.; Tanaka, J. XPS study for the microstructure development of hydroxyapatite–collagen nanocomposites cross-linked using glutaraldehyde. *Biomaterials* **2002**, *23*, 3879–3885. [CrossRef]
31. Montes, C.C.; Pereira, F.A.; Thome, G.; Alves, E.D.M.; Acedo, R.V.; de Souza, J.R.; Melo, A.C.M.; Trevilatto, P.C. Failing factors associated with osseointegrated dental implant loss. *Implant Dent.* **2007**, *16*, 404–412. [CrossRef]
32. Howe, M.-S.; Keys, W.; Richards, D. Long-term (10-year) dental implant survival: A systematic review and sensitivity meta-analysis. *J. Dent.* **2019**, *84*, 9–21. [CrossRef]
33. Walton, T.R. The up-to-14-year survival and complication burden of 256 TiUnite implants supporting one-piece cast abutment/metal-ceramic implant-supported single crowns. *Int. J. Oral Maxillofac. Implant.* **2016**, *31*, 1349–1358. [CrossRef]
34. Degidi, M.; Nardi, D.; Piattelli, A. 10-year follow-up of immediately loaded implants with TiUnite porous anodized surface. *Clin. Implants Dent. Relat. Res.* **2012**, *14*, 828–838. [CrossRef] [PubMed]
35. Cassetta, M. Immediate loading of implants inserted in edentulous arches using multiple mucosa-supported stereolithographic surgical templates: A 10-year prospective cohort study. *Int. J. Oral Maxillofac. Surg.* **2016**, *45*, 526–534. [CrossRef] [PubMed]
36. Fugazzotto, P.A. Success and failure rates of osseointegrated implants in function in regenerated bone for 72 to 133 months. *Int. J. Oral Maxillofac. Implant.* **2005**, *20*, 77–83.
37. Goodacre, C.J.; Kan, J.Y.; Rungcharassaeng, K. Clinical complications of osseointegrated implants. *J. Prosthet. Dent.* **1999**, *81*, 537–552. [CrossRef]
38. Kulkarni, M.; Mazare, A.; Schmuki, P.; Iglič, A. Biomaterial surface modification of titanium and titanium alloys for medical applications. *Nanomedicine* **2014**, *111*, 111.
39. Morra, M.; Cassinelli, C.; Cascardo, G.; Cahalan, P.; Cahalan, L.; Fini, M.; Giardino, R. Surface engineering of titanium by collagen immobilization. Surface characterization and in vitro and in vivo studies. *Biomaterials* **2003**, *24*, 4639–4654. [CrossRef]
40. Wu, X.; Liu, A.; Wang, W.; Ye, R. Improved mechanical properties and thermal stability of collagen fiber based film by crosslinking with casein, keratin or SPI: Effect of crosslinking process and concentrations of proteins. *Int. J. Biol. Macromol.* **2018**, *109*, 1319–1328. [CrossRef]

41. Jaikumar, D.; Baskaran, B.; Vaidyanathan, V. Effect of chromium (III) gallate complex on stabilization of collagen. *Int. J. Biol. Macromol.* **2017**, *96*, 429–435. [CrossRef]
42. Gu, L.; Shan, T.; Ma, Y.-x.; Tay, F.R.; Niu, L. Novel biomedical applications of crosslinked collagen. *Trends Biotechnol.* **2019**, *37*, 464–491. [CrossRef]
43. Casali, D.M.; Yost, M.J.; Matthews, M.A. Eliminating glutaraldehyde from crosslinked collagen films using supercritical CO_2. *J. Biomed. Mater. Res. A* **2018**, *106*, 86–94. [CrossRef] [PubMed]
44. Schliephake, H.; Aref, A.; Scharnweber, D.; Bierbaum, S.; Roessler, S.; Sewing, A. Effect of immobilized bone morphogenic protein 2 coating of titanium implants on peri-implant bone formation. *Clin. Oral Implants Res.* **2005**, *16*, 563–569. [CrossRef]
45. Sartori, M.; Giavaresi, G.; Parrilli, A.; Ferrari, A.; Aldini, N.N.; Morra, M.; Cassinelli, C.; Bollati, D.; Fini, M. Collagen type I coating stimulates bone regeneration and osseointegration of titanium implants in the osteopenic rat. *Int. Orthop.* **2015**, *39*, 2041–2052. [CrossRef] [PubMed]
46. Kilpadi, D.V.; Lemons, J.E.; Liu, J.; Raikar, G.N.; Weimer, J.J.; Vohra, Y. Cleaning and heat-treatment effects on unalloyed titanium implant surfaces. *Int. J. Oral Maxillofac. Implant.* **2000**, *15*, 219–230.
47. Massaro, C.; Rotolo, P.; De Riccardis, F.; Milella, E.; Napoli, A.; Wieland, M.; Textor, M.; Spencer, N.; Brunette, D. Comparative investigation of the surface properties of commercial titanium dental implants. Part I: Chemical composition. *J. Mater. Sci. Mater. Med.* **2002**, *13*, 535–548. [CrossRef]
48. Morra, M.; Cassinelli, C.; Bruzzone, G.; Carpi, A.; Santi, G.D.; Giardino, R.; Fini, M. Surface chemistry effects of topographic modification of titanium dental implant surfaces: 1. Surface analysis. *Int. J. Oral. Maxillofac. Implant.* **2003**, *18*, 40–45.
49. Ueno, T.; Yamada, M.; Hori, N.; Suzuki, T.; Ogawa, T. Effect of ultraviolet photoactivation of titanium on osseointegration in a rat model. *Int. J. Oral. Maxillofac. Implant.* **2010**, *25*, 287–294.
50. Crespi, R.; Capparè, P.; Gherlone, E. Sinus floor elevation by osteotome: Hand mallet versus electric mallet. A prospective clinical study. *Int. J. Oral. Maxillofac. Implant.* **2012**, *27*, 1144–1150.
51. D'Orto, B.; Tetè, G.; Polizzi, E. Osseointegrated dental implants supporting fixed prostheses in patients affected by Sjögren's Sindrome: A narrative review. *J. Biol. Regul. Homeost. Agents* **2020**, *34*, 89–91.
52. Misch, C.E. Occlusal considerations for implant supported prostheses. In *Comtemporary Implant Dentistry*; Elsevier: Amsterdam, The Netherlands, 1999; pp. 609–628.
53. Ciancaglini, R.; Gherlone, E.F.; Redaelli, S.; Radaelli, G. The distribution of occlusal contacts in the intercuspal position and temporomandibular disorder. *J. Oral Rehabil.* **2002**, *29*, 1082–1090. [CrossRef] [PubMed]

Article

Settable Polymeric Autograft Extenders in a Rabbit Radius Model of Bone Formation

Lauren A. Boller [1], Madison A.P. McGough [1], Stefanie M. Shiels [2], Craig L. Duvall [1], Joseph C. Wenke [2] and Scott A. Guelcher [1,3,4,*]

1. Department of Biomedical Engineering, Vanderbilt University, 2201 West End Ave, Nashville, TN 37235, USA; lauen.a.boller@vanderbilt.edu (L.A.B.); madison.a.mcgough@gmail.com (M.A.P.M.); craig.duvall@vanderbilt.edu (C.L.D.)
2. U.S. Army Institute of Surgical Research, 3698 Chambers Rd, San Antonio, TX 78234, USA; stefanie.m.shiels.ctr@mail.mil (S.M.S.); joseph.c.wenke.civ@mail.mil (J.C.W.)
3. Department of Chemical and Biomolecular Engineering, Vanderbilt University, 2201 West End Ave, Nashville, TN 37235, USA
4. Vanderbilt Center for Bone Biology, Vanderbilt University Medical Center, 1211 Medical Center Dr., Nashville, TN 37212, USA
* Correspondence: scott.guelcher@vanderbilt.edu

Abstract: Autograft (AG) is the gold standard for bone grafts, but limited quantities and patient morbidity are associated with its use. AG extenders have been proposed to minimize the volume of AG while maintaining the osteoinductive properties of the implant. In this study, poly(ester urethane) (PEUR) and poly(thioketal urethane) (PTKUR) AG extenders were implanted in a 20-mm rabbit radius defect model to evaluate new bone formation and graft remodeling. Outcomes including μCT and histomorphometry were measured at 12 weeks and compared to an AG (no polymer) control. AG control examples exhibited new bone formation, but inconsistent healing was observed. The implanted AG control was resorbed by 12 weeks, while AG extenders maintained implanted AG throughout the study. Bone growth from the defect interfaces was observed in both AG extenders, but residual polymer inhibited cellular infiltration and subsequent bone formation within the center of the implant. PEUR-AG extenders degraded more rapidly than PTKUR-AG extenders. These observations demonstrated that AG extenders supported new bone formation and that polymer composition did not have an effect on overall bone formation. Furthermore, the results indicated that early cellular infiltration is necessary for harnessing the osteoinductive capabilities of AG.

Keywords: autograft extender; bone; polyurethane

1. Introduction

Autograft (AG) bone is considered the gold standard in bone grafting. It is osteoinductive, osteoconductive, and osteogenic, and it does not pose a risk for disease transmission [1–3]. AG comes in various forms including both cancellous and cortical [3]. Cancellous AG is most often harvested from the iliac crest (IC); however, other donor sites such as the posterior superior iliac spine, femur, proximal tibia, and distal radius are utilized [4–7]. Cancellous AG contains mesenchymal stem cells (MSCs), osteoblasts, and growth factors including bone morphogenetic proteins (BMPs), which contribute to its osteoinductivity [3,8]. The trabeculae present within cancellous AG allow for enhanced cellular infiltration and vascularization in comparison to cortical AG [8]. Cortical AG is ideal for defects that require structural support as it offers superior mechanical properties compared with cancellous AG. However, cortical AG is less osteoinductive than cancellous AG, and its density results in slower revascularization and inhibits cellular infiltration [8,9]. Despite its osteogenic properties, AG is a scarce resource with multiple drawbacks including donor site morbidity (10–39% of patients), limited availability, the need for a second

surgical site [6], and rapid resorption dependent on the bone density and embryologic origin of the AG [10].

The use of allograft from donors is an alternative to AG. Allograft is more readily available than AG and provides structural support, but it does not possess the same osteoinductive capacity as AG due to its processing [2]. Furthermore, allograft faces potential immune rejection and slow osseointegration with host bone [11]. Synthetic materials such as recombinant human bone morphogenic proteins (rhBMPs) have emerged as substitutes for AG [12–14], but none of these alternatives has been shown to match all of the benefits provided by AG. Furthermore, the use of the FDA-approved rhBMP-2 treatment (INFUSE® bone graft, Medtronic) is limited to a few clinical indications [15–17].

To overcome the limitations in AG including availability and rapid resorption, various approaches to increase the overall volume of AG while maintaining its osteogenic and osteoinductive properties have been employed. Clinically, AG is typically blended with an 'extender' to reduce the volume of AG needed for implantation [18,19]. An early study demonstrated the utility of demineralized bone matrix as an AG extender [20] More recently, tissue engineered approaches to incorporate synthetic bone substitutes with AG have been investigated. Calcium phosphates (CaPs) such as β-tricalcium phosphate (β-TCP) and hydroxyapatite were evaluated as AG extenders for spinal applications [21–24]. Similarly, poly(propylene fumarate)- and poly(lactide-co-glycolide) (PLGA)-based polymer AG extenders have also been evaluated for spinal applications [25–28], while AG extenders utilizing bioactive glass particles have been investigated in the femur [29].

Lysine-based poly(ester urethanes) (PEURs) and poly(thioketal urethanes) (PTKURs) have been previously investigated in bone regeneration applications [30–33]. The mechanical properties of these materials can be easily altered, and the addition of ceramic particles, AG, and allograft supports new bone formation at various anatomic sites [34–37]. Previous work has demonstrated selective, cell-mediated, first-order degradation of PTKUR in vivo [38]. Furthermore, low-porosity PTKURs utilized in rabbit intertransverse processes [39] and femoral plugs [33] exhibited new bone formation, but minimal PTKUR degradation was observed. Slow degradation is advantageous in applications in which mechanical stability is required; however, in applications utilizing biologics, faster graft resorption is necessary to harness the osteoinductivity. In a previous study, PEUR was used to deliver rhBMP-2 and demonstrated balanced polymer resorption and new bone formation [34,40]. Therefore, we compared PEUR [41,42] with PTKUR [38] as an AG extender to test the hypothesis that faster degrading PEUR would support increased cellular infiltration and bone formation in a rabbit radius model.

Herein, settable and resorbable PTKUR-AG and PEUR-AG extenders were implanted into a 20 mm critical-sized segmental defect in the rabbit radius to investigate the effects of polymer composition on cellular infiltration, new bone formation, and polymer resorption. In this study, PTKUR or PEUR was blended with fresh IC AG and the resulting material subsequently molded to size and implanted in the defect. In vivo outcomes assessed post-operatively with X-ray, μCT, histology, and histomorphometry were compared to an AG control.

2. Materials and Methods
2.1. Materials

All chemicals were purchased from Sigma-Aldrich (St. Louis, MO, USA) with the exception of anhydrous diethyl ether purchased from Fisher Scientific. Lysine triisocyanate-polyethylene glycol (LTI-PEG) prepolymer (NCO = 21.7%) was obtained from Ricerca Biosciences LLC (Concord, OH, USA).

2.2. Polyester Triol and Thioketal Diol Synthesis

The polyester triol (molecular weight 450 g mol^{-1}) was synthesized utilizing a previously published method [43]. Briefly, glycerol, 70% ε-caprolactone, 20% glycolide, and 10% DL-lactide monomers were mixed for 40 h under argon at 140 °C. The resulting fluid was vacuum dried at 80 °C for 48 h. Thioketal (TK) diol was synthesized utilizing a previously published method [32]. Briefly, 2,2-dimethoxypropane and thioglycolic acid were reacted in the presence of bismuth (III) chloride at room temperature for 24 h. The resulting solution was filtered, dissolved in tetrahydrofuran, and added dropwise to LiAlH$_4$ under anhydrous conditions. The reaction was refluxed at 52 °C for 18 h and the product filtered and vacuum dried for 48 h.

2.3. AG Extender Fabrication

PTKUR- and PEUR-AG extenders were fabricated by adapted two-component reactive-liquid molding methods as previously described [39]. Briefly, polyisocyanate comprised of either TK diol or polyester triol, 10 pphp iron acetylacetonate (FeAA) catalyst in ε-caprolactone 0.5% (w/w), and LTI-PEG prepolymer were mixed together. Morselized AG (70 wt%) was added to the mixture and stirred by hand until homogeneous. The resulting mixture was injected as a viscous paste that was cured to form a solid implant in situ. The targeted index (NCO:OH) was 200.

2.4. AG Extenders in a Rabbit Radius Defect

Adult New Zealand White rabbits were used in this study (n = 12). The protocol was approved by the Ethics Committee of the U.S. Army Institute of Surgical Research (A-18-035). Animals were randomly assigned to PEUR-AG, PTKUR-AG, or AG control treatment groups (n = 4 per group). Assuming an effect size of 0.999 (determined from a previous study [41]) and alpha of 0.05, an a priori power analysis determined that a sample size of n = 3 would provide a power of 0.95. Thus, 4 animals per group were considered to provide sufficient power for this study. Animals were premedicated with slow-release buprenorphine (0.1 mg kg^{-1}) and anesthetized with isoflurane (1–3%). For all groups, the animal's left hindlimb and right forelimb were shaved and prepared for sterile surgery using alternating washes of alcohol and povidone-iodine. The left IC was exposed, and AG (0.6–0.7 g) was harvested using an oscillating saw. Excess soft tissue was removed and a bone mill (R. Quétin) was used to morselize the harvested bone. The IC harvest site was closed and the right radius exposed. An oscillating saw was used to create a 20 mm segmental defect in the radius. AG extenders were prepared as explained above and shaped to size (5 mm × 20 mm). AG control (morselized AG without PTKUR or PEUR) was molded to shape and carefully placed within the defect. A surgical elevator was used to place the AG extenders in the defects to ensure correct placement. AG extenders were allowed to cure in situ (Figure 1) after which the radial site was closed. Post-operative X-ray images (Faxitron X20) were taken throughout recovery and Calcein green and Xylenol orange fluorochromes were injected at 4 and 8 weeks post-operatively, respectively, to evaluate bone remodeling temporally. Animals were anesthetized and euthanized at 12 weeks. The radii were harvested and placed into formalin for further analysis.

2.5. µCT Analysis

µCT analysis was performed using a µCT50 (SCANCO, Brüttisellen, Switzerland). Radii were scanned at 70 kVp energy, 200 µA source current, 1000 projections per rotation, 800 ms integration time, and 17.2 µm voxel size. In order to spatially evaluate bone growth throughout the defect, bone area was calculated for each axial section (17.2 µm) totaling 20 mm. The area of interest (AOI) included the proximal onset of the defect and extended the length of the defect. It is not possible to distinguish AG from old or new bone utilizing µCT; thus, the ulna was included in analysis due to bone formation observed within the interosseous syndesmosis interfacing the ulna in some of the samples. The bone area was plotted as a function of defect length where 0 mm and 20 mm represented the proximal and distal ends of the defect, respectively. The bone volume (BV) and total volume (TV) within the AOI were measured to calculate the bone volume fraction (BV/TV). Additionally, trabecular thickening (Tb. Th.), trabecular spacing (Tb. Sp.), and trabecular number (Tb. N.) were evaluated.

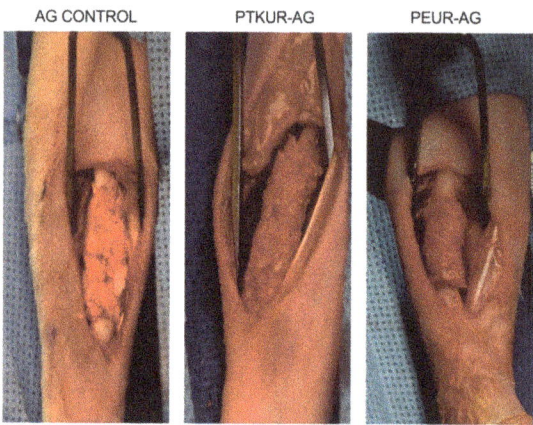

Figure 1. Surgical images. AG control, PTKUR-AG extender, and PEUR-AG extender in the 20 mm defect prior to closure.

2.6. Histological Evaluation

Non-decalcified histology was utilized to evaluate cellular infiltration, new bone formation, and residual polymer (n = 4 per treatment group) [38,42]. After formalin fixation, radii were dehydrated and embedded in poly (methyl methacrylate). Serial coronal sections were cut from the center of each defect with an Exakt band saw. Sections were polished and stained with Sanderson's Rapid Bone Stain to assess osteogenesis and remodeling. Safranin O staining was also performed to assess endochondral ossification. An unstained section was utilized to analyze fluorochrome binding. High-magnification histological images were obtained via bright-field and fluorescent microscopy (Olympus BX41, Tokyo, Japan).

For quantitative histomorphometry, images were taken at 4× via-bright field and fluorescent microscopy (Biotek Cytation). The AOI was defined as a 20 × 5 mm rectangular region that encompassed the entirety of the graft and defect. The ulna was excluded from the AOI. The same AOI was used for both Sanderson's Rapid stained and fluorescent sections. Quantification of new bone, infiltrating cells and tissue, and residual polymer was performed using Metamorph (Version 7.0.1). Bone was thresholded either as red (Sanderson's rapid) or green/orange (fluorochromes). Residual material was thresholded as black stain, and infiltrating cells were thresholded as blue/teal. The thresholded area was reported as an area percentage of the total AOI.

2.7. Statistical Analysis

Data were analyzed utilizing GraphPad Prism (Version 8.4.1) and reported as mean ± standard deviation. Treatment group outcomes at 12 weeks were evaluated using an ANOVA with a Tukey's multiple comparison test. Treatment group outcomes compared at 4 and 8 weeks were evaluated using a two-way ANOVA with Tukey's multiple comparison test. Statistical significance was set at $p < 0.05$.

3. Results

3.1. Surgical Outcomes

The surgical procedures and subsequent healing were uneventful. No fractures of the radii occurred. As shown in Figure 2, X-rays displayed healing progression from 0 to 12 weeks in all of the groups. The AG control presented challenges in implantation and shape maintenance during the surgical procedures due to the lack of a settable polymeric extender. However, AG control remained in place throughout the study and displayed at least partial bridging of the defect along the radial side of the defect within three of the four samples (Figure 2A). Both AG extenders were coherent throughout surgical placement and remained stable throughout the entirety of the study (Figure 2B,C). The AG extenders displayed new bone growth at the host bone/graft interfaces, and graft remodeling was observed in both AG extenders, specifically near the proximal and distal ends of the defect where new bone and decreasing residual graft were observed. Bridging of the defect was not observed in any of the AG extender samples. Both AG extenders exhibited increasing opacity within the grafts over the 12-week time course, and no qualitative differences in new bone formation within the graft were observed between PTKUR- and PEUR-AG groups.

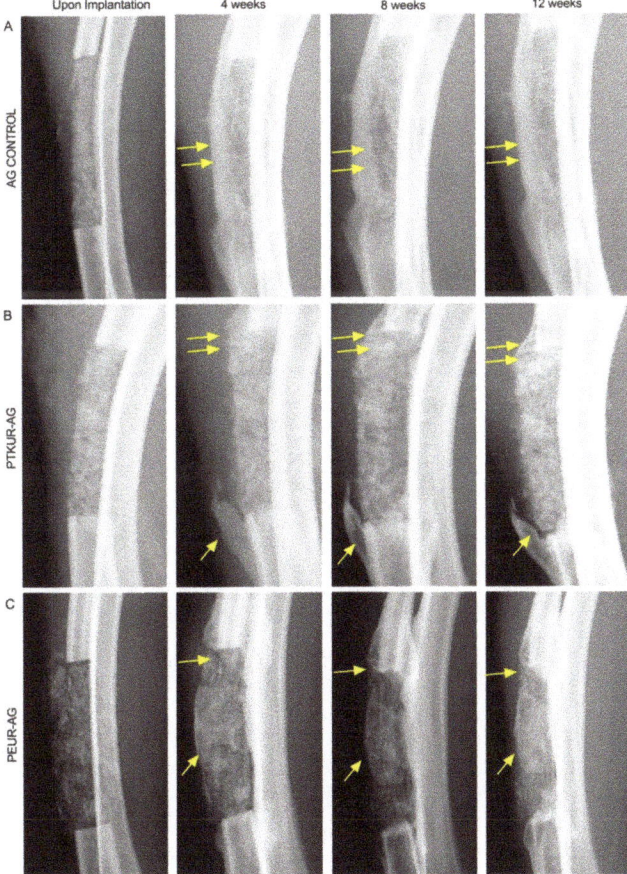

Figure 2. Representative X-ray images of (**A**) AG control, (**B**) PTKUR-AG, and (**C**) PEUR-AG acquired immediately after the surgical procedures and after 4, 8, and 12 weeks of healing. Areas of bone remodeling and formation are noted by yellow arrows.

3.2. In Vivo Bone Analysis

Representative μCT images revealed no significant difference in total bone (including new bone, residual AG, and host bone) between groups at 12 weeks (Figure 3A). New bone formation was observed in the interosseous membrane in the space between the radius and ulna. Bone area and volume were quantified by μCT analysis (Figure 3B,C). All groups displayed similar trends of increased bone area at the proximal and distal ends of the defect with a gradual decrease in bone area as the center of the defect was approached (Figure 3B). BV/TV in PEUR-AG extenders trended higher compared with PTKUR-AG (p = 0.070) and AG control (p = 0.337), but the differences were not significant (Figure 3C). Additional bone morphometric parameters including trabecular thickness (Tb.Th.), trabecular separation (Tb.Sp.), and trabecular number (Tb.N.) did not show significant differences between groups (Supplementary Figure S2).

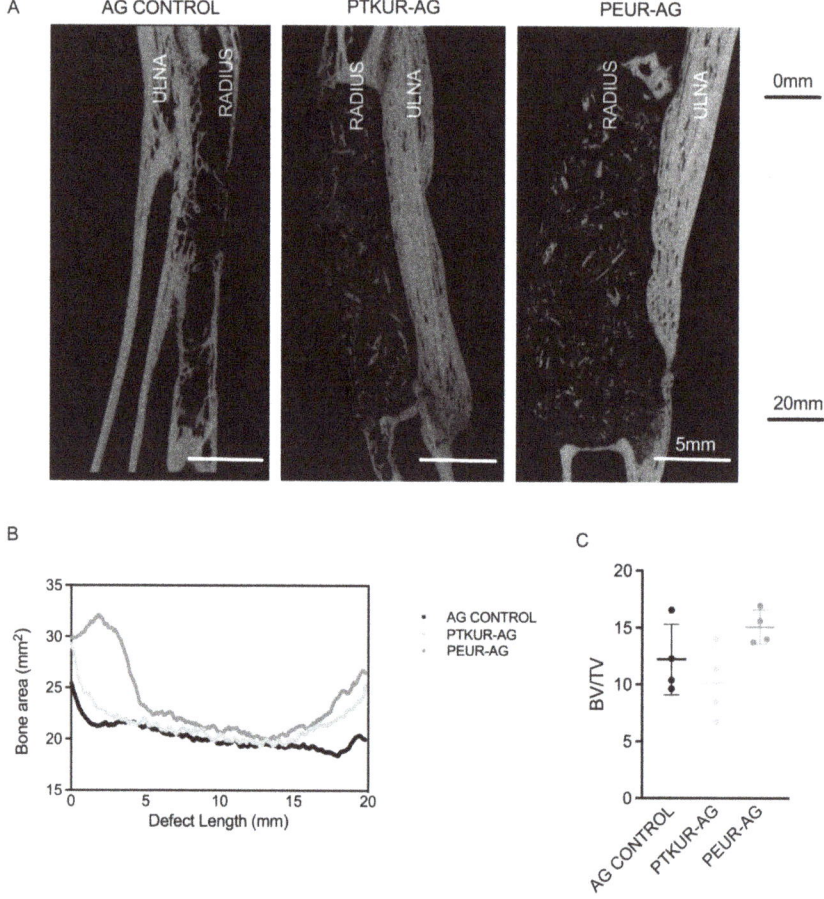

Figure 3. μCT analysis of bone remodeling. Representative μCT images of (**A**) AG control, PTKUR-AG, and PEUR-AG 12 weeks post-operatively. (**B**) Total bone area at 12 weeks measured as a function of defect length by μCT from the proximal to distal interfaces of the defect. Corresponding dotted lines representative standard deviation. (**C**) Bone volume/total volume (BV/TV) at 12 weeks for each treatment group.

New bone formation measured via histological analysis was observed within all groups (Figure 4A and Figure S1). Quantitative histomorphometric analysis at 12 weeks showed no significant difference in new bone formation between PTKUR- and PEUR-AG extenders (Figure 4B). While μCT analysis showed no significant difference in BV/TV between the AG control and extender groups (Figure 3C), histomorphometric analysis showed significantly higher new bone formation compared with both AG extenders (Figure 4B). This discrepancy can be explained in part by the different regions of interest used for μCT (entire defect including the ulna) and histomorphometry (center of the defect excluding the ulna). Although the AG control displayed greater new bone formation, healing within the samples appeared to be inconsistent (Figure S1).

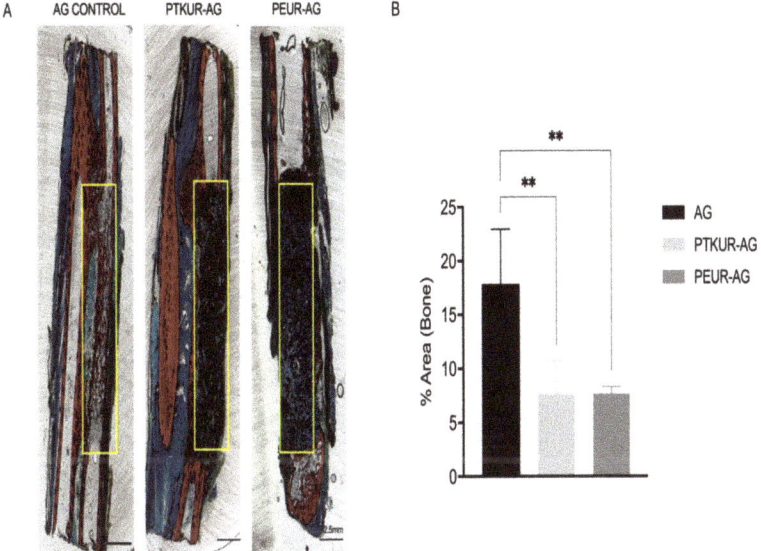

Figure 4. New bone formation in AG extenders. (**A**) Representative images of Sanderson's Rapid stained AG control, PTKUR-AG, and PEUR-AG histological sections. The AOI (20 mm × 5 mm) used for analysis is indicated by the yellow box. (**B**) Histomorphometric analysis of area percentage of new bone (red) at 12 weeks within the defect. Statistical significance determined using one-way ANOVA, ** $p < 0.01$.

Representative histological sections show the ingrowth of new bone at the graft interface indicating osseointegration in all groups (Figure 5A). While some specimens in the AG group showed increased adipogenesis in the marrow cavity compared with the extender groups (Figure 5A), images of histological sections from all AG specimens show variable adipogenesis (Supplementary Figure S1). Osteoblasts were observed around the perimeter of bone ingrowth, suggesting active ongoing remodeling (Figure 5B).

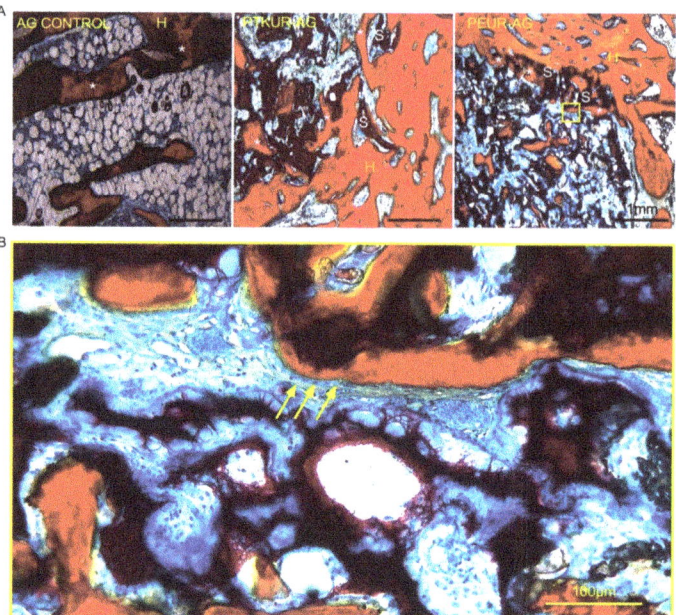

Figure 5. Osseointegration of AG extenders. (**A**) Histological images demonstrate osseointegration of the AG extenders at the host bone/material interface. H represents new bone, S represents scaffold, and * represents new bone growth. (Scale bar, 1 mm) (**B**) New bone growth occurring within the graft. Yellow arrows point to osteoblasts. (Scale bar, 100 µm).

Additionally, Safranin O staining revealed faint orange staining (cartilage) indicating that previous endochondral ossification had occurred within the AG control group (Figure 6A), while ongoing endochondral ossification was observed at 12 weeks in both AG extenders (Figure 6B,C).

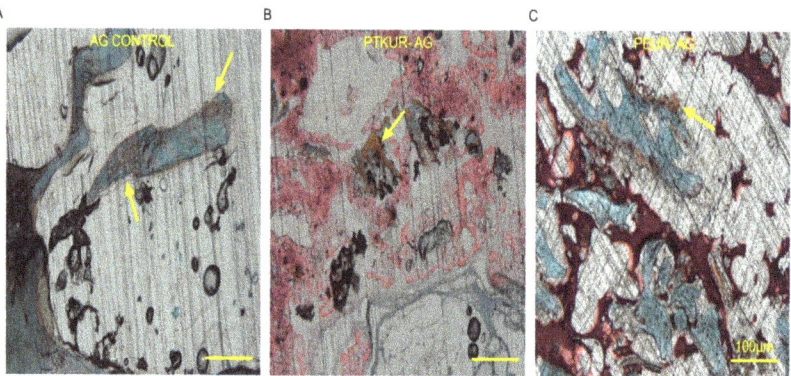

Figure 6. Endochondral ossification within AG extenders. Histological images demonstrate endochondral ossification with (**A**) AG control, (**B**) PTKUR-AG extenders, and (**C**) PEUR-AG extenders at 12 weeks. (Scale bar, 100 µm).

3.3. PTKUR and PEUR Graft Remodeling

Histological analysis revealed residual polymer (black) in AG extenders (Figure 7A and Figure S1). High-magnification images demonstrated that PTKUR-AG underwent slower resorption as evidenced by the higher amount of dense residual polymer in the PTKUR-AG sections compared with the extensive resorption evident in the PEUR-AG sections (Figure 7A). These findings were confirmed via histomorphometric analysis (AOI represented in Figure 4A) in which PTKUR-AG exhibited significantly more residual polymer compared with the PEUR-AG group (Figure 7B). All groups supported cellular and tissue infiltration (teal/blue), but significantly greater cellular and tissue infiltration was observed in the PEUR-AG group compared with PTKUR-AG and AG control (Figure 7B).

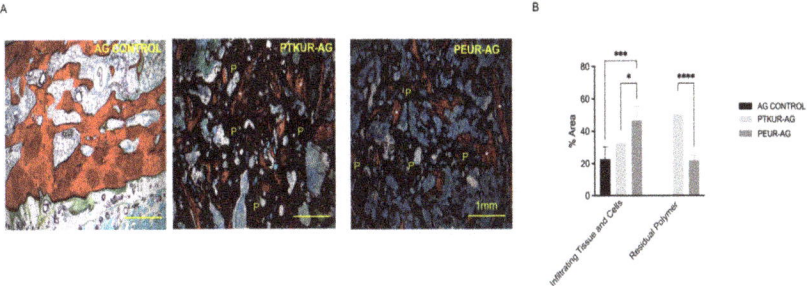

Figure 7. AG extender remodeling. (**A**) Representative images of residual polymer in AG control, PTKUR-AG, and PEUR-AG extenders. P denotes residual polymer and * denotes implanted AG. (Scale bar, 1 mm) (**B**) Histomorphometric analysis of area percentage of infiltrating cells and tissue and residual polymer within the defect after 12 weeks post implantation. Statistical significance determined using Two-way ANOVA, * $p < 0.05$, *** $p < 0.001$, **** $p < 0.0001$.

Bone remodeling throughout the healing process was observed in all groups, especially at the proximal and distal host bone/graft interfaces (Figure S1). Remodeling was observed within the PTKUR- and PEUR-AG grafts around the periphery of implanted AG at 4 and 8 weeks, indicating mineralization nucleating from implanted AG particles within the extenders (Figure 8A). PTKUR- and PEUR-AG extenders exhibited increased bone remodeling at 4 weeks (green) compared with 8 weeks (orange/red); however, these differences were not significant (Figure 8B). Additionally, increased bone remodeling was observed at the graft/host bone interface, indicating the osseointegration of both AG extenders (Figure 8C). The AG control demonstrated significantly greater bone remodeling compared with the AG extenders at both 4 and 8 weeks (Figure 8B); however, inconsistent healing was observed as only two of the four controls exhibited complete bridging along the lateral edge of the defect (Figure S1).

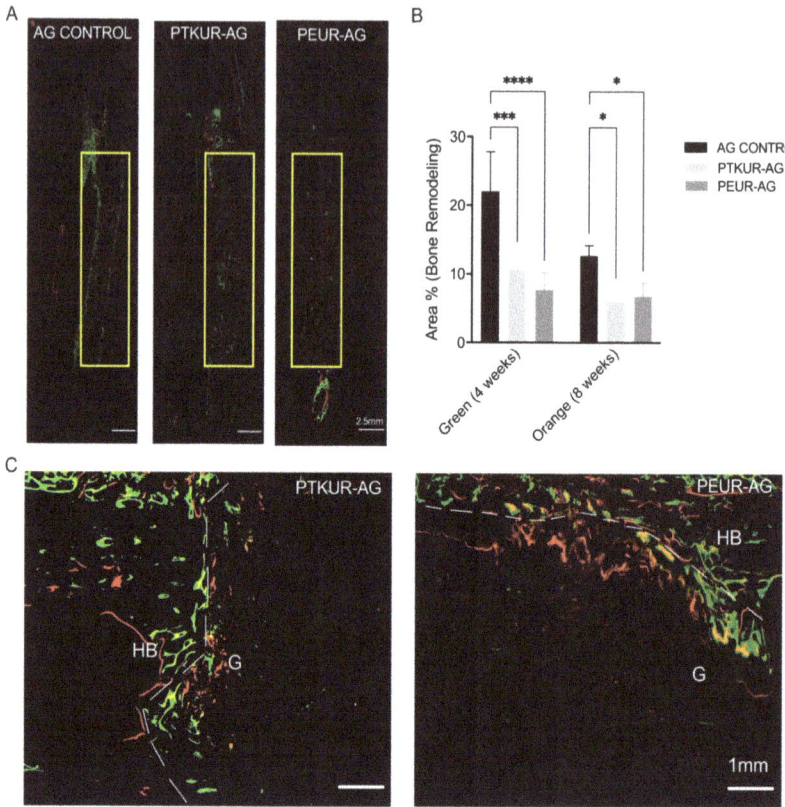

Figure 8. Dynamic histomorphometric analysis at 4 and 8 weeks. (**A**) Representative fluorescent images of AG, PTKUR-AG, and PEUR-AG groups. The AOI is indicated by the yellow box. (**B**) Histomorphometric analysis of area percentage of active bone remodeling at 4 (green) and 8 (orange/red) weeks within the defect. (**C**) Representative images of bone remodeling at the host bone–graft interface, demonstrating osseointegration in PTKUR-AG extenders and PEUR-AG extenders. HB indicates host bone and G indicates grafts. Statistical significance determined using two-way ANOVA, * $p < 0.05$, *** $p < 0.005$ **** $p < 0.001$.

4. Discussion

In this work, we implanted PTKUR-AG and PEUR-AG extenders in a 20 mm critical sized rabbit radial defect to evaluate the effects of polymer composition on both bone formation and graft remodeling in vivo. Both PTKUR- and PEUR-AG extenders supported new bone formation and utilized less AG than the AG control. Furthermore, the polymeric component of the AG extenders degraded and simultaneously maintained AG within the defect for 12 weeks. PEUR-AG extenders degraded more rapidly compared with PTKUR-AG extenders. However, new bone formation in both AG extenders was delayed compared with the AG control.

To understand the effect of polymer composition on bone formation and graft remodeling, AG extenders were implanted in a 20 mm critical-size radial defect in rabbits [44,45]. This model was selected as no external fixation was required [2]. No graft failure was observed in any of the groups throughout the 12 weeks, suggesting that AG extenders exhibited sufficient compression-resistant properties. Previous studies in the spine and mandible demonstrated that an elastic modulus >1 MPa provided compression-resistant properties [46,47]. We previously reported PTKUR-AG and PEUR-allograft moduli of 6.08 MPa and 4.38–9.47 MPa, respectively [39,48].

Previous studies performed in the rabbit radius have reported bone growth from the proximal end of the defect, the distal end of the defect, and the interosseus membrane [44,48–51]. Similarly, we observed bone formation in these directions. Due to the inability of µCT to distinguish between new bone, residual AG, host bone, and ossification extending from the ulna within the interosseous membrane to the radius, the ulna was included in µCT analysis. Interestingly, BV/TV trended higher in PEUR-AG compared with PTKUR-AG. These differences were not significant, but they were likely due in part to increased degradation of the PEUR, allowing for increased bone formation throughout the defect and within the interosseous membrane. Furthermore, µCT bone area quantification indicated increased bone at the proximal and distal end of the defect, indicating bone formation at the interfaces. Consistent with previous studies utilizing AG [52–54], new bone formation via creeping substitution at the host bone/graft interface was observed.

Histomorphometric analysis was performed to evaluate new bone formation specifically in the 5 mm × 20 mm defect space; thus, bone present in the interosseus membrane was excluded from analysis. Transverse sections were obtained from the center of the defect to evaluate bone formation at its most stringent point. Ultimately, no significant difference in bone between PTKUR- and PEUR-AG extender was observed via histomorphometry. The AG control demonstrated significantly increased new bone within the defect compared with AG extenders at 12 weeks via histomorphometric analysis, but new bone formation appeared to be variable throughout the defect. These differences were not observed in overall BV/TV between groups, suggesting that PTKUR- and PEUR-AG promoted bone formation, particularly in the interosseous membrane surrounding the defect while AG control promoted greater bone formation within the defect site itself.

In agreement with an earlier PTKUR-AG study in a biologically stringent intertransverse process defect [39], residual polymer was observed in PTKUR-AG extenders at 12 weeks. PEUR degradation occurred more rapidly than PTKUR degradation, as evidenced by significantly less residual polymer within the defect at 12 weeks. PTKUR degrades in response to specific cell types including osteoclasts, macrophages, and other ROS-secreting cells [38], while PEUR degrades via autocatalytic hydrolytic degradation [55]. Furthermore, AG control exhibited the least amount of cellular infiltration at 12 weeks, suggesting that cells recruited to the AG were osteoprogenitor cells that underwent direct differentiation. PTKUR-AG extenders exhibited less cellular infiltration than PEUR-AG extenders, demonstrating that cells were able to more readily infiltrate the graft as the polymer degraded. Early vascularization of cancellous AG begins at 2 days and is followed by the recruitment of MSCs in response to the osteoinductive signals of AG within the first weeks after implantation [56]. In contrast, AG particles were encapsulated in residual polymer in the AG extenders (Figure 5A), which delayed the rate of new bone formation. These findings are further confirmed by Safranin O staining in which faint positive Safranin O staining for cartilage was demonstrated within the AG control group, suggesting that new bone formation occurred via endochondral ossification and was near completion by 12 weeks. However, more intense positive Safranin O staining was observed in the AG extenders, suggesting that ongoing endochondral ossification was still occurring at 12 weeks. Thus, AG particles encapsulated in residual polymer retained their osteoinductivity beyond the first few weeks after implantation. Although the AG extenders showed delayed endochondral ossification suggesting a longer total healing time, the slower healing coupled with the observation that AG remains stable throughout the study suggests that PEUR and PTKUR extenders can reduce the risk for rapid resorption.

Dynamic bone histomorphometry is a widely utilized method for evaluating bone remodeling [57,58]. As mentioned above, ossification within the interosseus membrane was excluded from dynamic histomorphometric analysis. In agreement with our static histomorphometric findings, bone remodeling at 4 and 8 weeks was greater in AG controls compared with PTKUR- and PEUR-AG extenders within the defect. It is likely that the lack of polymer in AG controls allowed for more extensive cellular infiltration and new bone formation. Despite a smaller 10 mm defect size, a previous study utilizing highly porous cell-seeded hydroxyapatite scaffolds did not observe fluorochrome binding within the scaffold until six weeks post implantation [51]. Herein, fluorescent staining beginning at 4 weeks was apparent within the grafts in the AG extender groups, suggesting that embedded AG maintained bioactivity. Additionally, abundant osseointegration at the host bone/graft interface in both AG extenders was observed, further confirming bioactivity. AG is resorbed by osteoclasts and new bone is deposited by osteoblasts in a process known as creeping substitution [56].

The complete degradation of synthetic polymers requires from 4 to 24 months in vivo [59], which can delay the creeping substitution of AG particles encapsulated in polymer. However, while the encapsulation of AG particles within residual polymer delays new bone formation, it also protects AG from rapid resorption that can result in unpredictable healing [10]. While osteobiologics such as recombinant human bone morphogenetic protein-2 (rhBMP-2) have been shown to promote predictable bone healing at 12 weeks in the rabbit radius defect model [60], rhBMP-2 is approved by the FDA for only a limited number of indications, including lumbar fusion, ridge augmentation, and fresh fractures of the tibial shaft. Furthermore, rhBMP-2 delivered on a collagen sponge has weak mechanical properties, similar to AG. Thus, PEUR- and PTKUR-AG extenders may be most beneficial in clinical scenarios where long-term mechanical stability is required, such as posterolateral spine fusion and fractures of the mid-diaphysis. This study is limited by a single intermediate time point for endpoint outcomes (12 weeks), at which time bone healing and complete resorption of residual polymer and AG were not observed. Future studies should focus on optimizing the rate of polymer degradation to increase the rate of new bone formation while protecting the AG from excessive resorption.

5. Conclusions

In this work, PTKUR- and PEUR-AG extenders were compared with an AG control in a rabbit radius model of bone regeneration. PTKUR- and PEUR-AG extenders both maintained AG in the defect throughout the study and demonstrated bone formation along the host bone/graft interface comparable to AG control. Polymer resorption and subsequent cellular infiltration were observed within the defect space in both AG extenders but did not have an effect on overall bone formation. These results suggest that early polymer degradation and cellular infiltration are necessary for harnessing and maximizing the osteoinductive capabilities of AG.

Supplementary Materials: The following are available online at https://www.mdpi.com/article/10.3390/ma14143960/s1, Figure S1: Fluorescent and Sanderson's Rapid stained histological sections. Figure S2: Bone morphometric parameters.

Author Contributions: Conceptualization, L.A.B., M.A.P.M., C.L.D., J.C.W. and S.A.G.; methodology, L.A.B., M.A.P.M., S.A.G., S.M.S. and J.C.W.; formal analysis, L.A.B. and M.A.P.M.; investigation, L.A.B., M.A.P.M., S.M.S.; resources, L.A.B., M.A.P.M., S.M.S., J.C.W. and S.A.G.; data curation, L.A.B. and M.A.P.M.; writing—original draft preparation, L.A.B. and M.A.P.M.; writing—review and editing, L.A.B., S.M.S., C.L.D., J.C.W. and S.A.G.; visualization, L.A.B. and M.A.P.M.; supervision, J.C.W. and S.A.G.; project administration, S.M.S., J.C.W. and S.A.G.; funding acquisition, J.C.W. and S.A.G. All authors have read and agreed to the published version of the manuscript.

Funding: This work was funded by the National Institutes of Health (NIH) (R01AR064772 and T32DK101003) and the United States Army Institute of Surgical Research. Any opinions, findings, and conclusions or recommendations expressed in this material are those of the authors' and do not necessarily reflect the views of the National Institutes of Health or the Department of Defense.

Institutional Review Board Statement: All surgical and care procedures were approved by the Institutional Animal Care and Use Committee of the US Army Institute of Surgical Research, Fort Sam Houston, TX. Procedures were performed in compliance with the Animal Welfare Act, Animal Welfare Regulations, and the Guide for the Care and Use of Laboratory Animals.

Informed Consent Statement: Not applicable.

Data Availability Statement: The data presented in this study are available on request from the corresponding author. The data are not publicly available due to privacy restrictions.

Conflicts of Interest: The authors declare no conflict of interest. The funders had no role in the design of the study; in the collection, analyses, or interpretation of data; in the writing of the manuscript, or in the decision to publish the results.

References

1. Albrektsson, T.; Johansson, C. Osteoinduction, Osteoconduction and Osseointegration. *Eur. Spine J.* **2001**, *10*, S96–S101. [CrossRef]
2. Shafiei, Z.; Bigham, A.S.; Dehghani, S.N.; Torabi Nezhad, S. Fresh Cortical Autograft versus Fresh Cortical Allograft Effects on Experimental Bone Healing in Rabbits: Radiological, Histopathological and Biomechanical Evaluation. *Cell Tissue Bank.* **2009**, *10*, 19–26. [CrossRef] [PubMed]
3. Baldwin, P.; Li, D.J.; Auston, D.A.; Mir, H.S.; Yoon, R.S.; Koval, K.J. Autograft, Allograft, and Bone Graft Substitutes: Clinical Evidence and Indications for Use in the Setting of Orthopaedic Trauma Surgery. *J. Orthop. Trauma* **2019**, *33*, 203–213. [CrossRef]
4. Myeroff, C.; Archdeacon, M. Autogenous Bone Graft: Donor Sites and Techniques. *J. Bone Jt. Surg. Ser. A* **2011**, *7*, 2227–2236. [CrossRef]
5. Rogers, G.F.; Greene, A.K. Autogenous Bone Graft: Basic Science and Clinical Implications. *J. Craniofacial Surg.* **2012**, *3*, 323–327. [CrossRef]
6. Carlisle, E.R.; Fischgrund, J.S. Bone Graft and Fusion Enhancement. In *Surgical Management of Spinal Deformities*; Elsevier: Amsterdam, The Netherlands, 2009; ISBN 9781416033721.
7. Arrington, E.D.; Smith, W.J.; Chambers, H.G.; Bucknell, A.L.; Davino, N.A. Complications of Iliac Crest Bone Graft Harvesting. *Clin. Orthop. Relat. Res.* **1996**, *329*, 300–309. [CrossRef]
8. Roberts, T.T.; Rosenbaum, A.J. Bone Grafts, Bone Substitutes and Orthobiologics the Bridge between Basic Science and Clinical Advancements in Fracture Healing. *Organogenesis* **2012**, *8*, 114–124. [CrossRef] [PubMed]
9. Bae, D.S.; Waters, P.M. Free Vascularized Fibula Grafting: Principles, Techniques, and Applications in Pediatric Orthopaedics. *Orthop. J. Harv. Med. Sch.* **2006**, *8*, 86–89.
10. Lumetti, S.; Galli, C.; Manfredi, E.; Consolo, U.; Marchetti, C.; Ghiacci, G.; Toffoli, A.; Bonanini, M.; Salgarelli, A.; Macaluso, G.M. Correlation between Density and Resorption of Fresh-Frozen and Autogenous Bone Grafts. *BioMed Res. Int.* **2014**, *2014*, 508328. [CrossRef]
11. Aponte-Tinao, L.A.; Ayerza, M.A.; Muscolo, D.L.; Farfalli, G.L. What Are the Risk Factors and Management Options for Infection After Reconstruction With Massive Bone Allografts? *Clin. Orthop. Relat. Res.* **2016**. [CrossRef]
12. Urban, R.M.; Turner, T.M.; Hall, D.J.; Inoue, N.; Gitelis, S. Increased Bone Formation Using Calcium Sulfate-Calcium Phosphate Composite Graft. *Clin. Orthop. Related Res.* **2007**, *459*, 110–117. [CrossRef]
13. Parikh, S.N. Bone Graft Substitutes in Modern Orthopedics. *Orthopedics* **2002**, *25*, 1301–1309. [CrossRef]
14. Sohn, H.S.; Oh, J.K. Review of Bone Graft and Bone Substitutes with an Emphasis on Fracture Surgeries. *Biomater. Res.* **2019**, *23*, 9. [CrossRef] [PubMed]
15. McKay, W.F.; Peckham, S.M.; Badura, J.M. A Comprehensive Clinical Review of Recombinant Human Bone Morphogenetic Protein-2 (INFUSE® Bone Graft). *Int. Orthop.* **2007**, *31*, 729–734. [CrossRef] [PubMed]
16. Smucker, J.D.; Rhee, J.M.; Singh, K.; Yoon, S.T.; Heller, J.G. Increased Swelling Complications Associated with Off-Label Usage of RhBMP-2 in the Anterior Cervical Spine. *Spine* **2006**, *31*, 2813–2819. [CrossRef] [PubMed]
17. Stiel, N.; Hissnauer, T.N.; Rupprecht, M.; Babin, K.; Schlickewei, C.W.; Rueger, J.M.; Stuecker, R.; Spiro, A.S. Evaluation of Complications Associated with Off-Label Use of Recombinant Human Bone Morphogenetic Protein-2 (RhBMP-2) in Pediatric Orthopaedics. *J. Mater. Sci. Mater. Med.* **2016**, *27*, 184. [CrossRef]
18. Barrack, R.L. Bone Graft Extenders, Substitutes, and Osteogenic Proteins. *J. Arthroplas.* **2005**, *20*, 94–97. [CrossRef] [PubMed]
19. Girardi, F.P.; Cammisa, F.P. The Effect of Bone Graft Extenders to Enhance the Performance of Iliac Crest Bone Grafts in Instrumented Lumbar Spine Fusion. *Orthopedics* **2003**, *26*. [CrossRef]
20. Cammisa, F.P.; Lowery, G.; Garfin, S.R.; Geisler, F.H.; Klara, P.M.; McGuire, R.A.; Sassard, W.R.; Stubbs, H.; Block, J.E. Two-Year Fusion Rate Equivalency between Grafton® DBM Gel and Autograft in Posterolateral Spine Fusion: A Prospective Controlled Trial Employing a Side-by-Side Comparison in the Same Patient. *Spine* **2004**, *29*, 660–666. [CrossRef] [PubMed]
21. Smucker, J.D.; Petersen, E.B.; Nepola, J.V.; Fredericks, D.C. Assessment of Mastergraft(®) Strip with Bone Marrow Aspirate as a Graft Extender in a Rabbit Posterolateral Fusion Model. *Iowa Orthop. J.* **2012**, *32*, 61–68.
22. Smucker, J.D.; Petersen, E.B.; Fredericks, D.C. Assessment of MASTERGRAFT PUTTY as a Graft Extender in a Rabbit Posterolateral Fusion Model. *Spine* **2012**, *37*, 1017–1021. [CrossRef]
23. Dai, L.Y.; Jiang, L.S. Single-Level Instrumented Posterolateral Fusion of Lumbar Spine with β-Tricalcium Phosphate versus Autograft: A Prospective, Randomized Study with 3-Year Follow-Up. *Spine* **2008**, *33*, 1299–1304. [CrossRef]
24. Lerner, T.; Bullmann, V.; Schulte, T.L.; Schneider, M.; Liljenqvist, U. A Level-1 Pilot Study to Evaluate of Ultraporous β-Tricalcium Phosphate as a Graft Extender in the Posterior Correction of Adolescent Idiopathic Scoliosis. *Eur. Spine J.* **2009**, *18*, 170–179. [CrossRef] [PubMed]

25. Hile, D.D.; Kandziora, F.; Lewandrowski, K.U.; Doherty, S.A.; Kowaleski, M.P.; Trantolo, D.J. A Poly(Propylene Glycol-Co-Fumaric Acid) Based Bone Graft Extender for Lumbar Spinal Fusion: In Vivo Assessment in a Rabbit Model. *Eur. Spine J.* **2006**, *15*, 936–943. [CrossRef]
26. Chedid, M.K.; Tundo, K.M.; Block, J.E.; Muir, J.M. Hybrid Biosynthetic Autograft Extender for Use in Posterior Lumbar Interbody Fusion: Safety and Clinical Effectiveness. *Open Orthop. J.* **2015**, *9*, 218–225. [CrossRef]
27. Lewandrowski, K.U.; Bondre, S.; Gresser, J.D.; Silva, A.E.; Wise, D.L.; Trantolo, D.J. Augmentation of Osteoinduction with a Biodegradable Poly(Propylene Glycol-Co-Fumaric Acid) Bone Graft Extender. A Histologic and Histomorphometric Study in Rats. *Bio-Med Mater. Eng.* **1999**, *9*, 325–334.
28. Walsh, W.R.; Oliver, R.A.; Gage, G.; Yu, Y.; Bell, D.; Bellemore, J.; Adkisson, H.D. Application of Resorbable Poly(Lactide-Co-Glycolide) with Entangled Hyaluronic Acid as an Autograft Extender for Posterolateral Intertransverse Lumbar Fusion in Rabbits. *Tissue Eng. Part A* **2011**, *17*, 213–220. [CrossRef] [PubMed]
29. Keränen, P.; Itälä, A.; Koort, J.; Kohonen, I.; Dalstra, M.; Kommonen, B.; Aro, H.T. Bioactive Glass Granules as Extender of Autogenous Bone Grafting in Cementless Intercalary Implant of the Canine Femur. *Scand. J. Surg.* **2007**, *96*, 243–251. [CrossRef]
30. Fernando, S.; McEnery, M.; Guelcher, S.A. Polyurethanes for Bone Tissue Engineering. *Adv. Polyurethane Biomater.* **2016**, 481–501. [CrossRef]
31. Guelcher, S.A. Biodegradable Polyurethanes: Synthesis and Applications in Regenerative Medicine. *Tissue Eng. Part B Rev.* **2008**, *14*, 3–17. [CrossRef]
32. McEnery, M.A.P.; Lu, S.; Gupta, M.K.; Zienkiewicz, K.J.; Wenke, J.C.; Kalpakci, K.N.; Shimko, D.A.; Duvall, C.L.; Guelcher, S.A. Oxidatively Degradable Poly(Thioketal Urethane)/Ceramic Composite Bone Cements with Bone-like Strength. *RSC Adv.* **2016**, *6*, 109414–109424. [CrossRef]
33. McGough, M.A.P.; Boller, L.A.; Groff, D.M.; Schoenecker, J.G.; Nyman, J.S.; Wenke, J.C.; Rhodes, C.; Shimko, D.; Duvall, C.L.; Guelcher, S.A. Nanocrystalline Hydroxyapatite-Poly(Thioketal Urethane) Nanocomposites Stimulate a Combined Intramembranous and Endochondral Ossification Response in Rabbits. *ACS Biomater. Sci. Eng.* **2020**, *6*, 564–574. [CrossRef]
34. Talley, A.D.; Boller, L.A.; Kalpakci, K.N.; Shimko, D.A.; Guelcher, C.S.A. Injectable, Compression-Resistant Polymer/Ceramic Composite Bone Grafts Promote Lateral Ridge Augmentation without Protective Mesh in a Canine Model. *Clin. Oral Implant. Res.* **2018**, 592–602. [CrossRef] [PubMed]
35. Samavedi, S.; Whittington, A.R.; Goldstein, A.S. Calcium Phosphate Ceramics in Bone Tissue Engineering: A Review of Properties and Their Influence on Cell Behavior. *Acta Biomater.* **2013**, *9*, 8037–8045. [CrossRef]
36. Lu, S.; McGough, M.A.P.; Shiels, S.M.; Zienkiewicz, K.J.; Merkel, A.R.; Vanderburgh, J.P.; Nyman, J.S.; Sterling, J.A.; Tennent, D.J.; Wenke, J.C.; et al. Settable Polymer/Ceramic Composite Bone Grafts Stabilize Weight-Bearing Tibial Plateau Slot Defects and Integrate with Host Bone in an Ovine Model. *Biomaterials* **2018**, *179*, 29–45. [CrossRef] [PubMed]
37. Talley, A.D.; McEnery, M.A.; Kalpakci, K.N.; Zienkiewicz, K.J.; Shimko, D.A.; Guelcher, S.A. Remodeling of Injectable, Low-Viscosity Polymer/Ceramic Bone Grafts in a Sheep Femoral Defect Model. *J. Biomed. Mater. Res. Part B Appl. Biomater.* **2017**, *105*, 2333–2343. [CrossRef] [PubMed]
38. Martin, J.R.; Gupta, M.K.; Page, J.M.; Yu, F.; Davidson, J.M.; Guelcher, S.A.; Duvall, C.L. A Porous Tissue Engineering Scaffold Selectively Degraded by Cell-Generated Reactive Oxygen Species. *Biomaterials* **2014**. [CrossRef] [PubMed]
39. McGough, M.; Shiels, S.M.; Boller, L.A.; Zienkiewicz, K.J.; Duvall, C.L.; Wenke, J.C.; Guelcher, S.A. Poly(Thioketal Urethane) Autograft Extenders in an Intertransverse Process Model of Bone Formation. *Tissue Eng. Part A* **2018**, *25*, 949–963. [CrossRef] [PubMed]
40. Boller, L.; Jones, A.; Cochran, D.; Guelcher, S. Compression-Resistant Polymer/Ceramic Composite Scaffolds Augmented with RhBMP-2 Promote New Bone Formation in a Nonhuman Primate Mandibular Ridge Augmentation Model. *Int. J. Oral. Maxillofac. Implant.* **2020**, *35*, 616–624. [CrossRef]
41. Dumas, J.E.; Prieto, E.M.; Zienkiewicz, K.J.; Guda, T.; Wenke, J.C.; Bible, J.; Holt, G.E.; Guelcher, S. Balancing the Rates of New Bone Formation and Polymer Degradation Enhances Healing of Weight-Bearing Allograft/Polyurethane Composites in Rabbit Femoral Defects. *Tissue Eng. Part A* **2014**, *20*, 115–129. [CrossRef]
42. Hafeman, A.E.; Zienkiewicz, K.J.; Zachman, A.L.; Sung, H.J.; Nanney, L.B.; Davidson, J.M.; Guelcher, S.A. Characterization of the Degradation Mechanisms of Lysine-Derived Aliphatic Poly(Ester Urethane) Scaffolds. *Biomaterials* **2011**, *32*, 419–429. [CrossRef] [PubMed]
43. Guelcher, S.A.; Srinivasan, A.; Dumas, J.E.; Didier, J.E.; McBride, S.; Hollinger, J.O. Synthesis, Mechanical Properties, Biocompatibility, and Biodegradation of Polyurethane Networks from Lysine Polyisocyanates. *Biomaterials* **2008**, *29*, 1762–1775. [CrossRef] [PubMed]
44. Bodde, E.W.H.; Spauwen, P.H.M.; Mikos, A.G.; Jansen, J.A. Closing Capacity of Segmental Radius Defects in Rabbits. *J. Biomed. Mater. Res. Part A* **2008**, *85*, 206–217. [CrossRef]
45. Wheeler, D.L.; Stokes, K.E.; Park, H.M.; Hollinger, J.O. Evaluation of Particulate Bioglass in a Rabbit Radius Ostectomy Model. *J. Biomed. Mater. Res.* **1997**, *35*, 249–254. [CrossRef]
46. Talley, A.D.; Kalpakci, K.N.; Shimko, D.A.; Zienkiewicz, K.J.; Cochran, D.L.; Guelcher, S.A. Effects of Recombinant Human Bone Morphogenetic Protein-2 Dose and Ceramic Composition on New Bone Formation and Space Maintenance in a Canine Mandibular Ridge Saddle Defect Model. *Tissue Eng. Part A* **2016**, *22*, 469–479. [CrossRef]

47. Shiels, S.M.; Talley, A.D.; McGough, M.A.P.; Zienkiewicz, K.J.; Kalpakci, K.; Shimko, D.; Guelcher, S.A.; Wenke, J.C. Injectable and Compression-Resistant Low-Viscosity Polymer/Ceramic Composite Carriers for RhBMP-2 in a Rabbit Model of Posterolateral Fusion: A Pilot Study. *J. Orthop. Surg. Res.* **2017**, *12*, 107. [CrossRef]
48. Dumas, J.E.; Zienkiewicz, K.; Tanner, S.A.; Prieto, E.M.; Bhattacharyya, S.; Guelcher, S.A. Synthesis and Characterization of an Injectable Allograft Bone/Polymer Composite Bone Void Filler with Tunable Mechanical Properties. *Tissue Eng. Part A* **2010**, *16*, 2505–2518. [CrossRef]
49. Guda, T.; Walker, J.A.; Singleton, B.M.; Hernandez, J.W.; Son, J.S.; Kim, S.G.; Oh, D.S.; Appleford, M.R.; Ong, J.L.; Wenke, J.C. Guided Bone Regeneration in Long-Bone Defects with a Structural Hydroxyapatite Graft and Collagen Membrane. *Tissue Eng. Part A* **2013**, *9*, 1879–1888. [CrossRef] [PubMed]
50. Guda, T.; Walker, J.A.; Pollot, B.E.; Appleford, M.R.; Oh, S.; Ong, J.L.; Wenke, J.C. In Vivo Performance of Bilayer Hydroxyapatite Scaffolds for Bone Tissue Regeneration in the Rabbit Radius. *J. Mater. Sci. Mater. Med.* **2011**, *22*, 647–656. [CrossRef] [PubMed]
51. Rathbone, C.R.; Guda, T.; Singleton, B.M.; Oh, D.S.; Appleford, M.R.; Ong, J.L.; Wenke, J.C. Effect of Cell-Seeded Hydroxyapatite Scaffolds on Rabbit Radius Bone Regeneration. *J. Biomed. Mater. Res. Part A* **2014**, *102*, 1458–1466. [CrossRef]
52. Wang, W.; Yeung, K.W.K. Bone Grafts and Biomaterials Substitutes for Bone Defect Repair: A Review. *Bioact. Mater.* **2017**, *2*, 224–247. [CrossRef]
53. Ehrler, D.M.; Vaccaro, A.R. The Use of Allograft Bone in Lumbar Spine Surgery. *Clin. Orthop. Relat. Res.* **2000**, 38–45. [CrossRef] [PubMed]
54. Yazar, S. Onlay Bone Grafts in Head and Neck Reconstruction. *Semin. Plast. Surg.* **2010**, *24*, 255–261. [CrossRef]
55. Antheunis, H.; van der Meer, J.C.; de Geus, M.; Heise, A.; Koning, C.E. Autocatalytic Equation Describing the Change in Molecular Weight during Hydrolytic Degradation of Aliphatic Polyesters. *Biomacromolecules* **2010**, *11*, 1118–1124. [CrossRef]
56. Khan, S.N.; Cammisa, F.P.; Sandhu, H.S.; Diwan, A.D.; Girardi, F.P.; Lane, J.M. The Biology of Bone Grafting. *J. Am. Acad. Orthop. Surg.* **2005**, *13*, 77–86. [CrossRef] [PubMed]
57. Parfitt, A. The physiologic and clinical significance of bone histomorphometric data. In *Bone Histomorphometry: Techniques and Interpretations*; CRC Press: Boca Raton, FL, USA, 1983.
58. Compston, J.; Skingle, L.; Dempster, D.W. Bone Histomorphometry. In *Vitamin D*, 4th ed.; Elsevier: Amsterdam, The Netherlands, 2018; ISBN 9780128099667.
59. Sheikh, Z.; Najeeb, S.; Khurshid, Z.; Verma, V.; Rashid, H.; Glogauer, M. Biodegradable Materials for Bone Repair and Tissue Engineering Applications. *Materials* **2015**, *8*, 5744–5794. [CrossRef] [PubMed]
60. Hou, J.; Wang, J.; Cao, L.; Qian, X.; Xing, W.; Lu, J.; Liu, C. Segmental Bone Regeneration Using RhBMP-2-Loaded Collagen/Chitosan Microspheres Composite Scaffold in a Rabbit Model. *Biomed. Mater.* **2012**, *7*. [CrossRef]

Article

Promotion of Bone Regeneration Using Bioinspired PLGA/MH/ECM Scaffold Combined with Bioactive PDRN

Da-Seul Kim [1,2,†], Jun-Kyu Lee [1,†], Ji-Won Jung [1], Seung-Woon Baek [1,3,4], Jun Hyuk Kim [1], Yun Heo [1], Tae-Hyung Kim [2] and Dong Keun Han [1,*]

1. Department of Biomedical Science, CHA University, 335 Pangyo-ro, Bundang-gu, Seongnam-si 13488, Gyeonggi-do, Korea; dptmf4011@cau.ac.kr (D.-S.K.); jklee2020@chauniv.ac.kr (J.-K.L.); jeongjiwon97@gmail.com (J.-W.J.); baiksw830@g.skku.edu (S.-W.B.); 1016jeffrey@gmail.com (J.H.K.); yun.heo@icloud.com (Y.H.)
2. School of Integrative Engineering, Chung-Ang University, 84 Heukseok-ro, Dongjak-gu, Seoul 06974, Korea; thkim0512@cau.ac.kr
3. Department of Biomedical Engineering, SKKU Institute for Convergence, Sungkyunkwan University (SKKU), 2066 Seobu-ro, Jangan-gu, Suwon-si 16419, Gyeonggi-do, Korea
4. Department of Intelligent Precision Healthcare Convergence, SKKU Institute for Convergence, Sungkyunkwan University (SKKU), 2066 Seobu-ro, Jangan-gu, Suwon-si 16419, Gyeonggi-do, Korea
* Correspondence: dkhan@cha.ac.kr
† D.-S.K. and J.-K.L. contributed equally to this work.

Abstract: Current approaches of biomaterials for the repair of critical-sized bone defects still require immense effort to overcome numerous obstacles. The biodegradable polymer-based scaffolds have been required to expand further function for bone tissue engineering. Poly(lactic-co-glycolic) acid (PLGA) is one of the most common biopolymers owing to its biodegradability for tissue regenerations. However, there are major clinical challenges that the byproducts of the PLGA cause an acidic environment of implanting site. The critical processes in bone repair are osteogenesis, angiogenesis, and inhibition of excessive osteoclastogenesis. In this study, the porous PLGA (P) scaffold was combined with magnesium hydroxide (MH, M) and bone-extracellular matrix (bECM, E) to improve anti-inflammatory ability and osteoconductivity. Additionally, the bioactive polydeoxyribonucleotide (PDRN, P) was additionally incorporated in the existing PME scaffold. The prepared PMEP scaffold has pro-osteogenic and pro-angiogenic effects and inhibition of osteoclast due to the PDRN, which interacts with the adenosine A_{2A} receptor agonist that up-regulates expression of vascular endothelial growth factor (VEGF) and down-regulates inflammatory cytokines. The PMEP scaffold has superior biological properties for human bone-marrow mesenchymal stem cells (hBMSCs) adhesion, proliferation, and osteogenic differentiation in vitro. Moreover, the gene expressions related to osteogenesis and angiogenesis of hBMSCs increased and the inflammatory factors decreased on the PMEP scaffold. In conclusion, it provides a promising strategy and clinical potential candidate for bone tissue regeneration and repairing bone defects.

Keywords: bone regeneration; poly(lactide-co-glycolide); magnesium hydroxide; extracellular matrix; polydeoxyribonucleotide; porous scaffold

1. Introduction

Bone fracture is the most common injury, which has high healing efficiency by oneself, but critical-sized bone fraction indispensably requires orthopedic surgery. To enhance the bone repair rate, various methods has been used, including autograft, allograft, xenograft, and artificial bone graft materials (e.g., tricalcium phosphate, hydroxyapatite, and bioglass). Among these treatments, autograft, obtained from the patient's other position, is regarded as a 'gold standard' because of its high regeneration rate and superior osteoconductivity and osteoinductivity without any immune response [1–3]. Although autograft has numerous advantages including no risk of disease transfer, there are some limitations, such as

restricted bone supply, donor site morbidity, and poor capability to accommodate defects. To conquer these hurdles, scaffold implantation is considered as an ideal way for bone tissue regeneration. Current bone tissue engineering (BTE) approaches still have numerous limitations such as low biocompatibility, mechanical property, osteoinductivity, and osteoconductivity. The polymer-based scaffold has been studied because of its biodegradability and biocompatibility. Among them, Poly(lactic-co-glycolic) acid (PLGA) was approved by the Food and Drug Administration (FDA) for diverse types of bone implants. However, it has been reported that its degradation byproducts, lactic acid and glycolic acid cause an acidic microenvironment at implanting site [4,5]. In previous studies, magnesium hydroxide (MH) performed outstanding pH neutralization ability for diverse tissue regeneration [6–11], in particular, bone repair [12–14]. However, MH as a hydrophilic inorganic molecule is difficult to disperse evenly in the hydrophobic polymer-based scaffold. To disperse metal ion molecule, in the prior study, the MH modified with a ricinoleic acid (mMH) was attempted to PLGA porous scaffold [9]. Plus, the incorporation of mMH into PLGA implant would be used to attenuate acid-induced inflammation triggered by the degradation products from the polymer and to improve the hydrophilicity of the scaffold.

As noted above, to overcome the limitation of the BTE scaffold, the extracellular matrix (ECM) isolated from mammalian tissues has been attempted in the scaffold for enhancing biocompatibility [15–17] and mimicking the natural composition of bone tissue. Especially, bovine-derived decellularized bone extracellular matrix (bECM), comprising mostly of calcium and phosphate, can improve not only biocompatibility also osteoconductivity of the scaffold.

Aside from these improvements, because bone repairing takes a longer time than other tissue in general, the ideal BTE scaffold should have an osteoinductive property. In this respect, in order to enhance osteoinductivity and bioactive function of the scaffold, polydeoxyribonucleotide (PDRN) was applied in bone regeneration. PDRN is a natural bioactive molecule normally extracted from salmon trout (*Oncorhynchus mykiss*) gonads, which is a short DNA form (50 to 2000 base pairs). Recently, some studies reported that PDRN has great effects on improving tissue regeneration since it plays as an adenosine A_{2A} receptor agonist. Adenosine A_{2A} receptor is a member of the G protein-coupled receptor (GPCR) family that has been proven as effective in improving angiogenesis and reducing inflammation [18–20]. Additionally, PDRN provides building blocks, nucleotides, and nucleosides to produce nucleic acids using less energy via the salvage pathway [21].

In this study, we designed a bioinspired scaffold by integrating mMH (M), bECM (E), and PDRN (P) into a porous PLGA (P) scaffold. We hypothesized that mMH could suppress the detrimental effect caused by PLGA degradation, reduce osteoclastogenesis; bECM could mimic the natural bone tissue microenvironment, improve osteoconductivity; PDRN could promote angiogenesis during bone repair. Therefore, the functionalized biodegradable PMEP scaffold would be applicable for effective bone regeneration with synergistic effects from these bioactive molecules.

2. Materials and Methods
2.1. Materials

Poly(D,L-lactide-co-glycolide) (PLGA, lactide:glycolide = 75:25, I.V. = 0.8–1.2) was purchased from Evonik Ind. (Essen, Germany). Magnesium hydroxide (MH), L-ascorbic acid, dexamethasone, and β-glycerophosphate were purchased by Sigma Aldrich (St. Louis, MO, USA). Ricinoleic acid was purchased from TCI product (Tokyo, Japan). The bovine bone-derived extracellular matrix powder (bECM; InduCera) was supplied by Oscotec Inc. (Seongnam, Korea). Polydeoxyribonucleotide (PDRN) was obtained from Goldbio (St. Louis, MO, USA). D-Plus™ cell counting kit 8 (CCK-8) cell viability assay kit was obtained from Dongin LS (Seoul, Korea).

2.2. Scaffold Preperation

The modified $Mg(OH)_2$ was synthesized with ricinoleic acid (mMH) following the process with the previous study [9]. All scaffolds (PLGA, PME, and PMEP) were prepared by the freeze-drying method. In brief, the ice particles (200–300 μm) were prepared by spraying deionized water into liquid nitrogen as a porogen for the porous scaffold. A 20 wt% mMH and PDRN, and 50 wt% bECM (compared to PLGA) were mixed with 0.5 g of PLGA in 0.3 M dichloromethane solution. The mixtures and ice particles were stuffed into round PTFE mold (ø5 × 2 mm^2). The filled molds were freeze-dried for 2 days to remove the ice and remaining organic solvent, then the porous scaffolds were obtained.

2.3. Scaffold Characterization

The cross-section morphology of the scaffolds was observed using scanning electron microscopy (SEM; GENESIS-1000, Emcraft, Gwangju, Korea). The thermal property of the scaffolds was analyzed by a thermal gravimetric analyzer (TGA 4000, PerkinElmer, Waltham, MA, USA). To assess the neutralization capacity of the scaffolds, the mass and pH changes were measured in 500 μL phosphate-buffered saline (PBS) solution (pH 7.4) with 20 μg/mL protease K (Bioneer, Daejeon, Korea) for 14 days. The inorganic compositions of the scaffold were measured using inductively coupled plasma-optical emission spectroscopy (ICP-OES, Optima 8000, PerkinElmer, Waltham, MA, USA). The water contact angle (WCA) was analyzed using the sessile drop method at room temperature to evaluate the hydrophilicity and hydrophobicity of the scaffolds.

2.4. Cell and Cytotoxicity Assay

Human bone-marrow mesenchymal stem cells (hBMSCs) were cultured in DMEM/low glucose media supplemented with 10% FBS (Hyclone, Logan, UT, USA) and 1% antibiotic–antimycotic solution (Gibco, Thermo Scientific Inc., Waltham, MS, USA). The cells were maintained under a humidified atmosphere with 5% CO_2 at 37 °C. The viability and proliferation of the cells were determined using a Live-dead viability/cytotoxicity kit (Invitrogen, Thermo Scientific Inc., Waltham, MS, USA) and the fluorescence images were obtained using LSM880 (Zeiss, Jena, Germany) at 1, 3, and 7 days. The CCK-8 assay was conducted on the 3D scaffold at the same days.

2.5. Wound Healing Assay and Tubule Formation

The scratch wound healing assay was conducted to assess the migratory capacity of hBMSCs by PDRN. The cells were seeded into a 6-well culture plate at the density of 3×10^5 cells/well and cultured for 1 day. The confluent wells were scratched, then washed with PBS solution. Cells were cultured with DMEM/low glucose containing 1% (v/v) FBS and added 100 μg/mL of PDRN. After 12 and 24 h, the plates were photographed and quantified the healed area using Image J software. To assess the angiogenic effects of the PDRN, 250 μL of matrigel matrix (Corning, Brooklyn, NY, USA) was added to pre-cooled 24-well plate and then incubated at 37 °C for 1 h. The human umbilical vein endothelial cells (HUVECs) were seeded onto coated well at the density of 1.2×10^5 cells/well with EBM-2 (Lonza, Basel, Switzerland) containing 1% FBS, then added 100 μg/mL of PDRN. After 18 h, the cells were stained with calcein AM (C1430, Thermo Scientific Inc., Waltham, MS, USA), then photographed with a fluorescence microscope (U-RFL-T, Olympus, Tokyo, Japan). The tube length and branch point were quantified using Image J software.

2.6. RNA Extraction and Quantitative Real-Time PCR (qRT-PCR)

The RNA from scaffolds was extracted using Trizol reagent (15596018, Ambion, Invitrogen, Thermo Scientific Inc., Waltham, MS, USA) following the manufacturer's instructions. The RNA concentration and quality were measured by spectrophotometer (ND-1000; Thermo Scientific, Waltham, MA, USA). The cDNA was synthesized using PrimeScript RT Reagent Kit (Perfect Real Time, Takara, Tokyo, Japan). The qRT-PCR was performed using each primer and SYBR Green PCR Master Mix (Applied Biosystems, Thermo Scientific

Inc., Waltham, MS, USA). The expression of osteogenic, angiogenic, and inflammation-related genes was calculated with the 18S rRNA as a reference using the 2−ΔΔCt method. The primers used were as follows: 18S rRNA: forward, 5′-gcaattattccccatgaacg-3′ and reverse, 5′-gggacttaatcaacgcaagc-3′; IL6: forward, 5′-gatgagtacaaaagtcctgatcca-3′ and reverse, 5′-ctgcagccactggttctgt-3′; IL-1β: forward, 5′-tacctgtcctgcgtgttgaa-3′ and reverse, 5′-tctttgggtaatttttgggatct-3′; VEGF: forward, 5′-actggaccctggctttactg-3′ and reverse, 5′-tctgctcccttctgtcgt-3′; MMP2: forward, 5′-caccaccgaggattatgacc-3′ and reverse, 5′- cacccacagt ggacatagca-3′; ALP: forward, 5′-atgaaggaaaagccaagcag-3′ and reverse, 5′-ccaccaaatgt gaagacgtg-3′; RUNX2: forward, ggtcagatgcaggcggccc-3′ and reverse, 5′-tacgtgtggtagcgcgtgg c-3′; OCN: forward, 5′-cagcgaggtagtgaagagacc-3′ and reverse, 5′-tctggagtttatttgggagcag-3′.

2.7. Osteogenic Differentiation In Vitro

To assess the capacity of osteogenic differentiation on the 3D scaffold, hBMSCs were seeded onto the scaffold at the density of 5×10^5 cells/scaffold. After 1 day, the medium was replaced with an osteogenic differentiation medium, DMEM/low glucose, containing 50 μM L-ascorbic acid, 0.1 μM Dexamethasone, and 10 mM β-glycerophosphate. After 7 days of osteogenic differentiation, the scaffolds were fixed with 10% formalin for 20 min, rinsed with deionized water, and stained with an alkaline phosphatase (ALP) staining kit (MK300, Takara, Japan). The stained samples were incubated in 15 and 30% sucrose solution in order, and embedded with frozen section media (FSC 22, Leica Biosystems, Wetzlar, Germany). The frozen samples were sectioned with a cryostat microtome (CM3050S, Leica Biosystems, Wetzlar, Germany). For the quantification of ALP activity, the scaffolds were lysed using an ALP assay kit (MK301, Takara, Tokyo, Japan). The assay was conducted according to the produced protocol.

2.8. Tartrate-Resistant Acid Phosphatase Staining and Activity

The osteoclastogenesis was identified by Tartrate-resistant acid phosphatase (TRAP) staining (MK-300, Takara, Tokyo, Japan). The RAW264.7 cells, mouse macrophage cell line, were seeded into a 24-well culture plate at the density of 2×10^4 cells/well. After 1 day, 100 ng/mL of receptor activator of the NF-κB ligand (RANKL) was treated with RAW264.7 cells to induce differentiation into osteoclast. After 3 days, the scaffolds were put into trans-well inserts for co-culture with osteoclast. The TRAP staining was executed after 3 days.

2.9. Statistical Analysis

All experimental results were obtained through more than three independent experiments, and the values were described as mean ± standard deviation (SD). The statistical significance was analyzed by one-way ANOVA using Tukey's post hoc method in GraphPad Prism 7.0 software [12] (GraphPad Software, Inc., San Diego, CA, USA). The statistically significant difference was defined as the *p* value being less than 0.05. The differences were considered significant when * $p < 0.05$, ** $p < 0.01$, *** $p < 0.001$, and # $p < 0.0001$.

3. Results and Discussion

3.1. Scaffold Characterization

The biodegradable porous scaffolds containing PLGA, mMH, dECM, and PDRN were fabricated using the etching method with ice particles. In Figure 1A, the SEM images represent cross-section morphology that the pores of the scaffold were well-distributed and interconnected, so that the cells could easily attach and migrate in the scaffold during bone regeneration. Moreover, the 200–300 μm of porogens were used, that it is known as appropriate size for osteogenic differentiation in many other studies [22,23]. This size of porogen could be beneficial to cell ingrowth into the pore structures. The proportion of inorganic molecules in the scaffolds was analyzed by TGA (Figure 1B) and induced coupled plasma-optical emission spectroscopy (ICP-OES, Table 1). The PMEP scaffold consists of 195.44 ppm of magnesium, 265.74 ppm of calcium, and 151.20 ppm of phosphorus, respectively. The PME scaffold consists of 201.46 ppm of magnesium, 270.44 ppm of calcium, and 136.43 ppm of phosphorus, respectively. Interestingly, the amount of phosphorous in the PMEP was slightly higher than the PME because of the phosphate backbone in PDRN. The porous scaffolds with dissimilar surface roughness can cause different wettability and thus affect the permeability. The WCA was conducted to evaluate the wettability of the scaffold. The angles on PLGA, PME, and PMEP scaffold were 104.59, 93.99, and 77.12°, respectively. As bioactive molecules were added, the contact angles decreased. In other words, the PMEP scaffold has more hydrophilic property than the PLGA and PME ones.

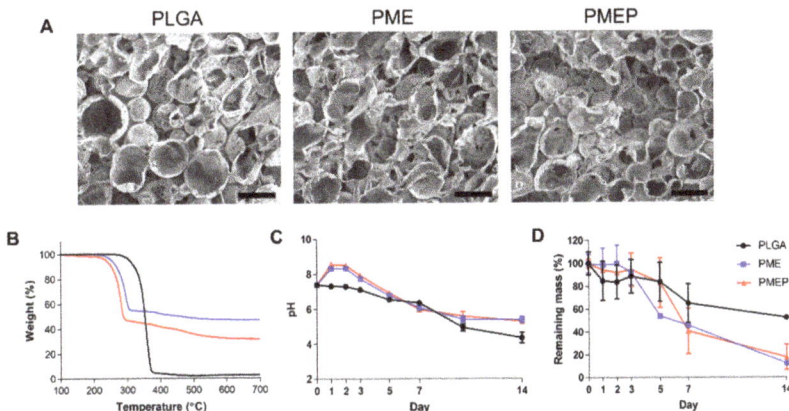

Figure 1. Scaffold characterization. (**A**) representative scanning electron microscopy (SEM) images of PLGA, PLGA/MH/ECM (PME), and PLGA/MH/ECM/PDRN (PMEP) scaffold. (Scale bars = 200 µm). (**B**) thermal gravimetric analysis (TGA) thermograms of each scaffold. Change of (**C**) pH and (**D**) mass during in vitro degradation in PBS solution with protease K at 37 °C for 14 days.

Table 1. ICP-OES and water contact angle.

Group	Mg (ppm)	Ca (ppm)	P (ppm)	Water Contact Angle (°)
PLGA	-	-	-	104.59 ± 4.24
PLGA/mMH/bECM (PME)	201.46 ± 0.93	270.44 ± 2.53	136.43 ± 0.94	93.99 ± 9.90
PLGA/mMH/bECM/PDRN (PMEP)	195.44 ± 3.03	265.74 ± 5.75	151.20 ± 1.60	77.12 ± 4.49

The degradation of porous PLGA, PME, and PMEP scaffolds was observed in the presence of 20 µg/mL protease K for 14 days at 37 °C. The accelerative condition was conducted using protease K due to relatively high molecular weight of PLGA. In Figure 1C, pH value of the PLGA in PBS solution drastically decreased to 4.3 after 14 days of degradation. However, pH value of the solution containing the PME and PMEP specimens initially reached 8.3 and 8.5 and dropped slowly for 14 days to reach 5.5 and 5.4, respectively due to neutralization ability of mMH. The PME and PMEP scaffolds showed fast degradation performance than the PLGA only scaffold since those were containing numerous soluble bioactive molecules (Figure 1D).

3.2. Biocompatibility of the Scaffold

To confirm cytotoxicity of the scaffolds in vitro, in Figure 2A, calcein AM and ethidium homodimer 1 (EthD-1) stainings were conducted with hBMSCs at 1, 3, and 7 days. Because of its well-known biocompatibility of PLGA, the EthD-1 positive cells indicating dead cells were observed rarely in all the scaffolds even the PLGA only group. What is more, the cells were observed evenly along with the pores of the scaffold. However, the population of calcein AM positive cells, the live cells, was getting increased in the PME and the PMEP than the PLGA at 1, 3, and 7 days, respectively. In Figure 2B, the cell viability was quantified using CCK-8 in 1, 3, and 7 days. The initial adhesion rate of hBMSCs on the PMEP significantly increased due to its surface hydrophilicity for cell recruitment ($p < 0.01$). Because a hydrophilic surface promotes the adhesion of the cells [24], the initial cell adhesion rate significantly increased on the PMEP scaffold.

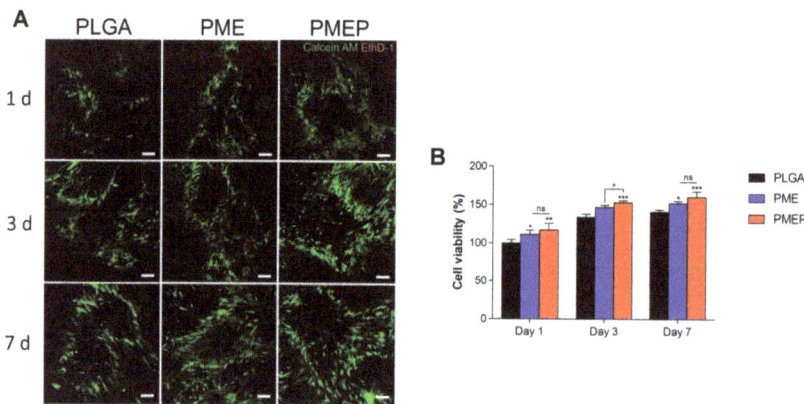

Figure 2. Biocompatibility of the scaffolds. (**A**) live-dead assay images on each scaffold at 1, 3, and 7 days (scale bar = 100 μm). (**B**) cell viability of the hBMSCs onto each scaffold at 1, 3, and 7 days in vitro. The differences were considered significant when ns = not significant ($p \geq 0.05$), * $p < 0.05$, ** $p < 0.01$, and *** $p < 0.001$ ($n \geq 3$).

As mentioned previously, the PLGA produced acidic byproducts during hydrolytic degradation, so that the cell slightly proliferated for 7 days. However, the cell viability on the PME and particularly, the PMEP scaffold was remarkably enhanced for 7 days, $p < 0.05$ and $p < 0.001$, respectively. Consequently, the incorporation of mMH, bECM, and PDRN could constrict the adverse effect on cell cytotoxicity caused by hydrolytic degradation of PLGA.

3.3. Confirmation of Angigenic Ability

Angiogenesis is a physiological process by which new blood vessels form from the pre-existing vascular network, allowing the delivery of oxygen and nutrients to the body's tissues. Angiogenesis has been studied as a therapeutic target in regenerative medicine. Bone is also richly vascularized tissue, so that new blood vessels play a critical role in maintaining the bone cells survival and stimulating their activity. However, in situ vascularized bone regeneration still remains in the extreme challenge [25–28].

As mentioned previously, PDRN has a pro-drug activity carried out through two different mechanisms. First, PDRN supplies purines and pyrimidines, promoting DNA synthesis or repair through the 'salvage pathway' [18,21]. Next, PDRN stimulates adenosine A_{2A} receptor, as suggested by Thellung et al. studied the effect of PDRN using 3,7-dimethyl-1-propargylxanthine (DMPX), a selective adenosine A_{2A} receptor antagonist [29]. Adenosine and adenosine A_{2A} receptor were considered clinically important to enhance angiogenesis. Wang et al. studied that adenosine enhances cell growth and induces tube formation in HUVECs in vitro [30]. In Figure 3, the biological ability of PDRN was investigated in angiogenesis and wound healing for effective bone tissue repair. When PDRN treated, HUVECs had formed a significant number of branch points and longer lengths of tubes. On the same side of Figure 2, because PDRN could enhance the growth and migratory ability of hBMSCs, the wound closure rates also highly increased to 34.8 and 31.9% in PDRN treated groups compared to control at 24 and 48 h, respectively. To conclude, these outstanding biological abilities of PDRN give a synergistic effect to achieving a novel strategy for bone regeneration.

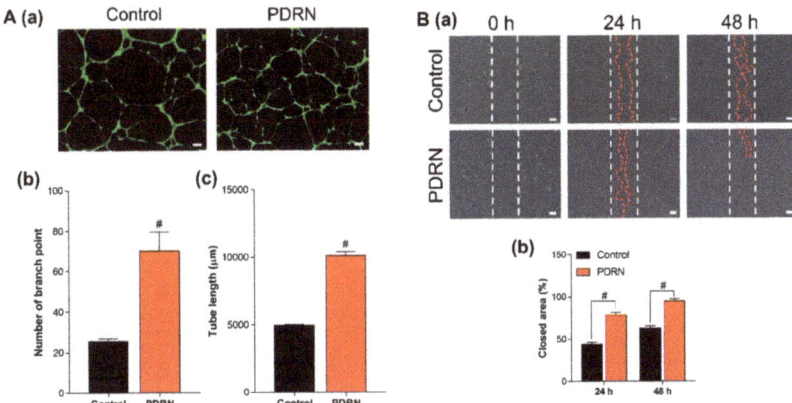

Figure 3. Biological effects of PDRN. (**A**) tubule-forming assay; images of HUVECs stained with calcein AM (scale bar = 200 µm) (**a**) and quantification of branch point (**b**) and tube length (**c**). (**B**) wound healing assay; optical images (scale bar = 200 µm) (**a**) and quantification of closed area at 24 and 48 h (**b**). The differences were considered significant when # $p < 0.0001$ ($n \geq 3$).

3.4. Biological Abilities of the PMEP Scaffold with hBMSCs: Anti-Inflammation and Angiogenesis

Since mineralization is affected by numerous mechanisms, biomaterials should have a variety of functions, such as vascularization, inhibition of inflammation, as well as osteogenesis to reach effective bone regeneration. The quantitative real-time PCR (qRT-PCR) was conducted to determine the expression of inflammation and angiogenesis-related genes on 3D scaffolds with hBMSCs. The effect of the scaffolds was assessed in osteogenic media at 7 and 21 days. As shown in Figure 4A, the PME scaffold restricted the expression of inflammatory genes, interleukin-6 (IL-6) and interleukin-1β (IL-1β), compared to the PLGA scaffold. Plus, the PMEP scaffold effectively suppressed the above-mentioned gene expression even in comparison with the PME one. In recent studies, the researchers demonstrated that the PDRN affects to increase expression of vascular endothelial growth factor (VEGF) and to suppress the production of pro-inflammatory cytokines by stimulating the A_{2A} receptor [18,21,31]. As a result, the PDRN could promote angiogenesis and inhibit inflammation during bone repair. In Figure 4B, the PME scaffold exhibited a negligible difference in the expression of angiogenesis-related genes, including VEGF and matrix metalloproteinase-2 (MMP2). Likewise, the PMEP scaffold promoted the highest gene expression of VEGF and MMP2 on both days. It is notable that the addition of PDRN on the scaffold has effects on not only reducing the inflammatory response but also significantly enhancing vascularization. In prior analysis (Figure 3), we confirmed the effectiveness of PDRN, treated directly in cells in the 2D environment. Further, incorporation of mMH, bECM, and PDRN in the biodegradable porous 3D scaffold displayed attenuating inflammatory response and enhancing angiogenesis, simultaneously. These results suggest that the PDRN effect is not only for angiogenesis but also may influence several factors containing the healing process.

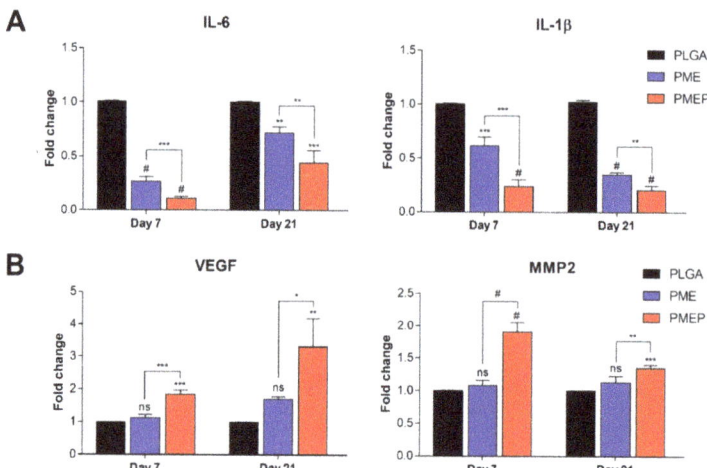

Figure 4. Anti-inflammatory and angiogenic effects on the scaffolds using hBMSCs. Gene expressions of hBMSCs onto the scaffolds related to (**A**) anti-inflammation: IL-6 and IL-1β, and (**B**) angiogenesis: VEGF and MMP2 at 7 and 21 days. The differences were considered significant when ns = not significant ($p \geq 0.05$), * $p < 0.05$, ** $p < 0.01$, *** $p < 0.001$, and # $p < 0.0001$ ($n \geq 3$).

3.5. Induction of Osteogenesis in 3D Scaffold

To identify the osteogenic capacity of the scaffolds, hBMSCs were seeded onto the scaffold. ALP is known as an early marker of osteogenesis. After 7 days of osteogenic differentiation, the ALP staining was conducted on each scaffold. Figure 5A showed that the PMEP scaffold formed more degrees of staining with less collapsing of internal structure compared to other scaffolds. Moreover, the PMEP scaffold enhanced ALP activity, which was even significantly higher than the PME one. These results implied that the PME scaffold could induce osteogenic differentiation of hBMSCs effectively, and by adding PDRN, the osteogenesis was more enhanced. Further investigation of cell differentiation was verified through gene expression analysis of the osteogenic markers. The expression of osteogenesis-related genes including ALP, runt-related transcription factor 2 (RUNX2), and osteocalcin (OCN) was also evaluated at 7 (Figure 5B) and 21 days (Figure 5C). In general, the RUNX2 and OCN are, respectively, used as mid- and late-responsive genes for bone formation. The results exhibited that the PMEP scaffold significantly up-regulated ALP, RUNX2, and OCN at all days. The mRNA expression of genes in the PMEP scaffold respectively increased by 2.48-, 2.05-, and 3.07-fold higher than the PLGA group at 7 days. The expressions on 21 days were also up-regulated by 2.08-, 1.75-, and 1.94-fold higher in the PMEP scaffold, respectively. These results indicated that the MH provides biocompatibility, bECM has osteoconductivity, and the PDRN promotes angiogenesis and osteogenesis by stimulating the A_{2A} receptor. In conclusion, the PMEP scaffold has the potentials that not only effectively induce early-stage of osteogenesis, but also affect the maturation of hBMSCs for bone regeneration.

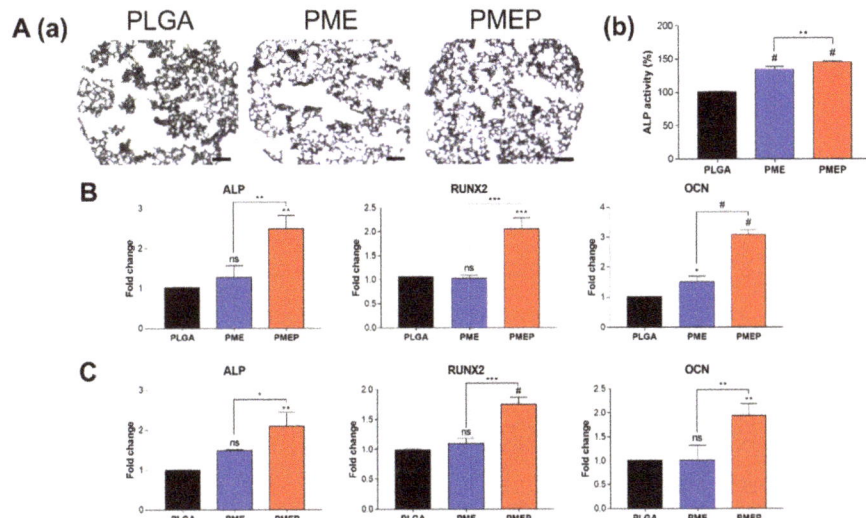

Figure 5. Osteogenic differentiation onto scaffolds using hBMSCs. (**A**) optical images of the scaffolds stained with ALP for 7 days in osteogenic medium (**a**). Scale bars indicate 400 μm. What is more, the quantification of ALP activity onto each scaffold (**b**). (**B**,**C**) gene expressions of hBMSCs related to osteogenesis onto the scaffolds; ALP, RUNX2, and OCN at 7 and 21 days of osteogenic differentiation. The differences were considered significant when ns = not significant ($p \geq 0.05$), * $p < 0.05$, ** $p < 0.01$, *** $p < 0.001$, and # $p < 0.0001$ ($n \geq 3$).

3.6. Attenuation of Osteoclastogenesis

Recently, to develop the bone repairing materials, one of the most important issues is bone homeostasis between osteoclasts and osteoblasts, because excessive differentiation of osteoclasts affects bone tissue resorption, which would occur metabolic bone-related diseases such as osteoporosis [32–34]. In general, when bone fracture occurs, both osteoblasts and osteoclasts are activated. Immoderate osteoclastogenesis and osteoblastogenesis cause eventually the delay of bone formation or nonunion. Thus, the BTE scaffold should control initial immoderate osteoclastogenesis, which is critical for promoting osteoblast activity and enhancing bone mineral density [35–37]. To evaluate the control ability of the 3D scaffold, we designed an indirect co-culture system (Figure 6) using a trans-well insert. In Figure 6A, optical images represent TRAP positive cells after 3 days of RANKL treatment. The stained cells (purple) significantly decreased in the PMEP group compared to the control, the PLGA and the PME. To quantify the osteoclast activity, TRAP activity was analyzed with the 3D scaffolds. In the PLGA group, the activity slightly increased in comparison to RANKL treated control group. However, the secreted bioactive molecules from the PME and PMEP scaffold attenuated RANKL-induced differentiation into osteoclast of RAW264.7 cells for 31.7 and 74.4%, respectively, than control. Overall, the PMEP scaffold has multifunctional abilities in inhibition of local inflammation, promotion of angiogenesis, and attenuation of osteoclastogenesis. The bioinspired PMEP scaffold would be clinically utilized as a bone grafting material for tissue regeneration of various sizes and shapes.

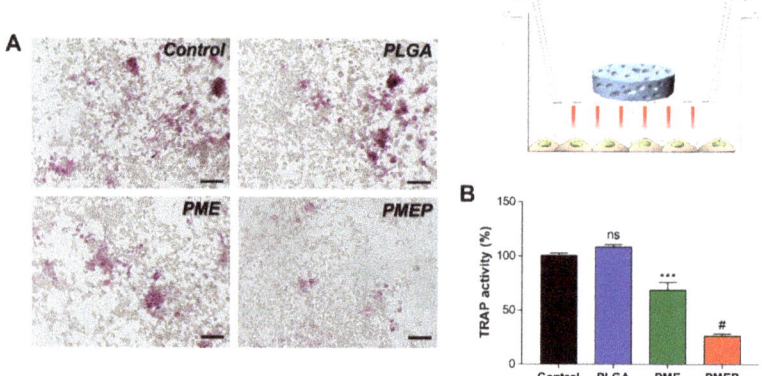

Figure 6. RANKL-induced osteoclastogenesis of RAW264.7 cells for 3 days. Experimental design of osteoclastogenesis using porous scaffold (right above). (**A**) optical images of TRAP+ cells (scale bar = 100 μm). (**B**) quantification of TRAP activity. The differences were considered significant when ns = not significant ($p \geq 0.05$), *** $p < 0.001$, and # $p < 0.0001$ ($n \geq 3$).

4. Conclusions

Because of their biodegradability and biocompatibility in physiological environments, biodegradable synthetic polymers are commonly used for a wide range of biomedical applications, especially bone repairing. Among them, PLGA has been clinically used as a bone grafting material. However, PLGA-based bioimplants often occur clinical failure due to low mechanical property and local acidification. Our findings proposed that mMH could enhance the mechanical property and neutralize acidification. The bECM was introduced to improve osteoconductivity by providing natural calcium and phosphate rich environment. Additionally, the DNA-derived bioactive molecule, PDRN facilitated biocompatibility and in situ vascularization during osteogenesis.

Taken together, we investigated that the synergistic interaction of mMH, bECM, and PDRN in the PMEP scaffold. The bioactive PMEP scaffold could inhibit osteoclastogenesis and promote adequate cell proliferation, angiogenesis, and osteogenesis in vitro. In the future study, we expect that the PMEP scaffold can regenerate the new bone in vivo by multifunctional abilities. This versatile biodegradable scaffold would apply to a novel bone tissue engineering as an advanced biomedical device.

Author Contributions: D.K.H. conceived and supervised the project. D.-S.K. and J.-K.L. contributed equally to this work. D.-S.K., J.-K.L., J.-W.J., S.-W.B. and J.H.K. performed the experiments and analyzed the data. The manuscript was written by D.-S.K., J.-K.L., Y.H. and T.-H.K. All authors have read and agreed to the published version of the manuscript.

Funding: This work was supported by Basic Science Research Program (2020R1A2B5B03002344) and Bio & Medical Technology Development Program (2018M3A9E2024579) through the National Research Foundation of Korea funded by the Ministry of Science and ICT (MSIT), and the Korea Medical Device Development Fund grant funded by the Korea government (the Ministry of Science and ICT, the Ministry of Trade, Industry and Energy, the Ministry of Health & Welfare, Republic of Korea, the Ministry of Food and Drug Safety (202011A05-05), and Korea Health Technology R&D Project through the Korea Health Industry Development Institute (KHIDI), funded by the Ministry of Health & Welfare, Republic of Korea (HR16C0002).

Institutional Review Board Statement: Not applicable.

Informed Consent Statement: Not applicable.

Data Availability Statement: The data presented in this study are available on request from the corresponding author.

Conflicts of Interest: The authors declare no conflict of interest. The funders had no role in the design of the study; in the collection, analyses, or interpretation of data; in the writing of the manuscript, or in the decision to publish the results.

References

1. Murphy, C.M.; O'Brien, F.J.; Little, D.G.; Schindeler, A. Cell-scaffold interactions in the bone tissue engineering triad. *Eur. Cell Mater.* **2013**, *26*, 120–132. [CrossRef]
2. Moussa, N.T.; Dym, H. Maxillofacial Bone Grafting Materials. *Dent. Clin. N. Am.* **2020**, *64*, 473–490. [CrossRef] [PubMed]
3. Bernardi, S.; Macchiarelli, G.; Bianchi, S. Autologous Materials in Regenerative Dentistry: Harvested Bone, Platelet Concentrates and Dentin Derivates. *Molecules* **2020**, *25*, 5330. [CrossRef] [PubMed]
4. Amini, A.R.; Wallace, J.S.; Nukavarapu, S.P. Short-term and long-term effects of orthopedic biodegradable implants. *J. Long. Term Eff. Med. Implants* **2011**, *21*. [CrossRef] [PubMed]
5. Koga, T.; Kumazawa, S.; Okimura, Y.; Zaitsu, Y.; Umeshita, K.; Asahina, I. Evaluation of PolyLactic-co-Glycolic Acid-Coated β-Tricalcium Phosphate Bone Substitute as a Graft Material for Ridge Preservation after Tooth Extraction in Dog Mandible: A Comparative Study with Conventional β-Tricalcium Phosphate Granules. *Materials* **2020**, *13*, 3452. [CrossRef] [PubMed]
6. Lih, E.; Park, K.W.; Chun, S.Y.; Kim, H.; Kwon, T.G.; Joung, Y.K.; Han, D.K. Biomimetic Porous PLGA Scaffolds Incorporating Decellularized Extracellular Matrix for Kidney Tissue Regeneration. *ACS Appl. Mater. Interfaces* **2016**, *8*, 21145–21154. [CrossRef] [PubMed]
7. Lih, E.; Park, W.; Park, K.W.; Chun, S.Y.; Kim, H.; Joung, Y.K.; Kwon, T.G.; Hubbell, J.A.; Han, D.K. A Bioinspired Scaffold with Anti-Inflammatory Magnesium Hydroxide and Decellularized Extracellular Matrix for Renal Tissue Regeneration. *ACS Cent. Sci.* **2019**, *5*, 458–467. [CrossRef]
8. Ko, K.-W.; Park, S.-Y.; Lee, E.H.; Yoo, Y.-I.; Kim, D.-S.; Kim, J.Y.; Kwon, T.G.; Han, D.K. Integrated Bioactive Scaffold with Polydeoxyribonucleotide and Stem-Cell-Derived Extracellular Vesicles for Kidney Regeneration. *ACS Nano* **2021**, *15*, 7575–7585. [CrossRef]
9. Lih, E.; Kum, C.H.; Park, W.; Chun, S.Y.; Cho, Y.; Joung, Y.K.; Park, K.S.; Hong, Y.J.; Ahn, D.J.; Kim, B.S.; et al. Modified Magnesium Hydroxide Nanoparticles Inhibit the Inflammatory Response to Biodegradable Poly(lactide-co-glycolide) Implants. *ACS Nano* **2018**, *12*, 6917–6925. [CrossRef]
10. Park, K.S.; Kim, B.J.; Lih, E.; Park, W.; Lee, S.H.; Joung, Y.K.; Han, D.K. Versatile effects of magnesium hydroxide nanoparticles in PLGA scaffold-mediated chondrogenesis. *Acta Biomater.* **2018**, *73*, 204–216. [CrossRef]
11. Ko, K.W.; Choi, B.; Kang, E.Y.; Shin, S.W.; Baek, S.W.; Han, D.K. The antagonistic effect of magnesium hydroxide particles on vascular endothelial activation induced by acidic PLGA degradation products. *Biomater. Sci.* **2020**, *9*, 892–907. [CrossRef]
12. Bedair, T.M.; Lee, C.K.; Kim, D.S.; Baek, S.W.; Bedair, H.M.; Joshi, H.P.; Choi, U.Y.; Park, K.H.; Park, W.; Han, I.; et al. Magnesium hydroxide-incorporated PLGA composite attenuates inflammation and promotes BMP2-induced bone formation in spinal fusion. *J. Tissue Eng.* **2020**, *11*, 2041731420967591. [CrossRef]
13. Go, E.J.; Kang, E.Y.; Lee, S.K.; Park, S.; Kim, J.H.; Park, W.; Kim, I.H.; Choi, B.; Han, D.K. An osteoconductive PLGA scaffold with bioactive β-TCP and anti-inflammatory Mg(OH)$_2$ to improve in vivo bone regeneration. *Biomater. Sci.* **2020**, *8*, 937–948. [CrossRef] [PubMed]
14. Lee, S.K.; Han, C.M.; Park, W.; Kim, I.H.; Joung, Y.K.; Han, D.K. Synergistically enhanced osteoconductivity and anti-inflammation of PLGA/β-TCP/Mg(OH)$_2$ composite for orthopedic applications. *Mater. Sci. Eng. C Mater. Biol. Appl.* **2019**, *94*, 65–75. [CrossRef] [PubMed]
15. Bianco, J.E.R.; Rosa, R.G.; Congrains-Castillo, A.; Joazeiro, P.P.; Waldman, S.D.; Weber, J.F.; Saad, S.T.O. Characterization of a novel decellularized bone marrow scaffold as an inductive environment for hematopoietic stem cells. *Biomater. Sci.* **2019**, *7*, 1516–1528. [CrossRef]
16. Chun, S.Y.; Lim, J.O.; Lee, E.H.; Han, M.-H.; Ha, Y.-S.; Lee, J.N.; Kim, B.S.; Park, M.J.; Yeo, M.; Jung, B.; et al. Preparation and Characterization of Human Adipose Tissue-Derived Extracellular Matrix, Growth Factors, and Stem Cells: A Concise Review. *Tissue Eng. Regen. Med.* **2019**, *16*, 385–393. [CrossRef]
17. Munir, A.; Døskeland, A.; Avery, S.J.; Fuoco, T.; Mohamed-Ahmed, S.; Lygre, H.; Finne-Wistrand, A.; Sloan, A.J.; Waddington, R.J.; Mustafa, K.; et al. Efficacy of copolymer scaffolds delivering human demineralised dentine matrix for bone regeneration. *J. Tissue Eng.* **2019**, *10*, 2041731419852703. [CrossRef]
18. Baek, A.; Kim, Y.; Lee, J.W.; Lee, S.C.; Cho, S.R. Effect of Polydeoxyribonucleotide on Angiogenesis and Wound Healing in an In Vitro Model of Osteoarthritis. *Cell Transplant.* **2018**, *27*, 1623–1633. [CrossRef] [PubMed]
19. Bahreyni, A.; Khazaei, M.; Rajabian, M.; Ryzhikov, M.; Avan, A.; Hassanian, S.M. Therapeutic potency of pharmacological adenosine receptor agonist/antagonist in angiogenesis, current status and perspectives. *J. Pharm. Pharmacol.* **2018**, *70*, 191–196. [CrossRef]
20. Squadrito, F.; Bitto, A.; Altavilla, D.; Arcoraci, V.; de Caridi, G.; de Feo, M.E.; Corrao, S.; Pallio, G.; Sterrantino, C.; Minutoli, L. The effect of PDRN, an adenosine receptor A2A agonist, on the healing of chronic diabetic foot ulcers: Results of a clinical trial. *J. Clin. Endocrinol. Metab.* **2014**, *99*, E746–E753.
21. Squadrito, F.; Bitto, A.; Irrera, N.; Pizzino, G.; Pallio, G.; Minutoli, L.; Altavilla, D. Pharmacological Activity and Clinical Use of PDRN. *Front Pharmacol.* **2017**, *8*, 224. [CrossRef] [PubMed]

22. Chen, Z.; Yan, X.; Yin, S.; Liu, L.; Liu, X.; Zhao, G.; Ma, W.; Qi, W.; Ren, Z.; Liao, H.; et al. Influence of the pore size and porosity of selective laser melted Ti6Al4V ELI porous scaffold on cell proliferation, osteogenesis and bone ingrowth. *Mater. Sci. Eng. C* **2020**, *106*, 110289. [CrossRef] [PubMed]
23. Ouyang, P.; Dong, H.; He, X.; Cai, X.; Wang, Y.; Li, J.; Li, H.; Jin, Z. Hydromechanical mechanism behind the effect of pore size of porous titanium scaffolds on osteoblast response and bone ingrowth. *Mater. Des.* **2019**, *183*, 108151. [CrossRef]
24. Gianfreda, F.; Antonacci, D.; Raffone, C.; Muzzi, M.; Pistilli, V.; Bollero, P. Microscopic Characterization of Bioactivate Implant Surfaces: Increasing Wettability Using Salts and Dry Technology. *Materials* **2021**, *14*, 2608. [CrossRef] [PubMed]
25. Subbiah, R.; Hwang, M.P.; Van, S.Y.; Do, S.H.; Park, H.; Lee, K.; Kim, S.H.; Yun, K.; Park, K. Osteogenic/angiogenic dual growth factor delivery microcapsules for regeneration of vascularized bone tissue. *Adv. Healthc. Mater.* **2015**, *4*, 1982–1992. [CrossRef]
26. Filipowska, J.; Tomaszewski, K.A.; Niedźwiedzki, Ł.; Walocha, J.A.; Niedźwiedzki, T. The role of vasculature in bone development, regeneration and proper systemic functioning. *Angiogenesis* **2017**, *20*, 291–302. [CrossRef]
27. Khojasteh, A.; Behnia, H.; Naghdi, N.; Esmaeelinejad, M.; Alikhassy, Z.; Stevens, M. Effects of different growth factors and carriers on bone regeneration: A systematic review. *Oral Surg. Oral Med. Oral Pathol. Oral Radiol* **2013**, *116*, e405–e423. [CrossRef]
28. Palomino-Durand, C.; Lopez, M.; Marchandise, P.; Martel, B.; Blanchemain, N.; Chai, F. Chitosan/Polycyclodextrin (CHT/PCD)-Based Sponges Delivering VEGF to Enhance Angiogenesis for Bone Regeneration. *Pharmaceutics* **2020**, *12*, 784. [CrossRef]
29. Thellung, S.; Florio, T.; Maragliano, A.; Cattarini, G.; Schettini, G. Polydeoxyribonucleotides enhance the proliferation of human skin fibroblasts: Involvement of A2 purinergic receptor subtypes. *Life Sci.* **1999**, *64*, 1661–1674. [CrossRef]
30. Wang, Y.; Shao, J.-H. Effect of adenosine on three dimensional tube formation and angiogenesis of human umbilical vein endothelial cell (HUVEC) *in vitro*. *Zhongguo Ying Yong Sheng Li Xue Za Zhi* **2005**, *21*, 160–162.
31. Galeano, M.; Bitto, A.; Altavilla, D.; Minutoli, L.; Polito, F.; Calò, M.; Lo Cascio, P.; Stagno d'Alcontres, F.; Squadrito, F. Polydeoxyribonucleotide stimulates angiogenesis and wound healing in the genetically diabetic mouse. *Wound Repair Reg.* **2008**, *16*, 208–217. [CrossRef]
32. Kampleitner, C.; Changi, K.; Felfel, R.M.; Scotchford, C.A.; Sottile, V.; Kluger, R.; Hoffmann, O.; Grant, D.M.; Epstein, M.M. Preclinical biological and physicochemical evaluation of two-photon engineered 3D biomimetic copolymer scaffolds for bone healing. *Biomater. Sci.* **2020**, *8*, 1683–1694. [CrossRef]
33. Dong, R.; Bai, Y.; Dai, J.; Deng, M.; Zhao, C.; Tian, Z.; Zeng, F.; Liang, W.; Liu, L.; Dong, S. Engineered scaffolds based on mesenchymal stem cells/preosteoclasts extracellular matrix promote bone regeneration. *J. Tissue Eng.* **2020**, *11*, 2041731420926918. [CrossRef]
34. Smieszek, A.; Seweryn, A.; Marcinkowska, K.; Sikora, M.; Lawniczak-Jablonska, K.; Witkowski, B.S.; Kuzmiuk, P.; Godlewski, M.; Marycz, K. Titanium Dioxide Thin Films Obtained by Atomic Layer Deposition Promotes Osteoblasts' Viability and Differentiation Potential While Inhibiting Osteoclast Activity—Potential Application for Osteoporotic Bone Regeneration. *Materials* **2020**, *13*, 4817. [CrossRef]
35. Blair, H.C. How the osteoclast degrades bone. *Bioessays* **1998**, *20*, 837–846. [CrossRef]
36. Lin, K.; Xia, L.; Li, H.; Jiang, X.; Pan, H.; Xu, Y.; Lu, W.W.; Zhang, Z.; Chang, J. Enhanced osteoporotic bone regeneration by strontium-substituted calcium silicate bioactive ceramics. *Biomaterials* **2013**, *34*, 10028–10042. [CrossRef] [PubMed]
37. Tian, X.; Yuan, X.; Feng, D.; Wu, M.; Yuan, Y.; Ma, C.; Xie, D.; Guo, J.; Liu, C.; Lu, Z. In vivo study of polyurethane and tannin-modified hydroxyapatite composites for calvarial regeneration. *J. Tissue Eng.* **2020**, *11*, 2041731420968030. [CrossRef] [PubMed]

Article

Preliminary Animal Study on Bone Formation Ability of Commercialized Particle-Type Bone Graft with Increased Operability by Hydrogel

So-Yeun Kim [1,†], You-Jin Lee [2,†], Won-Tak Cho [2], Su-Hyun Hwang [2], Soon-Chul Heo [3], Hyung-Joon Kim [3,*] and Jung-Bo Huh [2,*]

1. Department of Prosthodontics, Kyungpook National University Dental Hospital, Daegu 41940, Korea; soyeunkim179@gmail.com
2. Department of Prosthodontics, Dental and Life Sciences Institute, Education and Research Team for Life Science on Dentistry, School of Dentistry, Pusan National University, Yangsan 50612, Korea; nicejin17@naver.com (Y.-J.L.); joonetak@hanmail.net (W.-T.C.); hsh2942@hanmail.net (S.-H.H.)
3. Department of Oral Physiology, Periodontal Diseases Signaling Network Research Center, Dental and Life Science Institute, School of Dentistry, Pusan National University, Yangsan 50612, Korea; snchlheo@gmail.com
* Correspondence: hjoonkim@pusan.ac.kr (H.-J.K.); neoplasia96@hanmail.net (J.-B.H.); Tel.: +82-10-6326-4189 (H.-J.K.); +82-10-8007-9099 (J.-B.H.); Fax: +82-55-510-8208 (H.-J.K.); +82-55-360-5134 (J.-B.H.)
† These authors contributed equally to this work.

Abstract: The purpose of this study was to evaluate the bone-generating ability of a new bovine-derived xenograft (S1-XB) containing hydrogel. For control purposes, we used Bio-Oss and Bone-XB bovine-derived xenografts. S1-XB was produced by mixing Bone-XB and hydrogel. Cell proliferation and differentiation studies were performed to assess cytotoxicities and cell responses. For in vivo study, 8 mm-sized cranial defects were formed in 16 rats, and then the bone substitutes were transplanted into defect sites in the four study groups, that is, a Bio-Oss group, a Bone-XB group, an S1-XB group, and a control (all $n = 4$); in the control group defects were left empty. Eight weeks after surgery, new bone formation areas were measured histomorphometrically. In the cell study, extracts of Bio-Oss, Bone-XB, and S1-XB showed good results in terms of the osteogenic differentiation of human mesenchymal stem cells (hMSCs) and no cytotoxic reaction was evident. No significant difference was observed between mean new bone areas in the Bio-Oss (36.93 ± 4.27%), Bone-XB (35.07 ± 3.23%), and S1-XB (30.80 ± 6.41%) groups, but new bone area was significantly smaller in the control group (18.73 ± 5.59%) ($p < 0.05$). Bovine-derived bone graft material containing hydrogel (S1-XB) had a better cellular response and an osteogenic effect similar to Bio-Oss.

Keywords: bone regeneration; bone substitute; bovine-derived xenograft; hydrogel

1. Introduction

Currently, various types of graft materials are used for guided bone regeneration (GBR). Graft materials are classified as autografts, allografts, or xenografts, which have different osteoblast adhesion, proliferation, and differentiation properties [1]. Autografts are considered ideal graft material as regards osteogenesis, osteoinduction, and osteoconduction [2], but they require an additional surgical site for bone harvesting, which increases patient discomfort and risks of complications [3,4]. For this reason, development is being actively conducted on other types of bone graft materials [5,6].

Synthetic bone material and xenograft have advantages in terms of supply and cost. Synthetic bone materials, including ceramic-based hydroxyl-apatite (HA) and tricalcium phosphate (β-TCP), but are limited in terms of biocompatibility, osteoconduction, and osteoinduction [7,8]. In addition, since synthetic bone materials are generally stronger than surrounding bone stress, shielding problems may occur [9]. Xenografts obtained from animals (e.g., bovine and porcine sourced) are similar to human bones [10], and because xenografts are biocompatible and have high porosities, they provide space for osteoinduction and osteoconduction [11,12]. Bovine-derived xenografts are the most widely used graft materials and have been reported on many occasions to produce good results [13,14].

Graft materials are made in the form of particles of various sizes and are applied using the conventional GBR technique by compaction in transplant areas and covering the compacted materials with a membrane. However, when defects are irregularly shaped or large, it is difficult to maintain particle-type bone graft materials stably, and prognoses are relatively poor [15]. In addition, particle-type bone graft materials do not compact well due to a lack of adhesion between particles, and thus, handling properties deteriorate during surgery, and sometimes graft particles are malpositioned or lost [16,17]. In order to solve these problems, several types of bone graft materials have recently appeared on the market. These products improve handling properties by preventing particle separation by interposing organic substances between bone graft particles [18].

Hyaluronic acid, glycerol, chitosan, and others have been used as carrier materials to increase the handling of bone graft powder [19]. Recently, moldable putties were developed to fill bone defects and are being widely used because they hold their shapes and are not dispersed by irrigation or blood [20]. In previous studies, various hydrogels were applied to bone grafts in combination with bioactive molecules or cells [21]. In particular, in a recent review article, it was reported that hydrogels improve the stabilities of graft materials in injection or graft sites [22], and in a recent study, alginate hydrogel was reported to improve bone graft stability during surgery [23].

The use of hydrogels for preparing bone grafts confers several advantages. Hydrogels have several potential benefits for bone repair because they have swollen network structures that can contain biologically active agents and excellent biocompatibility. Furthermore, the three-dimensional hydrophilic characteristics of hydrogels provide mechanical strength and nutritional environments suitable for endogenous cell growth. The hydrogel reduces inflammatory responses, has soft textures, and its viscoelastic properties increase operability [21,24,25]. It has also been reported that hydrogels promote the spreading, proliferation, and differentiation of mesenchymal stem cells [26]. The use of hydrogels by mixing with bone graft materials such as ceramics (hydroxyapatite, tricalcium phosphate) and bioglass particles has been studied [27].

Currently, many products for bone graft are being developed, and attempts to improve the operability of the widely used bovine-derived xenograft materials are continuing. The development of injectables, putties, cements, pastes, gels, etc. is attractive in the field of bone regeneration because of its formability and stability, and products with improved operability using various materials are on the market [28]. However, there are few studies evaluating the clinical application of hydrogel-added xenograft materials among various materials. The purpose of this study is to evaluate the clinical applicability of a commercially available xenograft material developed to be manipulated in a shape suitable for the defect area by applying particle xenograft materials and hydrogels at the cellular and small animal level.

2. Materials and Methods

2.1. Experimental Xenogenic Bone Substitutes

In this study, two commercial, powder-type bovine-derived bone grafts were used as experimental groups. Bio-Oss group (Bio-Oss®, Geistlich Pharma AG, Wolhusen, Switzerland) is the most widely used product in the market and is generally used as a positive control because of its considerable scientific basis. Bone-XB group (Bone-XB®, Medpark, Busan, Korea) is also derived from bovine cancellous bone, but its granule size is smaller and its Ca/P ratio is greater than that of Bio-Oss. S1-XB group (S1-XB®, Medpark, Busan, Korea) is prepared by mixing Bone-XB with hydrogel. The company did not disclose patent or specific manufacturing details. Specific porosity, Ca/P ratio, particle sizes, and manufacturers' details are provided in Table 1 [16].

Table 1. Characteristics of bone graft materials.

Product Name	Type	Derivation	Granule Size (μm)	Porosity (%)	Ca/P (%)	Manufacturer
Bio-Oss®	Powder	Bovine cancellous	250–1000	70.5	1.54	Geistlich Pharma AG, Wolhusen, Switzerland
Bone-XB®	Powder	Bovine cancellous	200–1000	70.20	1.7063	Medpark, Busan, Korea
S1-XB®	Powder	Bovine cancellous/hydrogel	200–1000	70.20	1.7063	Medpark, Busan, Korea

2.2. In Vitro Study

2.2.1. Scanning Electron Microscope (SEM) Image Analysis

The surface morphology of the xenogenic bone substitutes was observed by scanning electron microscopy (SEM, Hitachi S3500N, Hitachi, Tokyo, Japan). Specimens were coated with Au using a sputter coater (SCD 005, BAL-TEC, Balzers, Liechtenstein), and SEM was conducted at an accelerating voltage of 15kV. Surface compositional analysis was performed by onboard energy-dispersive X-ray spectroscopy (EDX, Apollo X, Ametek EDAX, Mahwah, NJ, USA). EDX dot scans were performed three times for each group of specimens.

2.2.2. Preparation of Extracts

We mixed 250 mg of each xenogeneic bone substitute (Bio-Oss, Bone-XB, and S1-XB) with 10 mL of alpha-modified Eagle's medium (α-MEM; Welgene, Deagu, Korea) and incubated suspensions at 37 °C for 24 h. Control medium was prepared in the same manner without bone substitute. Suspensions were centrifuged at 1500 rpm for 5 min and filtered through 0.2 μm membranes (Sartorius, Göttingen, Germany).

2.2.3. Culture of Human Bone Marrow Mesenchymal Stem Cells (hMSCs)

Human mesenchymal stem cells (hMSCs) were purchased from Lonza (Walkersville, MD, USA) and cultured in α-MEM supplemented with 10% fetal bovine serum (FBS; Gibco, Carlsbad, CA, USA), 100 U/mL penicillin, and 100 μg/mL streptomycin (Gibco™, Thermo Fisher Scientific, Waltham, MA, USA), at 37 °C in a 5% CO_2 atmosphere. Media were refreshed every other day, and the hMSCs used in all experiments were passaged less than seven times.

2.2.4. Cell Viability and Proliferation Assay

hMSCs were seeded into 48-well plates (Nunc, Roskilde, Denmark) at a density of 10^4 cells/well in medium containing 20% or 50% of the bone substitute extracts and cultured for 0, 1, 2, or 3 days. Cell proliferations were assessed by adding 20 μL of CCK-8 solution (Dojindo, Rockville, MD, USA). Absorbances were measured at 450 nm using an Opsys MR micro-plate reader (DYNEX Technologies Inc., Denkendorf, Germany). Cell viabilities were determined with 24 h absorbance data.

2.2.5. Osteoblasts Differentiation and Alkaline Phosphatase (ALP) Activity Assay

hMSCs were seeded into a 48-well plate at a density of 10^4 cells/well and cultured for 24 h. Osteoblast differentiation was induced by adding 10 mM of β-glycerophosphate and 50 μg/mL of ascorbic acid (Sigma-Aldrich, Milan, Italy) to α-MEM supplemented with 10% FBS. The medium was refreshed every other day. Bone substitute extracts constituted 50% of the total osteogenic induction medium. On day 12, alkaline phosphatase (ALP) activities were estimated using the ALP staining kit (86R-1KT, Sigma-Aldrich, Milan, Italy). Intensities of alkaline phosphatase staining were quantified by imaging culture plates (Nikon, Eclipse Ts2, Melville, NY, USA) and measuring integrated densities using the ImageJ program (version 1.52p, National Institutes of Health, Bethesda, MD, USA).

2.2.6. Quantitative Real-Time Polymerase Chain Reaction (qPCR) Analysis

The qPCR analysis was performed to examine the expressions of osteogenic differentiation markers. Five days after osteogenic induction, total RNA was extracted from hMSCs using TRIzol (Life Technologies, Grand Island, NY, USA), and 2 µg of RNA was reverse transcribed using Superscript II (Invitrogen) (Gibco^{TM}, Thermo Fisher Scientific, Waltham, MA, USA) according to the manufacturer's instructions. For qPCR analysis, 50 ng of cDNA was mixed with SYBR Green PCR Master Mix (Applied Biosystems, Forster, CA, USA) and amplified for 40 cycles using an AB7500 unit (Applied Biosystems). Experiments were performed in triplicate, and the data were normalized versus β-actin mRNA. Besides GAPDH, beta-actin is one of the most widely used genes as an internal reference gene for quantitative real-time PCR. The stable expression of beta-actin has been validated in human mesenchymal stem cell as well as osteoblast [29,30]. The analysis was performed using the $\Delta(\Delta CT)$ method. The primer sequences used in qPCR were as follows: Runx2: F(5'-TGCTTTGGTCTTGAAATCACA-3'), R(5'-TCTTAGAACAAATTCTGCCCTTT-3'); ALP: F(5'-ATCCAGAATGTTCCACGGAGGCTT-3'), R(5'-AGACACATATGATGGCCGAGG); OCN: F(5'-CAGCGAGGTAGTGAAGAGAC-3'), R(5'-TGAAAGCCGATGTGGTCAG-3'); OSX: F(5'-TCCCTGCTTGAGGAG GAAG-3'), R(5'-AGTTGTTGAGTCCCGCAGAG-3'); OPN: F(5'-AGACACATATGATGGCCGAGG-3'), R(5'-GGCCTTGTATGCACCATTCAA-3'); β-actin: F(5'-ACTCTTCCAGCCTTCCTTCC-3'), R(5'-TGTTGGCGTACAGGTCTTTG-3') (Table 2).

Table 2. Primer sequences used for real-time PCR.

Target Genes	Sequences
Runx2	F: 5'-TGCTTTGGTCTTGAAATCACA-3' R: 5'-TCTTAGAACAAATTCTGCCCTTT-3'
ALP	F: 5'-ATCCAGAATGTTCCACGGAGGCTT-3 R: 5'-AGACACATATGATGGCCGAGG
OCN	F: 5'-CAGCGAGGTAGTGAAGAGAC-3' R: 5'-TGAAAGCCGATGTGGTCAG-3'
OSX	F: 5'-TCCCTGCTTGAGGAG GAAG-3' R: 5'-AGTTGTTGAGTCCCGCAGAG-3'
OPN	F: 5'-AGACACATATGATGGCCGAGG-3' R: 5'-GGCCTTGTATGCACCATTCAA-3'
β-actin	F: 5'-ACTCTTCCAGCCTTCCTTCC-3' R: 5'-TGTTGGCGTACAGGTCTTTG-3'

2.3. In Vivo Experiment

2.3.1. Animals and Operative Procedures

The experimental protocol was approved beforehand by the IACUC (Institutional Animal Care and Use Committee), the Animal Experimental Ethics Committee of Pusan National University (PNU-2020-2618). Sixteen Sprague-Dawley rats (male, 13 weeks old, Koatech, Pyeongtaek, Korea) were used for in vivo experiments. Animals were acclimated to a pellet diet with free access to water in individual cages for 2 weeks. For surgery, general anesthesia was achieved by administering an intraperitoneal injection of avertin. In each animal, the skull was exposed through a sagittal incision using a #15 surgical blade, and a trephine bur was used to make a circular defect of diameter 8 mm under saline irrigation. Bone graft materials were applied to defects according to the manufacturer's instructions (Figure 1), and surgery was completed by suturing periosteum and skin with 3-0 Vicryl sutures without a membrane.

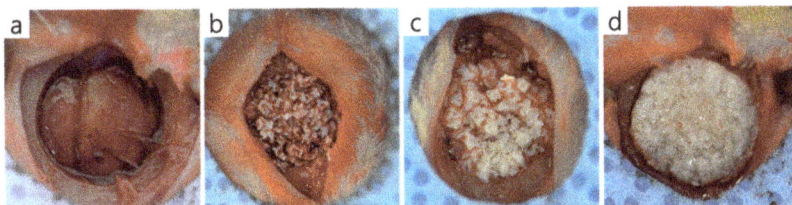

Figure 1. The bone graft materials applied to rat calvaria defects. (**a**) Control group, (**b**) Bio-Oss group, (**c**) Bone-XB group, and (**d**) S1-XB group.

Eight weeks after surgery, rats were sacrificed by CO_2 inhalation. Tissue samples, which included peri-defect samples, were carefully secured and fixed in 10% neutral buffered formalin (Sigma-Aldrich) for 2 weeks.

2.3.2. Histologic Analysis

Specimens were decalcified using decalcifying Solution (Calci-Clear Rapid®, National Diagnostics, Atlanta, GA, USA) and dehydrated in an alcohol series (70%, 80%, 90%, and 100%) before they were embedded in paraffin. After that, the paraffin block was sectioned 3–4 μm thickness using a microtome(Leica® RM2255, Leica Biosystems Inc., Buffalo Grove, IL, USA). Slides were stained with hematoxylin-eosin and Masson's trichrome and imaged under an optical microscope (Olympus BX51, Olympus Co., Ltd., Tokyo, Japan). New bone areas (%) were determined using an image analysis program (iSolution, IMT, Vancouver, Canada) (Figure 2).

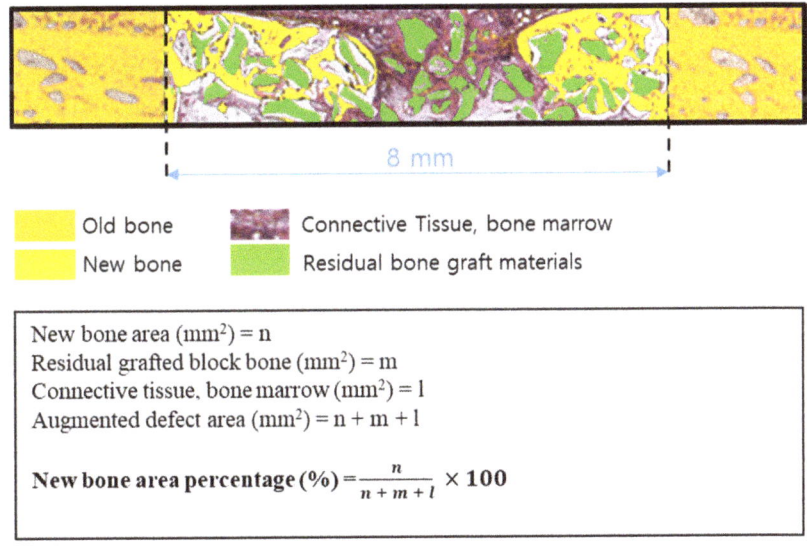

New bone area (mm^2) = n
Residual grafted block bone (mm^2) = m
Connective tissue, bone marrow (mm^2) = l
Augmented defect area (mm^2) = n + m + l

New bone area percentage (%) $= \dfrac{n}{n + m + l} \times 100$

Figure 2. Schematic of histometric analysis.

2.3.3. Statistical Analysis

The analyses were performed using SPSS ver. 25.0 (SPSS, Chicago, IL, USA). In vitro data were obtained from at least three independent experiments conducted in triplicate. Results of multiple observations are presented as means ± SEMs. The significances of intergroup differences were determined by one-way or two-way ANOVA followed by the Bonferroni post hoc test. In vivo results were analyzed using the Kruskal-Wallis test followed by the Mann-Whitney U post hoc test. Statistical significance was accepted for p values < 0.05.

3. Results
3.1. In Vitro Findings
3.1.1. Scanning Electron Microscope Surface Analysis

Images were taken at ×60, ×500, and ×3000 magnification to investigate the surface morphologies of bone graft materials. Similar particle sizes and macro-porous structures were observed for Bio-Oss, Bone-XB, and S1-XB. In S1-XB, distinct hydrogel layers were not visible on the SEM image (Figure 3).

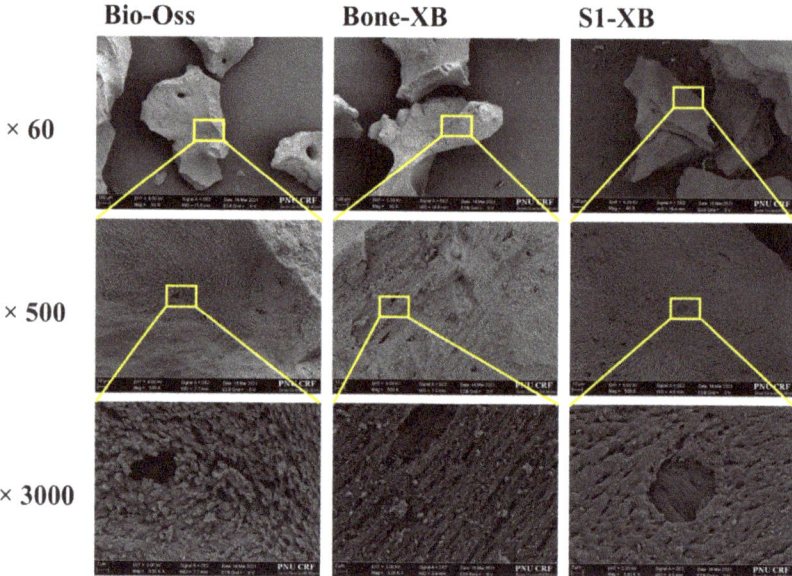

Figure 3. Scanning electron microscope (SEM) images of each group. [Original magnification: ×60, ×500, ×3000.

3.1.2. Energy-Dispersive X-ray Spectroscopy (EDX) Findings

The elements in bone graft materials and Ca/P ratios were determined by EDX (Table 3). Ca/P ratios were similar for Bio-Oss and Bone-XP, the ratio was slightly lower for S1-XB. For all bone substitutes, the stoichiometric ratio of HA was less than 1.67. In addition, S1-XB had higher C and O contents than Bio-Oss or Bone-XB, which attributed to the hydrogel.

Table 3. EDS results of the bone substitutes (Atomic %; Mean ± SD).

Elements	Bio-Oss	Bone-XB	S1-XB
C	2.947 ± 1.034	2.29 ± 0.611	6.837 ± 6.433
O	28.19 ± 9.45	27.047 ± 6.339	34.78 ± 5.699
P	35.91 ± 1.192	37.04 ± 2.581	33.79 ± 7.053
Ca	54.78 ± 4.996	57.40 ± 3.789	48.44 ± 8.954
Ca/P	1.526	1.550	1.434

3.1.3. CCK-8 Assays of Cell Viability and Proliferation

Liquid extraction method has been recommended as a standard procedure in testing cytotoxicity of biocompatible materials by the ISO 10993. After treating hMSCs with extracts of bone substitutes, cell viability and proliferation were investigated using the CCK-8 assay. Extracts were collected by incubating Bio-Oss, Bone-XB, or S1-XB for 24 h. No significant intergroup viability differences were observed after incubation for 24 h with 20% or 50% extracts of the bone substitute versus non-treated controls. hMSCs proliferation was measured every 24 h for 72 h after supplementation with 20% or 50% extracts. Supplementation with 20% extracts derived from Bio-Oss significantly increased hMSCs proliferation at 48 h. Notably, supplementation with 20% and 50% extracts derived from S1-XB increased hMSCs proliferation most at 48 h (Figure 4).

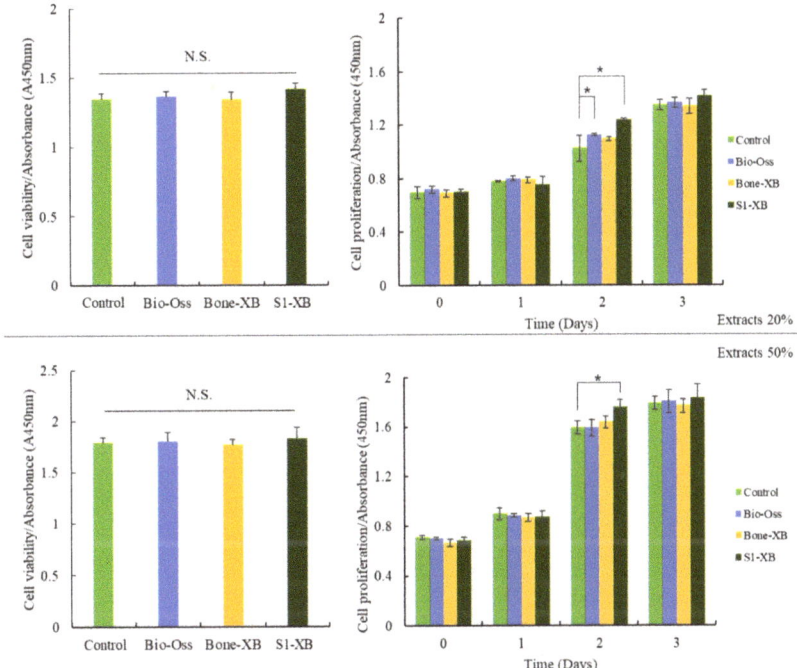

Figure 4. Cell viability and proliferation of human mesenchymal stem cells (hMSCs) exposed to the different extracts. The letter "*" indicates significant differences ($p < 0.05$); N.S.—No significant.

3.1.4. Alkaline Phosphatase (ALP) Staining for Osteogenic Differentiation Analysis

Since we could not find significant differences between 20% and 50% extracts treatment on hMSCs proliferation at the final time point in our system, only 50% extracts were used in the subsequent experiments. To determine the effects of bone substitutes on osteogenic differentiation, ALP staining was conducted 12 days after the induction of osteogenic differentiation by culturing hMSCs in the presence of 50% extracts. No obvious differences were observed between the ALP positive areas of cells treated with Bio-Oss extracts and non-treated controls. However, treatments with Bone-XB and S1-XB extracts significantly increased ALP positive areas versus controls, and the greatest increase was observed for cells treated with S1-XB extracts (Figure 5).

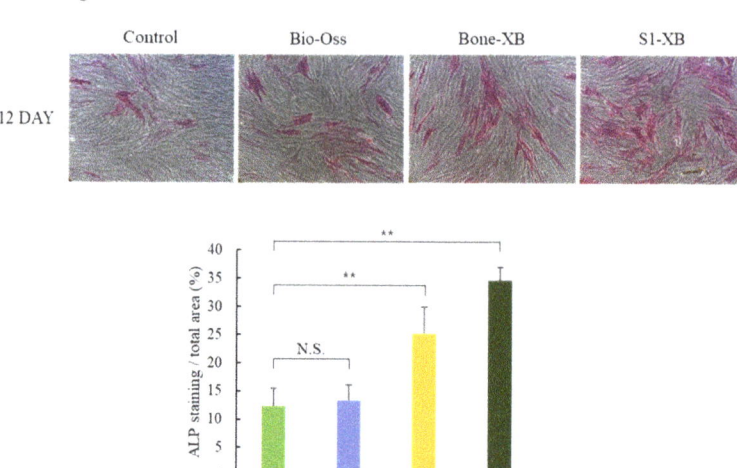

Figure 5. Cell osteogenic differentiation assay (** $p < 0.01$). The letter "**" indicates significant differences ($p < 0.05$); N.S.—No significant.

3.1.5. Quantitative Real-Time Polymerase Chain Reaction (qPCR) for Osteogenic Differentiation Analysis

To evaluate the effects of the three bone substitutes on osteogenic differentiation further, qPCR was performed to examine the expressions of osteogenesis-related genes. Consistent with ALP staining results, supplementation of extracts derived from Bio-Oss did not induce any osteogenesis-related genes versus non-treated controls. However, supplementation of extracts derived from Bone-XB and S1-XB greatly induced expressions of the key osteogenic transcription factors Runx2 and OSX and those of genes involved in osteoblast differentiation and matrix mineralization; ALP, OCN, and OPN (Figure 6).

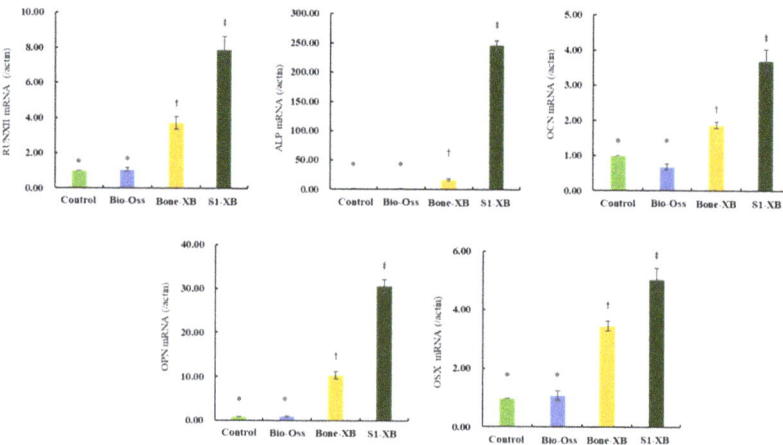

Figure 6. The effects of the three bone substitute extracts on osteogenesis-related genes. hMSCs were cultured in osteogenic induction media supplemented with 50% extracts for 12 days. The mRNA expressions of osteogenic differentiation-related genes, that is, Runx2, OSX, ALP, OCN, and OPN were determined by qPCR. Normalization was performed versus β-actin expression. "*", "†", "‡" Different letters indicate significant differences ($p < 0.05$).

3.2. In Vivo Findings

3.2.1. Clinical Findings

All rats survived during the experimental periods, and no side effects such as inflammation or specimen exposure were observed.

3.2.2. Histologic Findings

Bio-Oss, Bone-XB, and S1-XB bone substitutes were retained stably at defect sites, and no inflammatory cells or unexpected reactions were observed (Figure 7).

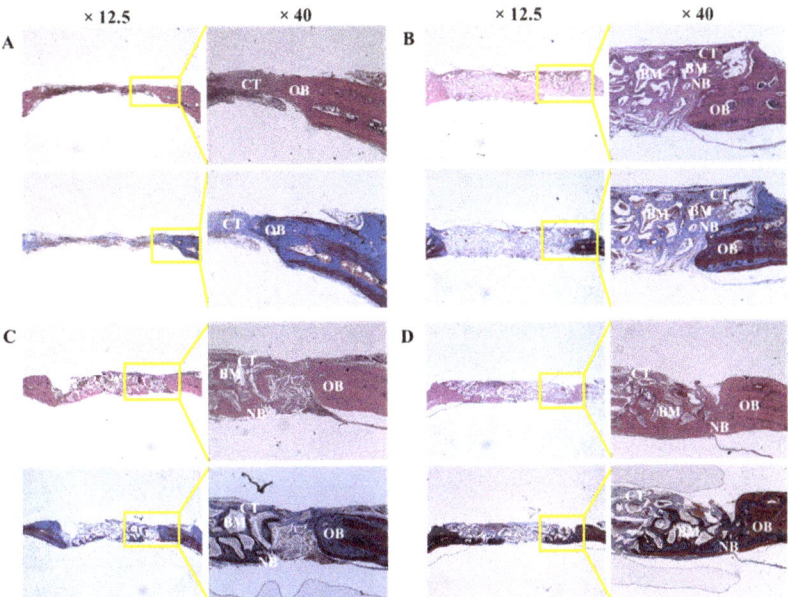

Figure 7. Histologic sections of defect sites at 8 weeks after surgery. (**A**) Control, (**B**) Bio-Oss, (**C**) Bone-XB, (**D**) S1-XB. Note: NB; new bone, CT; connective tissue, GM; graft materials, OB; old bone.

3.2.3. Histomorphometric Findings

New bone area results are shown in Table 4 and Figure 8. New bone area was significantly less in the control group than in the Bio-Oss, Bone-XB, or S1-XB groups, but no significant difference was observed between the three bone graft materials ($p > 0.05$).

Table 4. New bone area within areas of interest (AOIs).

	Group	Mean	SD	*p*-Value
New bone area (%)	Control	18.73	5.59	0.026 *
	Bio-Oss	36.93	4.27	
	Bone-XB	35.07	3.23	
	S1-XB	30.80	6.41	

The letter "*" indicates significant differences ($p < 0.05$).

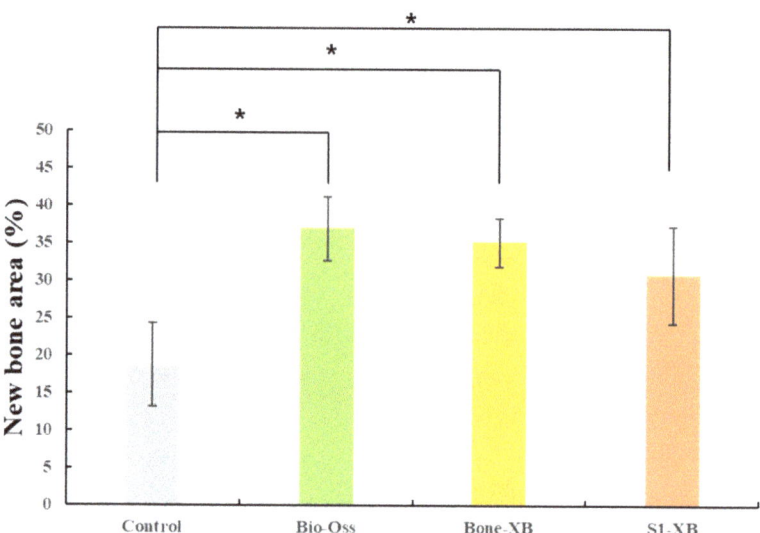

Figure 8. Percentages of new bone area within areas of interest (AOIs). The letter "*" indicates significant differences ($p < 0.05$).

4. Discussion

In this study, we compared three commercially available xenograft materials that have been officially confirmed to be safe. The first one, Bio-Oss is well-known for its effectiveness, safety, bone formation quality and quantity, and high success rate [31–33]. The second one, Bone-XB is a new bovine bone graft material developed in Korea. It has a particle size of 0.2–1.0 mm and a macro/micropore structure similar to Bio-Oss. The third one, S1-XB is a product from the same manufacturer as Bone-XB, a mixture of the xenograft particle and hydroxypropyl methylcellulose that is the polymer making the hydrogel. After the xenograft particle adsorbed with osteogenic protein and hydroxypropyl methyl cellulose powder are mixed and stirred to obtain a viscous gel, it is prepared under vacuum and freeze-drying conditions. Cause of the viscous condition, S1-XB had a putty-like consistency that allowed it to be shaped easily by hand, and it also showed an ability to retain its shape. In Figure 1b,c, the particles of the material are together in the defect, but the particles are seen separately one by one. The graft material in Figure 1d looks like a single, round lump with a rough surface.

The electron microscopy (SEM) images of Bio-Oss, Bone-XB, and S1-XB groups showed similar pore morphologies, presumably because they are all derived from bovine bone. Macro pores contribute to the formation of blood vessels, whereas micro pores contribute to osteoblast adhesion, and thus, Bio-Oss and Bone-XB can be considered to provide environmentally similar osteoblast scaffolds [34]. In the S1-XB group, the hydrogel coating layer was not well-visualized by SEM, which implied it was evenly and thinly distributed. a is useful for studying the degree of bone mineralization [35], and the theoretical value (1.67) of stoichiometric hydroxyapatite, the main component of human bone, and the Ca/P ratio of the three bone substitutes were different. Bone-XB was 1.550, Bio-Oss was 1.526, and S1-XB was 1.434, all smaller than 1.67 [36,37]. In the previous study, when the Ca/P ratio was 2.0 or higher, it decreased osteoblast viability after 72 h; in contrast, when the Ca/P ratio was less than 2.0, it optimized osteoblast viability and promoted osteoblast alkaline phosphatase activity after 72 h. In addition, although HA with a Ca/P ratio of 1.67 has excellent biocompatibility, it is known to interrupt bone re-generation due to slow absorption in the body. However, TCP with Ca/P 1.5, a commercially available bone graft material, has high biodegradability and is mixed with HA in a certain ratio to compensate for each disadvantage. The materials used in this study showed results close to TCP with Ca/P of 1.4 to 1.5. Compared to Bio-Oss and Bone-XB, the C and O levels were much higher in S1-XB, and thus, Ca and P levels were reduced, presumably due to the presence of the hydrogel [24]. The hydrogel is made of various polymers, and since there are many C and O in the polymer, these results were expected to be shown in S1-XB [38].

In previous studies, Bio-Oss was reported to have a positive effect on the osteoblast differentiation of hMSCs [39]. In this study, S1-XB and Bone-XB were compared with Bio-Oss. In the cell study, the toxicity evaluation was performed by measured CCK-8 assay, areas of differentiated osteoblastic cells were compared by ALP staining, and levels of bone remodeling-related mRNAs were assessed by real-time PCR. All three bone substitutes had viabilities and cell proliferations similar to those of non-treated controls. Degrees of osteoblast differentiation followed the order S1-XB > Bone-XB > Bio-Oss and were similar for Bio-Oss and non-treated controls group but higher for S1-XB and Bone-XB than controls. ALP activity was investigated by real-time PCR as a surrogate of osteoblast activity because ALP is a marker of the change from pre- to mature osteoblasts. Our results show that S1-XB extract significantly promoted osteoblast differentiation. In addition, levels of the transcription factors of osteoblast differentiation Runx2 and osterix (OSX) were measured. OSX and Runx2 levels were significantly higher for S1-XB followed by Bone-XB. Extracellular matrix proteins, osteopontin (OPN) and osteocalcin (OCN), which are expressed during the formation of osteoblast or woven bone formation, were also assessed by PCR. The results showed expressions were significantly higher for S1-XB than Bone-XB and significantly higher for Bone-XB than Bio-Oss [40–42]. The CCK-8 assay revealed that the extracts from Bio-Oss, Bone-XB, and S1-XB showed no significant cytotoxic effect on hMSCs viability under these experimental conditions. The modest proliferation of hMSC was observed in response to supplementation with Bio-Oss and S1-XB at 48 h, but no detectable differences at 72 h. Based on these results, it appears that the osteoinductive effects of bone graft extracts on hMSCs are mainly derived by inducing osteogic differentiation capacity, not the stimulation of proliferation.

Histological analysis showed osteoblasts were distributed around transplanted xenografts in the Bio-Oss, Bone-XB, and S1-XB groups, indicating that all three xenografts used in this study had good bone conductivity [43]. Histomorphometric analysis showed new bone areas were not significantly different for the three bone substitutes, though all had significantly larger new bone areas than in the control group. Summarizing, S1-XB produced better results in cell studies and similar results in animal studies than Bio-Oss or Bone-XB, which shows S1-XB is biocompatible and osteoblast differentiation characteristics similar to that of Bio-Oss.

A wide variety of products are now in production, requiring clinicians to evaluate new products and validate their clinical applicability. It is meaningful that a product with a new function, such as S1-XB, has no significant difference in animal test results from existing products that are acknowledged and used. The main problems of previous bone graft material particles were manipulation, transportation, and molding [44,45]. The original particle-type bone graft material is less stable due to its intrinsic physical properties, so a space-maintaining barrier such as a membrane must be used [46]. S1-XB with the addition of hydrogel was not inferior to the existing materials in terms of bone formation ability in this study, even though no membrane was used. Products with similar bone formation performance and superior operability are valuable in clinical practice.

There are several limitations in the current study that are worth considering. First limitation, since this study was designed as a pilot, the number of animals is limited, so results from animal studies may be inferior to cell studies. Second limitation, as the graft materials have cohesiveness, we did not cover the graft areas with collagen membrane, although we doubt whether the absence of a membrane caused xenograft losses or displacements. The barrier membrane can prevent bone graft material loss by ensuring the stability of the bone graft material [47]. Third limitation, we have not investigated whether the increased cohesiveness of S1-XB affects passages and porosity of graft materials for new bone growth [48]. Further research is required on these topics. Despite the limitations of this study, the surface properties, cell activities, and bone regeneration ability of S1-XB were not inferior to Bio-Oss, and it was increased operability by hydrogel, which indicated its suitability as a bone graft material. Further research is required on these materials and topics.

5. Conclusions

Within the limitations of this preliminary study, the osteogenic capacity and biocompatibility of S1-XB, a commercially available particle-type bone graft with increased manipulability by hydrogel, were not inferior to those of currently commercially available bovine-derived xenograft materials, and showed favorable operability clinically.

Author Contributions: Conceptualization, J.-B.H.; methodology, J.-B.H., H.-J.K., S.-Y.K., and Y.-J.L.; formal analysis, W.-T.C. and S.-H.H.; investigation, W.-T.C. and S.-H.H.; data curation, W.-T.C., S.-H.H., S.-Y.K., Y.-J.L., and S.-C.H.; writing—original draft preparation, J.-B.H., S.-Y.K., and Y.-J.L.; writing—review and editing, H.-J.K., S.-Y.K., Y.-J.L., S.-C.H., and J.-B.H.; supervision, J.-B.H. All authors have read and agreed to the published version of the manuscript.

Funding: This research was supported by the National Research Foundation of Korea (NRF; Daejeon, Korea) and funded by the Korean government (MSIT; grant no. NRF-2018R1A5A2023879), and supported by the National Research Foundation of Korea (NRF) funded by the Korean government (MSIT; grant no. NRF-2020R1A2C1004927).

Institutional Review Board Statement: The study was approved by the Institutional Animal Care and Use Committee (IACUC) of Pusan National University (PNU-2020-2618).

Informed Consent Statement: Not applicable.

Data Availability Statement: Data sharing not applicable.

Conflicts of Interest: The authors have no conflict of interest to declare.

References

1. Yip, I.; Ma, L.; Mattheos, N.; Dard, M.; Lang, N.P. Defect healing with various bone substitutes. *Clin. Oral Implants Res.* **2015**, *26*, 606–614. [CrossRef] [PubMed]
2. Borstlap, W.A.; Heidbuchel, K.L.; Freihofer, H.P.M.; Kuijpers-Jagtman, A.M. Early secondary bone grafting of alveolar cleft defects: A comparison between chin and rib grafts. *J. Cranio-Maxillofac. Surg.* **1990**, *18*, 201–205. [CrossRef]
3. Misch, C.M. Comparison of intraoral donor sites for onlay-grafting prior to implant placement. *Int. J. Oral Maxillofac. Implants* **1997**, *12*, 767–776.
4. Raghoebar, G.M.; Louwerse, C.; Kalk, W.W.; Vissink, A. Morbidity of chin bone harvesting. *Clin. Oral Implants Res.* **2001**, *12*, 503–507. [CrossRef]
5. Piattelli, M.; Favero, G.F.; Scarano, A.; Orsini, G.; Piattelli, A. Bone reactions to anorganic bovine bone used in sinus lifting procedure: A histologic long-term report of 20 cases in man. *Int. J. Oral Maxillofac. Implants* **1999**, *14*, 835–840.
6. Norton, M.R.; Odell, E.W.; Thompson, I.D.; Cook, R.J. Efficacy of bovine bone mineral for alveolar augmentation: A human histologic study. *Clin. Oral Implants Res.* **2003**, *14*, 775–783. [CrossRef]
7. Suh, H.; Park, J.C.; Han, D.W.; Lee, D.H.; Han, C.D. A bone replaceable artificial bone substitute: Cytotoxicity, cell adhesion, proliferation, and alkaline phosphatase activity. *Artif. Organs* **2001**, *25*, 14–21. [CrossRef]
8. Liu, J.; Kerns, D.G. Suppl 1: Mechanisms of guided bone regeneration: A review. *Open Dent. J.* **2014**, *8*, 56–65. [CrossRef]
9. Park, S.-A.; Shin, J.-W.; Yang, Y.-I.; Kim, Y.-K.; Park, K.-D.; Lee, J.-W.; Jo, I.-H.; Kim, Y.-J. In vitro study of osteogenic differentiation of bone marrow stromal cells on heat-treated porcine trabecular bone blocks. *Biomaterials* **2004**, *25*, 527–535. [CrossRef]
10. Poumarat, G.; Squire, P. Comparison of mechanical properties of human, bovine bone and a new processed bone xenograft. *Biomaterials* **1993**, *14*, 337–340. [CrossRef]
11. Wetzel, A.; Stich, H.; Caffesse, R. Bone apposition onto oral implants in the sinus area filled with different grafting materials. A histological study in beagle dogs. *Clin. Oral Implants Res.* **1995**, *6*, 155–163. [CrossRef]
12. Klinge, B.; Alberius, P.; Isaksson, S.; Jönsson, J. Osseous response to implanted natural bone mineral and synthetic hydroxylapatite ceramic in the repair of experimental skull bone defects. *J. Oral Maxillofac. Surg.* **1992**, *50*, 241–249. [CrossRef]
13. Buser, D.; Brägger, U.; Lang, N.; Nyman, S. Regeneration and enlargement of jaw bone using guided tissue regeneration. *Clin. Oral Implants Res.* **1990**, *1*, 22–32. [CrossRef]
14. Hämmerle, C.; Olah, A.; Schmid, J.; Fluckiger, L.; Gogolewski, S.; Winkler, J.; Lang, N. The biological effect of deproteinized bovine bone on bone neoformation on the rabbit skull. *Clin. Oral Implants Res.* **1997**, *8*, 198–207. [CrossRef] [PubMed]

15. Buser, D.; Dula, K.; Belser, U.; Hirt, H.-P.; Berthold, H. Localized ridge augmentation using guided bone regeneration. I. Surgical procedure in the maxilla. *Int. J. Periodontics Restor. Dent.* **1993**, *13*, 29–45.
16. Le, B.T.; Borzabadi-Farahani, A. Simultaneous implant placement and bone grafting with particulate mineralized allograft in sites with buccal wall defects, a three-year follow-up and review of literature. *J. Cranio-Maxillofac. Surg.* **2014**, *42*, 552–559. [CrossRef] [PubMed]
17. Wong, R.; Rabie, A. Effect of Bio-Oss Collagen and Collagen matrix on bone formation. *Open Biomed. Eng. J.* **2010**, *4*, 71–76. [CrossRef]
18. Kato, E.; Lemler, J.; Sakurai, K.; Yamada, M. Biodegradation property of beta-tricalcium phosphate-collagen composite in accordance with bone formation: A comparative study with Bio-Oss Collagen®in a rat critical-size defect model. *Clin. Implants Dent. Relat. Res.* **2014**, *16*, 202–211. [CrossRef]
19. Kinard, L.A.; Dahlin, R.L.; Lam, J.; Lu, S.; Lee, E.J.; Kasper, F.K.; Mikos, A.G. Synthetic biodegradable hydrogel delivery of demineralized bone matrix for bone augmentation in a rat model. *Acta Biomater.* **2014**, *10*, 4574–4582. [CrossRef]
20. Schallenberger, M.; Lovick, H.; Locke, J.; Meyer, T.; Juda, G. The effect of temperature exposure during shipment on a commercially available demineralized bone matrix putty. *Cell Tissue Bank.* **2016**, *17*, 677–687. [CrossRef]
21. Wu, G.; Feng, C.; Quan, J.; Wang, Z.; Wei, W.; Zang, S.; Kang, S.; Hui, G.; Chen, X.; Wang, Q. In situ controlled release of stromal cell-derived factor-1α and antimiR-138 for on-demand cranial bone regeneration. *Carbohydr. Polym.* **2018**, *182*, 215–224. [CrossRef]
22. Gibbs, D.M.; Black, C.R.; Dawson, J.I.; Oreffo, R.O. A review of hydrogel use in fracture healing and bone regeneration. *J. Tissue Eng. Regen. Med.* **2016**, *10*, 187–198. [CrossRef]
23. He, B.; Ou, Y.; Zhou, A.; Chen, S.; Zhao, W.; Zhao, J.; Li, H.; Zhu, Y.; Zhao, Z.; Jiang, D. Functionalized d-form self-assembling peptide hydrogels for bone regeneration. *Drug Des. Dev. Ther.* **2016**, *10*, 1379–1388. [CrossRef]
24. Bai, X.; Gao, M.; Syed, S.; Zhuang, J.; Xu, X.; Zhang, X.-Q. Bioactive hydrogels for bone regeneration. *Bioact. Mater.* **2018**, *3*, 401–417. [CrossRef]
25. Buwalda, S.J.; Vermonden, T.; Hennink, W.E. Hydrogels for therapeutic delivery: Current developments and future directions. *Biomacromolecules* **2017**, *18*, 316–330. [CrossRef] [PubMed]
26. Chaudhuri, O. Viscoelastic hydrogels for 3D cell culture. *Biomater. Sci.* **2017**, *5*, 1480–1490. [CrossRef] [PubMed]
27. Tozzi, G.; Mori, A.D.; Oliveira, A.; Roldo, M. Composite Hydrogels for Bone Regeneration. *Materials* **2016**, *9*, 267. [CrossRef] [PubMed]
28. Lopera-Echavarría, A.M.; Medrano-David, D.; Lema-Perez, A.M.; Araque-Marín, P.; Londono, M.E. In vitro evaluation of confinement, bioactivity, and degradation of a putty type bone substitute. *Mater. Today Commun.* **2021**, *26*, 102105.
29. Li, X.; Yang, Q.; Bai, J.; Xuan, Y.; Wang, Y. Identification of appropriate reference genes for human mesenchymal stem cell analysis by quantitative real-time PCR. *Biotechnol. Lett.* **2015**, *37*, 1–67. [CrossRef]
30. Stephens, A.S.; Stephens, S.R.; Morrison, N.A. Internal control genes for quantitative RT-PCR expression analysis in mouse osteoblasts, osteoclasts and macrophages. *BMC Res. Notes* **2011**, *4*, 410. [CrossRef]
31. Yildirim, M.; Spiekermann, H.; Handt, S.; Edelhoff, D. Maxillary sinus augmentation with the xenograft Bio-Oss and autogenous intraoral bone for qualitative improvement of the implant site: A histologic and histomorphometric clinical study in humans. *Int. J. Oral Maxillofac. Implants* **2001**, *16*, 22–33.
32. Meijndert, L.; Raghoebar, G.M.; Schüpbach, P.; Meijer, H.J.A.; Vissink, A. Bone quality at the implant site after reconstruction of a local defect of the maxillary anterior ridge with chin bone or deproteinised cancellous bovine bone. *Int. J. Oral Maxillofac. Surg.* **2005**, *34*, 877–884. [CrossRef]
33. Artzi, Z.; Nemcovsky, C.E.; Tal, H. Efficacy of porous bovine bone mineral in various types of osseous deficiencies: Clinical observations and literature review. *Int. J. Periodontics Restor. Dent.* **2001**, *21*, 395–405.
34. Chan, O.; Coathup, M.J.; Nesbitt, A.; Ho, C.Y.; Hing, K.A.; Buckland, T.; Campion, C.; Blunn, G.W. The effects of microporosity on osteoinduction of calcium phosphate bone graft substitute biomaterials. *Acta Biomater.* **2012**, *8*, 2788–2794. [CrossRef]
35. Prati, C.; Zamparini, F.; Botticelli, D.; Ferri, M.; Yonezawa, D.; Piattelli, A.; Gandolfi, M.G. The Use of ESEM-EDX as an Innovative Tool to Analyze the Mineral Structure of Peri-Implant Human Bone. *Materials* **2020**, *13*, 1671. [CrossRef] [PubMed]
36. Kusrini, E.; Sontang, M. Characterization of x-ray diffraction and electron spin resonance: Effects of sintering time and temperature on bovine hydroxyapatite. *Radiat. Phys. Chem.* **2012**, *81*, 118–125. [CrossRef]
37. Mostafa, N.Y. Characterization, thermal stability and sintering of hydroxyapatite powders prepared by different routes. *Mater. Chem. Phys.* **2005**, *94*, 333–341. [CrossRef]
38. Rinaudo, M. Chitin and chitosan: Properties and applications. *Prog. Polym.Sci.* **2006**, *31*, 603–632. [CrossRef]
39. Berglundh, T.; Lindhe, J. Healing around implants placed in bone defects treated with Bio-Oss®. An experimental study in the dog. *Clin. Oral Implants Res.* **1997**, *8*, 117–124. [CrossRef]
40. Amanso, A.M.; Kamalakar, A.; Bitarafan, S.; Abramowicz, S.; Drissi, H.; Barnett, J.V.; Wood, L.B.; Goudy, S. Osteoinductive effect of soluble transforming growth factor beta receptor 3 on human osteoblast lineage. *J. Cell Biochem.* **2021**, *122*, 538–548. [CrossRef]
41. Kangwannarongkul, T.; Subbalekha, K.; Vivatbutsiri, P.; Suwanwela, J. Gene Expression and Microcomputed Tomography Analysis of Grafted Bone Using Deproteinized Bovine Bone and Freeze-Dried Human Bone. *Int. J. Oral Maxillofac. Implants* **2018**, *33*, 541–548. [CrossRef]
42. Safari, B.; Aghanejad, A.; Roshangar, L.; Davaran, S. Osteogenic effects of the bioactive small molecules and minerals in the scaffold-based bone tissue engineering. *Colloids Surf. B Biointerfaces* **2021**, *198*, 111462. [CrossRef]

43. Caetano Uetanabaro, L.; Claudino, M.; Mobile, R.Z.; Cesar Zielak, J.; Pompermaier Garlet, G.; Rodrigues de Araujo, M. Osteoconductivity of Biphasic Calcium Phosphate Ceramic Improves New Bone Formation: A Histologic, Histomorphometric, Gene Expression, and Microcomputed Tomography Study. *Int. J. Oral Maxillofac. Implants* **2020**, *35*, 70–78. [CrossRef] [PubMed]
44. Banjar, A.A.; Mealey, B.L. A clinical investigation of demineralized bone matrix putty for treatment of periodontal bony defects in humans. *Int. J. Periodontics Restor. Dent.* **2013**, *33*, 567–573. [CrossRef] [PubMed]
45. Gruskin, E.; Doll, B.A.; Futrell, F.W.; Schmitz, J.P.; Hollinger, J.O. Demineralized bone matrix in bone repair: History and use. *Adv. Drug Deliv. Rev.* **2012**, *64*, 1063–1077. [CrossRef] [PubMed]
46. Rocchietta, I.; Simion, M.; Hoffmann, M.; Trisciuoglio, D.; Benigni, M.; Dahlin, C. Vertical bone augmentation with an autogenous block or particles in combination with guided bone regeneration: A clinical and histological preliminary study in humans. *Clin. Implants Dent. Relat. Res.* **2016**, *18*, 19–29. [CrossRef]
47. Pang, C.; Ding, Y.; Zhou, H.; Qin, R.; Hou, R.; Zhang, G.; Hu, K. Alveolar ridge preservation with deproteinized bovine bone graft and collagen membrane and delayed implants. *J. Craniofacial Surg.* **2014**, *25*, 1698–1702. [CrossRef] [PubMed]
48. Hoare, T.R.; Kohane, D.S. Hydrogels in drug delivery: Progress and challenges. *Polymer* **2008**, *49*, 1993–2007. [CrossRef]

Article

Sustainable Surface Modification of Polyetheretherketone (PEEK) Implants by Hydroxyapatite/Silica Coating—An In Vivo Animal Study

Thomas Frankenberger [1], Constantin Leon Graw [2], Nadja Engel [2], Thomas Gerber [1], Bernhard Frerich [2] and Michael Dau [2,*]

1. Institute of Physics, Rostock University, 18057 Rostock, Germany; thomas.frankenberger@uni-rostock.de (T.F.); thomas.gerber@uni-rostock.de (T.G.)
2. Department of Oral, Maxillofacial and Plastic Surgery, Rostock University Medical Center, 18057 Rostock, Germany; constantin.graw@uni-rostock.de (C.L.G.); nadja.engel@med.uni-rostock.de (N.E.); bernhard.frerich@med.uni-rostock.de (B.F.)
* Correspondence: michael.dau@med.uni-rostock.de; Tel.: +49-381-494-6688

Abstract: Polyetheretherketone (PEEK) has the potential to overcome some of the disadvantages of titanium interbody implants in anterior cervical and discectomy and fusion (ACDF). However, PEEK shows an inferior biological behavior regarding osseointegration and bioactivity. Therefore, the aim of the study was to create a bioactive surface coating on PEEK implants with a unique nanopore structure enabling the generation of a long-lasting interfacial composite layer between coating material and implant. Seventy-two PEEK implants—each thirty-six pure PEEK implants (PI) and thirty-six PEEK implants with a sprayed coating consisting of nanocrystalline hydroxyapatite (ncHA) embedded in a silica matrix and interfacial composite layer (SPI)—were inserted in the femoral condyles of adult rats using a split-side model. After 2, 4 and 8 weeks, the femur bones were harvested. Half of the femur bones were used in histological and histomorphometrical analyses. Additionally, pull-out tests were performed in the second half. Postoperative healing was uneventful for all animals, and no postoperative complications were observed. Considerable crestal and medullary bone remodeling could be found around all implants, with faster bone formation around the SPI and fewer regions with fibrous tissue barriers between implant and bone. Histomorphometrical analyses showed a higher bone to implant contact (BIC) in SPI after 4 and 8 weeks ($p < 0.05$). Pull-out tests revealed higher pull-out forces in SPI at all time points ($p < 0.01$). The presented findings demonstrate that a combination of a bioactive coating and the permanent chemical and structural modified interfacial composite layer can improve bone formation at the implant surface by creating a sustainable bone-implant interface. This might be a promising way to overcome the bioinert surface property of PEEK-based implants.

Keywords: polyetheretherketone (PEEK) implants; silicon dioxide; hydroxyapatite (HA); animal rat model; mechanical testing; tissue integration; implant interface; osteoconductive modification; pull-out

1. Introduction

Interbody cage implants filled with bone graft substitutes are one of the main methods in anterior cervical discectomy and fusion (ACDF) and a safe, as well as efficient, alternative to iliac crest bone autografts [1]. These implants are mainly made of titanium (Ti), polyetheretherketone (PEEK) and carbon fiber PEEK. While Ti supports the osseointegration of the implants [2], PEEK offers the advantage of radiolucency and an elastic modulus closer to the bone, resulting in theoretically reduced levels of subsidence [3]. However, the bioinert property of PEEK limits the interaction with surrounding tissue, making PEEK inferior to titanium as an implant material [4]. Numerous animal studies have histologi-

cally and histomorphometrically proven that this limitation leads to a fibrous tissue barrier between the PEEK implant and newly formed bone, resulting in poor osseointegration [5].

Modifying the surface properties of PEEK implants might be a way to increase osseointegration. Therefore, various modifications for improving the biological behavior regarding osseointegration were tested [6–15].

The typical hydrophobic surface can be turned into more hydrophilic by a combination of immersions with H_2SO_4 and NaOH. This ambient temperature sulfonation, which reduced the PEEK water contact angle from 78 to 37 degrees, seems to be more effective than conventional plasma treatments [9]. Another technique to change the surface is by using a treatment with accelerated neutral atom beam (ANAB) technology. The human osteoblast-like cells seeded on the modified surface showed enhanced proliferation and increased expression of ALPL (alkaline phosphatase-like proteins), RUNX2 (runt-related transcription factor 2), COL1A (collagen type I alpha), IBSP (integrin-binding sialoprotein) and BMP2 (bone morphogenetic protein 2) [6]. These effects on the cellular level were connected with an observed increased bone contact and push-out force in a sheep animal model [16].

One of the key factors for biological integration is the early adsorption of proteins and cell colonization at the surface of the implants. Improving cell adhesion can be achieved by modifying the implant surface [8]. Typically, PEEK implants have a smooth surface, which is generally known to limit osseointegration [17]. One simple approach to overcome this disadvantage of PEEK implants is roughening the surface [11]. Micro roughened PEEK implant surfaces showed a higher cell proliferation accompanied by an increased osteocalcin (OC) expression and bone-like nodule formation on the surface [17], as well as higher pull-out force compared to mirror-polished PEEK [17]. Whereas introducing nano roughness to the PEEK surface promotes especially the adsorption of growth factors and proteins, which are known to play a crucial role in the initial cellular mechanisms of osseointegration [18,19].

Another approach to turn the bioinert PEEK surface into a more bioactive one is the surface coating with biomaterials such as alumina [12], calcium phosphate (CaP) [14], especially hydroxyapatite (HA) [13]. Materials such as HA have good osteoconductive properties [20] due to their similar composition to the mineralized part of the bone. HA applied as a coating to implant surfaces was found to increase the rate of bone formation around the implant and enhance the osseointegration [21]. While there is a whole range of different HA materials, nanocrystalline hydroxyapatite is known to deliver good clinical results [22]. Regardless of the good biological behavior of HA applied as a coating, mechanical ablation during the implantation process and delamination of the coating results only in a time-limited improvement of the implant surface properties regarding bioactivity. In order to utilize the bioactive properties of HA coating and to ensure a stable long-term osseointegration, the PEEK surface itself needs to be modified as well.

Achieving an adequately adhered coating on PEEK is a challenging task on its own and a necessary requirement for the success of the medical application. The chemical structure of PEEK limits the bonding strengths of bioactive coatings, e.g., HA, because of the absent chemical bond between the materials. One way to create a bonding is surface modifications of PEEK, which lead to functionalization of the repeat unit and establish dipole–dipole interactions with polar groups of the coating material [23]. Another point of view to achieve a strong bond is a firmly anchored coating on the substrate by creating interphase of both materials. Especially porous bioactive materials are well suited for melt infiltration of polymers in order to form a composite. Such a composite as an intermediate layer between a porous coating material and bulk PEEK provides a structural interlock and may prevent the delamination of the coating material.

Following our previous research on the bioactive coating of titan implants [24], the purpose of the study at hand was to histologically and histomorphometrically investigate the osseointegration of PEEK implants coated with a nanostructured SiO_2/HA biomaterial with an intermediate composite layer as interface. We hypothesized that (i) the interfacial composite layer between the implant and the bioactive coating might enhance the bioactivity of PEEK and (ii) provide a long-term modification of the PEEK surface due to the increased bone conductive properties of the composite layer.

2. Materials and Methods

2.1. Study Materials

The used custom frustum-like shape PEEK implants were made of machined medical grade PEEK (VESTAKEEP i4, Evonik Degussa GmbH, Essen, Germany) with a length of 4.00 mm and a diameter of 3.10 mm at the top and 2.90 mm at the bottom (Figure 1A). The frustum shape was chosen to fulfill the needs of the planed shear force-free pull-out test within the study. All implants were ultrasonically cleaned in deionized alkaline water and afterward generously rinsed with ethanol. Half of the implants (n = 36) were coated with the biomaterial (SPI). The remaining implants (n = 36) were left uncoated (PI). The coating was based upon the principle of the sol gel technique creating a highly porous silica matrix with embedded nanocrystalline hydroxyapatite as platelets with a length of 50–70 nm, 20–25 width and 3–4 nm thickness [24]. The silica sol, derived by acidic hydrolysis of tetraethyl orthosilicate (Sigma Aldrich, St. Louis, MO, USA) and HA (Artoss GmbH, Rostock, Germany), were dispersed in pure ethanol with a ratio HA:SiO_2 of 61:39 wt.%. The dispersion was spin-spray-coated onto the implant surface and subsequently dried under dry oil-free airflow; see [5,24] for a thorough description of the complete process. After coating, the implants were heat-treated with a hot air stream above the melting temperature of PEEK in order to form an interface between the polymer and the coating material. During the heat treatment at 400 °C, the implant surface was locally molten in the range of viscous flow behavior leading to the melt infiltration of the polymer into the unique interconnected pore structure of the coating material. Both groups, coated and uncoated implants, were sterilized in a vacuum oven at 160 °C for 24 h [25].

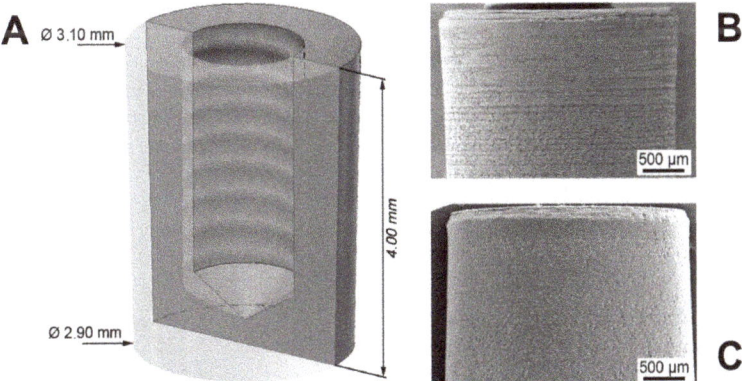

Figure 1. (**A**) Dimensions of PEEK implants. (**B**) SEM images of SiO_2/HA-coated PEEK implant. (**C**) SEM images of pure PEEK implant.

2.2. Material Characterization

Scanning electron microscopy (Merlin VP compact, Zeiss, Germany) analyses were performed on uncoated and coated implants before (Figure 1B,C) and after the in vivo experiments. In order to render the surface conductive, all implants were sputtered with a Cu layer of approximately 5 nm. Images were taken at different magnifications with an acceleration voltage of 10 keV. The local chemical composition was determined by energy-dispersive X-ray analysis (EDX).

The nanostructure of cross-sectional slices was investigated by transmission electron microscopy (EM912, Zeiss, Germany) with an acceleration voltage of 120 keV. Samples were cut into thin sections with a thickness of about 20 to 50 nm by using an ultramicrotome (Leica EM UC6, Leica, Germany) and subsequently were placed on copper grids.

2.3. Animal Model and Procedures

A well-established animal rat model was used in the study at hand [26,27]. All animal experiments were approved by the State Office of Agriculture, Food Safety and Fishery Mecklenburg-Vorpommern, Germany (LALLF M-V/TSD/7221.3-1-040/18) and followed the National Institutes of Health guidelines for the care and use of laboratory animals, as well the ARRIVE Guidelines for Reporting Animal Research [28].

Thirty-six male adult Wistar rats with an average mean body weight of 578.4 g ± 55.6 g were included in the study. The animals were kept at the laboratory with dry food and water ad libitum under standard conditions, including an artificial light–dark cycle of 12 h. All performed surgical interventions were made under general anesthesia with isoflurane (Forene®, AbbVie Germany GmbH and Co. KG, Ludwigshafen, Germany) and intramuscular analgesia via 0.5 mg carprofenum (Rimadyl®, Zoetis Germany GmbH, Berlin, Germany). Following anesthesia, the lateral epicondylus of the femur was exposed via longitudinal incisions at the lateral aspect of the femur. After drilling the implant hole (diameter 3.0 mm, Figure 2) using a sodium chloride cooled implant burr (pilot drill bur G1001 RAXL, Hager and Meisinger GmbH, Neuss, Germany), the implants—randomly chosen by using a computer-generated list—were inserted. In each animal, one implant was inserted on each side, resulting in a total of seventy-two implants (SPI: n = 36; PI: n = 36). After closing periosteum, muscle and skin in layers using absorbable suture material (Vicryl®, Ethicon, Norderstedt, Germany), the animals received intramuscular analgesia via 0.5 mg carprofenum (Rimadyl®, Zoetis Germany GmbH, Berlin, Germany). Additionally, metamizol natrium (Novaminsulfon-ratiopharm®, ratio-pharm GmbH, Ulm, Germany) was applied in the drinking water on a daily basis for the first 5 days postoperatively.

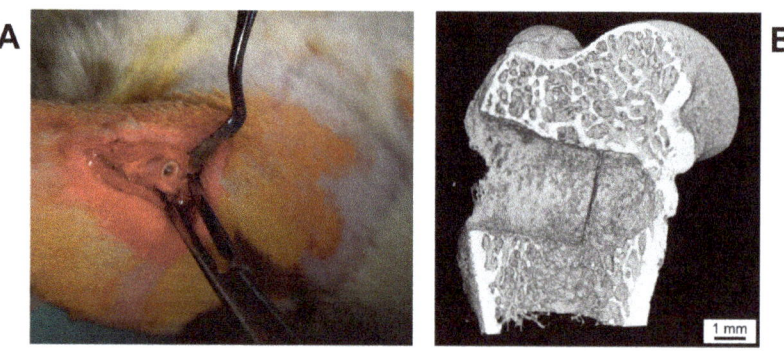

Figure 2. (**A**) Intraoperative picture of inserted implant in lateral aspect of femur. (**B**) μCT image of the bony defect in lateral femur condyle with SiO$_2$/HA-coated PEEK implant (SPI), 4 weeks postoperative. The PEEK implant is radiolucent and thus not visible.

The animals were sacrificed after 2, 4 and 8 weeks (24 implants each time point) by carbon dioxide gas inhalation and bleeding dry. The femurs were harvested for further histological examination (n = 36) and pull-out test (n = 36).

2.4. Pull-Out Tests

After cleaning the harvested femurs from residual soft tissue, the distal portion of the distal part of the femora was separated. The gathered bone samples with the implants were wrapped in gauze soaked with saline and stored at −40 °C between 4 and 8 weeks. Before performing the pull-out tests, the samples were thawed at 6 °C overnight in saline solution for the sake of rehydration. Subsequently, the samples were prepared for the pull-out analyses by embedding in a polyurethane-based resin (RenCast FC 52/53, FC 52 Polyol with a ratio of 50:50 wt.%, Huntsman, Belgium) for fixation in a custom-made tool to align the implant parallel to the vector of the applied force. After curing and alignment, implants were pulled out with a displacement speed of 0.5 mm/min using a universal testing machine (BT1-FR1.0Tn.140, Zwick GmbH and Co. KG, Pforzheim, Germany) equipped with a 200 N measuring box. The maximal load, which is defined as the load of implant loosening, was recorded for n = 6 implants of each group and time point.

2.5. Histological Procedures and Histomorphometrical Measurement

Following fixation in 4% phosphate-buffered formaldehyde for two weeks, the femur bone was dehydrated with alcohol in ascending order (70%–80%–90%–96%–100% each for 24 h and 100% for 48 h) for 7 days. Afterward, xylene was used twice (for 24 h and for 48 h). The specimens were embedded in combination (Technovit 9100® VLC; Heraeus Kulzer, Hanau, Germany) of methyl methacrylate (MMA) and ultraviolet light-activated polymethyl methacrylate (PMMA) in sequential steps (stabilized MMA + Xylol for 48 h, stabilized MMA for 72 h, destabilized MMA for 72 h and destabilized MMA + PMMA). The slices were carefully photopolymerized and processed, applying the sawing and grinding technique [29] using a micro-grinding system (Exact, Norderstedt, Germany). The specimens were ground to a final thickness of 30–40 µm and stained with Giemsa and toluidine blue.

Histological and histomorphometrical analyses were performed for n = 6 implants of each group and time point in the Giemsa toluidine blue-stained sections using a light microscope (Zeiss Axioskop 40, Oberkochen, Germany). The bone to implant contact (BIC) along both sides of the implant was measured with the imaging software (Adobe Photoshop CS® V12.0, Adobe Systems Software Ireland Ltd., Dublin, Ireland).

2.6. Statistics

All statistical analyses were performed with SPSS statistical package version 20.0 (SPSS Inc., Chicago, IL, USA). The results in the study at hand were expressed as arithmetic means ± standard deviation (SD). Before testing for statistical significances, all variables were evaluated for normal distribution via Shapiro–Wilk test. Depending on the presence or the absence of normal distribution, additional analyses were conducted using Mann–Whitney and Student's t-test, respectively. p-values < 0.05 were considered as significant.

3. Results

3.1. Coating Characterization

The coating process was successfully applied to the PEEK implants. Before coating, the bare PEEK implants offered a smooth surface with irregularities due to the machining process, see Figure 3A. These features were balanced by the coating material, which has fully covered the surface with a uniform layer. The thickness of the as-deposited coating was approximately 4–5 µm. An SEM image with high magnification in Figure 3C reveals the nanostructure of the coating material. HA nanocrystals are embedded in a highly porous silica matrix and form together bigger aggregates, creating pores in a size range of a few nanometers up to several hundred. Based on the wet-chemistry route, all pores are interconnected and generate a high surface of 146 m^2/g with a mean pore size of 26.9 nm. The topography of the coating is characterized by open pores due to the morphology of the HA crystals and the silica matrix.

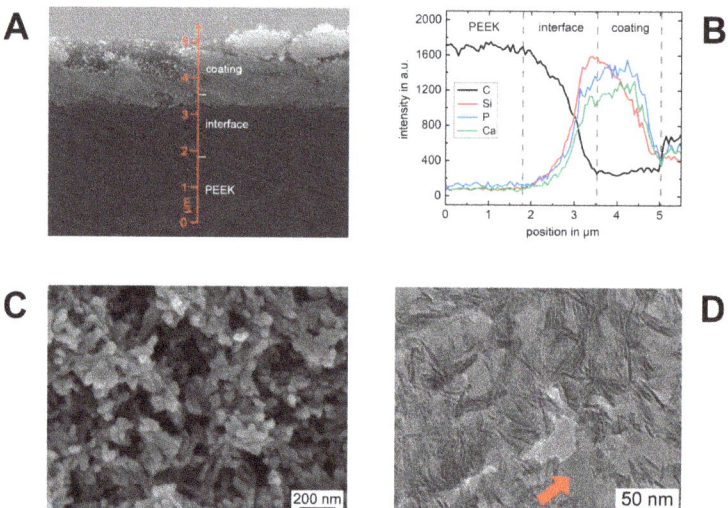

Figure 3. (**A**) SEM image of the cross-section of a coated and heat-treated implant. Three distinct areas are distinguishable: bulk PEEK, interface and coating material. (**B**) Chemical composition of the elements carbon, silicium, phosphor and calcium along the arrow in the SEM image (**A**). (**C**) SEM image of the nanostructure of the coating material with interconnecting pores. (**D**) TEM image of the interface cross-section. All pores of the coating material (darker contrast) are filled with PEEK (brighter contrast).

3.2. Interface Characterization

In Figure 3A, the cross-section of a coated implant with an interface is shown in an SEM image. The as-deposited coating material can be seen as a bright layer with a high electron intensity on top of the PEEK implant with less intensity. Locally molten PEEK from the implant surface partially infiltrated into the pore structure of the as-deposited coating material and filled the free pore volume. In the interface region, a dense composite layer arose which secondary electron intensity cannot be distinguished from the pure bulk PEEK. An analysis of the chemical composition perpendicular to the surface (Figure 3B) indicates at the beginning only a carbon signal belonging to bulk PEEK. As soon as the intensity of silicon, phosphorous and calcium rise, the interface starts as an overlap region of the elements occurring in the coating material (Si, P and Ca) and PEEK (C). Further, towards the top, the carbon signal drops and reaches a minimum, whereas the elements of the coating material follow an inverse behavior. Therefore, PEEK penetrated not the whole coating thickness. On top of the interface exists a residual portion of the coating material, which was not altered by melt infiltration, thus being unmodified. According to the SEM investigation, the interface has characteristics of a uniform layer with a thickness of approximately 2 μm and connects the bulk PEEK of the implant and the residual coating material as an intermediate layer.

The nanostructure of the interface is shown in the TEM image in Figure 3D. HA crystals, which are orientated parallel to the electron beam, were visible as acicular lines. The silica matrix around the HA crystals is not visible due to a similar scattering absorption contrast to PEEK. Together with the randomized HA crystals within the coating, the nanostructure of the coating material is specifiable through differences in the contrast. As can be seen, regardless of the size, all pores within the coating material were completely filled with PEEK, indicated as structureless areas with a brighter contrast (arrow).

3.3. In Vivo Experiments

For all animals, postoperative healing was uneventful, and no postoperative complications were observed.

3.4. Pull-Out Tests

The conducted pull-out tests ran smoothly without any complications. Analyses of the mean applied forces revealed an increase over time in both groups PI and SPI. The mean applied pull-out forces in SPI were about twice as high in PI at each point in time (2 weeks: 7.3 ± 3.7 versus 2.1 ± 1.5 N; 4 weeks: 10.4 ± 1.6 versus 3.2 ± 1.4 N; 8 weeks: 13.6 ± 7.0 versus 5.7 ± 2.5 N) and significantly increased compared to the PI at the same time points ($p < 0.01$). The results of the pull-out test are illustrated in Figure 4.

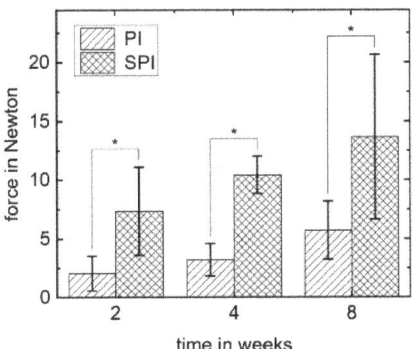

Figure 4. Results of the pull-out test for PI and SPI groups. The pull-out force (mean values in Newton) is pictured with standard deviation. The mean values for SPI were significantly higher for each time point (* $p < 0.01$).

3.5. Histomorphometrical Analysis

Histomorphometrical analysis focusing on the bone to implant contact (BIC) showed a significant increase of the mean values over time in both groups, PI and SPI ($p < 0.05$). The BIC of SPI and PI after 2 weeks were not statistically significant (26.6 ± 6.6% versus 15.0 ± 9.8%). With increasing healing time, the BIC of the SPI was significantly higher than the corresponding value of the PI after 4 weeks (35.7 ± 4.9% versus 25.0 ± 6.6%) and 8 weeks (53.0 ± 7.8% versus 38.5 ± 4.9%). The results of the histomorphometrical analyses are illustrated in Figure 5.

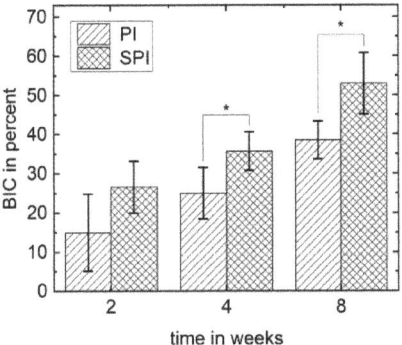

Figure 5. Bone to implant contact (BIC) of the PI and SPI group with standard deviation. The mean BIC of both groups increases as a function of time. After 4 and 8 weeks, the BIC is significantly higher for SPI (* $p < 0.05$).

3.6. Histological Analysis

All implants of the PI and SPI groups showed osseointegration without any sign of inflammation or encapsulation, e.g., Figure 6A. Due to the frustum-like shape of the implants, the press-fit was established only in the cortical portion. Different amounts of newly formed bone were visible around the implants and in the residual drilling channel. Generally, bone formation continued over time, and implants were increasingly enframed of matured lamellae bone. All slices also showed healthy bone marrow with a regular number of hematopoietic cells and adipocytes. At low magnifications, there was no qualitative difference between PI and SPI.

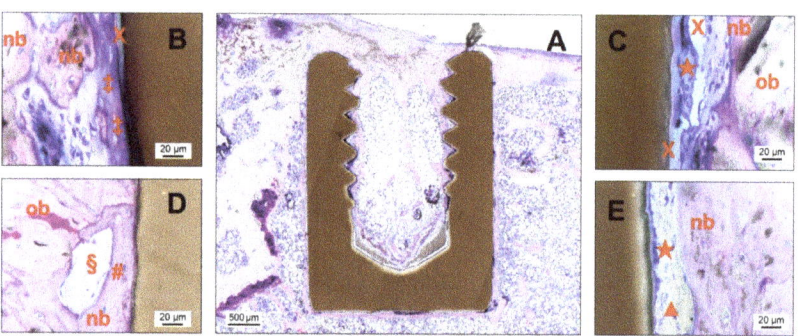

Figure 6. Images of undecalcified cut and ground sections of the PI and SPI group after different time points postoperatively. The sections were histologically stained with a mixture of Giemsa and toluidine blue. (**A**) Overview image of an uncoated pure PEEK implant (PI) after 8 weeks showing new bone formation around the implant outline. (**B–E**) Additional detail images with higher magnification focusing on the direct bone-implant contact area for SPI (2 weeks (**B**) and 8 weeks (**D**)) and PI (2 weeks (**C**) and 8 weeks (**E**)). Note: nb—new bone, ob—old bone, X—preparation artifact, ★—elongated irregular-shaped giant cell, ‡—osteoid, #—osteocyte in contact to the implant surface, §—blood vessel, ▲—fibroblast.

Higher magnifications of the interface region of PI and SPI after 2 and 8 weeks are shown in Figure 6B–E. For both groups, the bone formation took place in the direction of the implant surface. Darker stained new bone could be seen next to the lighter stained old bone. Especially in the PI group, a soft tissue barrier of several micrometers between new bone and the implant surface occurred (Figure 6C,E). The implant surface in these regions was covered with a formation of elongated irregular-shaped giant cells. Other areas showed a one and two-layered stack of smaller and darker stained cells in close proximity to the implant surface. Further, towards the soft tissue, lighter stained cells were visible, among other fibroblasts.

Conversely, more bone-forming regions with direct bone contact could be observed for implants of the SPI group (Figure 6B,D). A pronounced portion of non-mineralized tissue was evident around the implant after 2 weeks (Figure 6B). It appeared as osteoid with included osteocytes in contact with the implant surface. Therefore, osteogenesis occurred in the SPI group in both directions, from the old bone, and from the implant surface. However, osteoblastic rims at osteoid formation sides were difficult to detect. After 8 weeks, the osteoid at the implant surface has mineralized and matured into the lamellar bone with blood vessels in the vicinity of the implant. Osteocytes arranged parallel to the surface showed a vital contact to the implant through cellular extensions.

3.7. Implant Surface Characterization

After finishing the biomechanical investigation, pulled-out implants (PI and SPI) were supercritically dried for electron microscopic characterization of the interface. SEM analyses confirmed that all implants were removed almost without any residual tissue at the surface. When comparing the implant surface topography of the PI group before implantation and after the in vivo study, no specific change can be seen. This situation is representatively pictured in the SEM image Figure 7A for implant post-production and after 8 weeks in vivo, Figure 7B. Neither bone nor coverage with soft tissue was found at the implant surface. These results are in accordance with EDX investigations, which confirm the absence of calcium and phosphor as elements occurring in the inorganic portion of bone.

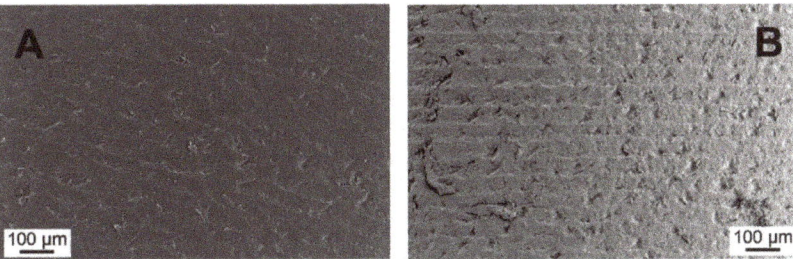

Figure 7. SEM images of the unmodified implant surface (PI) (**A**) post-production and (**B**) after 8 weeks in vivo. Almost no tissue remains on the surface after pull-out test.

In contrast to the PI group, the surface of the SPI implants showed an altered surface topography, pictured in the SEM image Figure 8A for an implant after 8 weeks in vivo. Regardless of the time point, the surface was covered with remains of soft tissue. A qualitative analysis of the surface composition is pictured for the elements calcium, carbon and phosphor as EDX-mapping in Figure 8A. Areas with soft tissue were mainly composed of carbon without the bone-specific elements calcium and phosphor. In contrast, these elements were abundant in regions without soft tissue. These findings applied over the whole test period until 8 weeks.

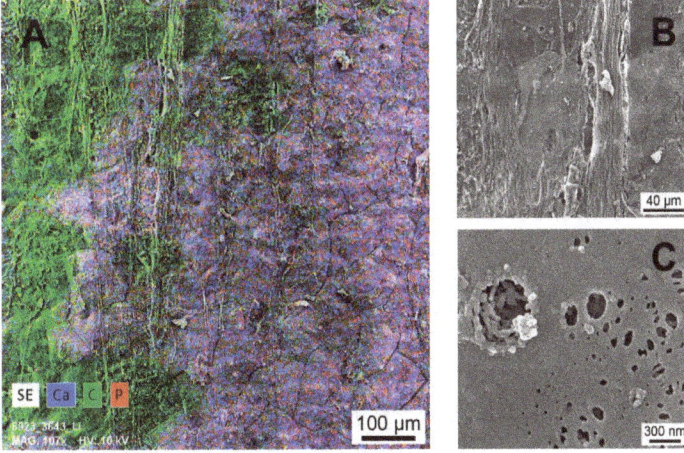

Figure 8. (**A**) EDX mapping of the implant surface (SPI) 8 weeks postoperative. The positions of the elements calcium, carbon and phosphor are indicated in blue, green and red, respectively. Areas covered with soft tissue are correlated with carbon. The residual surface shows higher concentration of calcium and phosphor. (**B**) Higher magnification of a surface portion with fibrous soft tissue coverage. (**C**) Smooth extracellular matrix covering areas with Ca and P occurrence—irregularities in this matrix exhibit an underlying nanostructure.

Higher magnifications of areas with tissue coverage and without are pictured in Figure 8A,C, respectively. The structure of the detected regions with soft tissue was mainly governed by fibrous tissue. Regions without fibrous tissue were covered with a smooth extracellular matrix. Irregularities in the coverage of this matrix exposed an underlying structure with agglomerates of plate-like particles in the nanometer range.

4. Discussion

Pure HA, with its strong bond to native bone, is supposed to alter the bioinert surface of PEEK in a more favorable way regarding osseointegration [13,30,31]. However, a pure composite material will change the tensile strength of the bulk in an unfavorable way. Therefore, the aim of the study was to create a bioactive surface modification on PEEK implants, investigating the influence of its properties after 2, 4 and 8 weeks in a femur rat animal model. The unique coating is a compound of a pure biomaterial layer with an underlying interfacial nanocomposite layer based on the biomaterial and PEEK. As a transition zone between coating and implant, the interface is supposed to firmly connect the coating material with the implant surface. Moreover, the interface is hypothesized to prevent the complete remodeling of the biomaterial in it to enhance sustainable bone conductive properties of the implant surface.

Histological and histomorphometrical results of the study at hand demonstrate that the coated implants with interfacial composite layer (SPI) sustainably improve osseointegration compared to the uncoated pure PEEK implants (PI). The observed mean BIC of the SiO_2/HA-coated PEEK implants after eight weeks is comparable to the earlier reported BIC values of PEEK implants with a plasma-sprayed titanium coating [32]. Similar to titanium surfaces, which are known for high osteoprogenitor cell adhesions [33], HA surfaces are also favorable for the adhesion of osteoprogenitor cells [34]. While at the moment, high numbers of osteoprogenitor cells are only proven for titanium-coated PEEK implant surfaces, the noted HA surface osteoprogenitor cell connection might be the reason for the improved BIC values in the presented study with SiO_2/HA-coated PEEK implants. Generally, these findings support the known approach of adding HA and HA-based biomaterials to PEEK surfaces for better osseointegration [13,35,36]. In contrast to studies focusing on surface modifications but lacking pull-out or push-out tests [36,37], in the presented study, the fixed bone–implant connection between the coated PEEK implants and the surrounding bone is underlined by the two-times higher pull-out forces needed in SPI. Taking the pure cylindrical shape of the used implants in mind, factors such as the screw design of implants do not have to be considered as additional interfering factors. Therefore, the higher pull-out forces can be seen as direct proof of the increased bone–implant connection and underpin the improved osseointegration of the coated implants.

The applied coating consists of HA nanocrystals being morphologically identical to biological HA [38], which are embedded in a silica gel matrix. Due to the wet chemical route, the resulting silica gel is a weakly cross-linked network characterized by numerous open bonds (Si–O$^-$ and Si–OH groups) and high porosity. The composition of HA and SiO_2 forms a matrix with interconnecting nanopores generating a high surface area (Figure 3A). Since the structural properties of the coating material and the original bone graft are based on the same technology, results concerning the biological degradation and the effects on osteogenesis are transferable. Several in vivo studies have shown high biocompatibility of the fully degradable synthetic bone graft in different shapes (e.g., as granulate or microspheres) and a fast formation of new bone [39]. The early substitution of the silica matrix within view days by autologous molecules and bone-specific proteins with known functions in attraction, adhesion and differentiation of bone cells were regarded as the key mechanism of the bone graft's function [40,41]. Despite this autologous self-coating process, the released silica is in general known to play a biochemical role in bone formation and mineralization [42,43]. Adam et al., who coated titan implants with a similar nanostructured SiO_2/HA combination, observed in vitro a drastic decrease in the silica within the coating material shortly after immersion of the coated implants in human blood [24].

This behavior seems similar to the earlier described exchange in the silica gel matrix with an autologous extracellular matrix in the pure nanostructured SiO_2/HA bone substitute material [41]. In the presented in vivo study, this known mechanism of the SiO_2 within the coating material might additionally induce a considerable contribution to the enhanced BIC and generally direct bone formation on the implant surface.

Typically, temporary coatings or damages in the coating's integrity expose the underlying implant surface. In the case of the PEEK surface, the exposure will lead to a non-connection between bone and implant material and might also decrease bone to implant contact in the long run. In the study at hand, even after 8 weeks, SEM analyses still revealed fractions of calcium, carbon and phosphor in SPI, while in the group of the pure PEEK implants (PI), an absence of calcium and phosphor was found. This finding, in combination with the observed underlying nanostructure (Figure 8B), might be caused by the interfacial composite layer. Osteoclast with their known size in the micrometer range is three magnitudes bigger than the primary pore size in the coating material, which might delay or even prevent the ability of osteoclasts to degrade the SiO_2/HA-based coating within the nanocomposite layer. In the end, the applied coating leads to both a chemical and structural modification of the PEEK implant surface that enabled a vital, long-lasting connection between the modified PEEK implant surface and the surrounding bone. The promising results on the interface support the hypothesis that the combination of the nanostructured surface and the biochemical imitation of organic and inorganic components of bone [44] in the interface increases the direct apposition of bone and its vital bond with the implant surface.

Focusing on the question of long-term stability of the used implant surface modification, increased BIC and pull-out force values were found still in SPI (coated implants) after 8 weeks. Based on the observation time to rat lifetime ratio in the experiment, the observed data suggest sustainable stability of the nanostructured SiO_2/HA coating modification. In relation to the human-to-rat age ratio [45], our results theoretically cover even the lifespan of dental implants in humans.

In conclusion, the HA/SiO_2-based bioactive coating with interfacial composite on PEEK implants created a lasting bone-implant interface and might be a promising way to alter the bioinert surface property of PEEK-based implants permanently. Despite the presented results in the study at hand, further research regarding the underlying chemical and physical structure of the interfacial composite layer in vivo is needed. Particularly the exact mechanisms on cellular levels enabling the connection between the SiO_2/HA coating and the surrounding bone tissue are unknown and should be subject to future research. Towards clinical applications, more complex implants modified with the nanostructured SiO_2/HA need to prove the implant stability in vivo.

5. Patents

The coating process described in the study at hand is connected with the US Patent for "Osteoconductive coating of implants made of plastic" (Patent # 9,833,319).

Author Contributions: M.D., B.F., T.G. and T.F. carried out the conception and design of the presented experiments. M.D., T.F. and C.L.G. organized and supervised animal experiments, the measurements and the collection of the data. T.F. produced the modified PEEK implants used in the experiment and conducted the material analyses. M.D. and C.L.G. performed animal experiments and took the measurements of the histopathologic specimens. M.D., T.F., N.E. and C.L.G. analyzed the data, performed the statistics and drafted the manuscript. M.D., T.F., C.L.G., N.E., B.F. and T.G. conducted the interpretation of the data. All authors have read and agreed to the published version of the manuscript.

Funding: The presented research was funded by European Regional Development Fund, grant number TBI-V-1-054-VBW-018.

Institutional Review Board Statement: The experiment was approved by the State Office of Agriculture, Food Safety and Fishery Mecklenburg-Vorpommern, Germany (LALLF M-V/TSD/7221.3-1-040/18) and followed the National Institutes of Health guidelines for the care and use of laboratory animals as well the ARRIVE Guidelines for Reporting Animal Research [28].

Informed Consent Statement: Not applicable.

Data Availability Statement: Data is contained within the article.

Acknowledgments: We thank Daniel Wolter (Laboratory Technical Assistant, Department of Oral, Maxillofacial Plastic Surgery, University Medical Center Rostock, Rostock, Germany) for excellent technical assistance and Werner Götz (Head of laboratory for oral biological research, Department of Orthodontics, Center of Dento-Maxillo-Facial Medicine, Faculty of Medicine, University of Bonn, Bonn, Germany) for support with difficult histological questions.

Conflicts of Interest: Thomas Gerber is CEO of Artoss GmbH, which developed the bone graft material. All other authors declare that they have no conflict of interests and financial interests related to any products involved in this study.

References

1. Shao, M.H.; Zhang, F.; Yin, J.; Xu, H.C.; Lyu, F.Z. Titanium cages versus autogenous iliac crest bone grafts in anterior cervical discectomy and fusion treatment of patients with cervical degenerative diseases: A systematic review and meta-analysis. *Curr. Med. Res. Opin.* **2017**, *33*, 803–811. [CrossRef]
2. Svehla, M.; Morberg, P.; Zicat, B.; Bruce, W.; Sonnabend, D.; Walsh, W.R. Morphometric and mechanical evaluation of titanium implant integration: Comparison of five surface structures. *J. Biomed. Mater. Res.* **2000**, *51*, 15–22. [CrossRef]
3. Rao, P.J.; Pelletier, M.H.; Walsh, W.R.; Mobbs, R.J. Spine interbody implants: Material selection and modification, functionalization and bioactivation of surfaces to improve osseointegration. *Orthop. Surg.* **2014**, *6*, 81–89. [CrossRef] [PubMed]
4. Najeeb, S.; Zafar, M.S.; Khurshid, Z.; Siddiqui, F. Applications of polyetheretherketone (PEEK) in oral implantology and prosthodontics. *J. Prosthodont. Res.* **2016**, *60*, 12–19. [CrossRef] [PubMed]
5. Adam, M.; Ganz, C.; Xu, W.G.; Sarajian, H.R.; Frerich, B.; Gerber, T. How to enhance osseointegration—roughness, hydrophilicity or bioactive coating. *Key Eng. Mater.* **2012**, *493–494*, 467–472. [CrossRef]
6. Khoury, J.; Maxwell, M.; Cherian, R.E.; Bachand, J.; Kurz, A.C.; Walsh, M.; Assad, M.; Svrluga, R.C. Enhanced bioactivity and osseointegration of PEEK with accelerated neutral atom beam technology. *J. Biomed. Mater. Res. B Appl. Biomater.* **2017**, *105*, 531–543. [CrossRef]
7. Mobbs, R.J.; Phan, K.; Assem, Y.; Pelletier, M.; Walsh, W.R. Combination Ti/PEEK ALIF cage for anterior lumbar interbody fusion: Early clinical and radiological results. *J. Clin. Neurosci.* **2016**, *34*, 94–99. [CrossRef]
8. Ramenzoni, L.L.; Attin, T.; Schmidlin, P.R. In vitro effect of modified polyetheretherketone (PEEK) implant abutments on human gingival epithelial keratinocytes migration and proliferation. *Materials* **2019**, *12*, 1401. [CrossRef]
9. Wang, W.; Luo, C.J.; Huang, J.; Edirisinghe, M. PEEK surface modification by fast ambient-temperature sulfonation for bone implant applications. *J. R. Soc. Interface* **2019**, *16*, 20180955. [CrossRef]
10. Walsh, W.R.; Bertollo, N.; Christou, C.; Schaffner, D.; Mobbs, R.J. Plasma-sprayed titanium coating to polyetheretherketone improves the bone-implant interface. *Spine J.* **2015**, *15*, 1041–1049. [CrossRef]
11. Henriques, B.; Fabris, D.; Mesquita-Guimaraes, J.; Sousa, A.C.; Hammes, N.; Souza, J.C.M.; Silva, F.S.; Fredel, M.C. Influence of laser structuring of PEEK, PEEK-GF30 and PEEK-CF30 surfaces on the shear bond strength to a resin cement. *J. Mech. Behav. Biomed. Mater.* **2018**, *84*, 225–234. [CrossRef] [PubMed]
12. Akkan, C.K.; Hammadeh, M.E.; May, A.; Park, H.W.; Abdul-Khaliq, H.; Strunskus, T.; Aktas, O.C. Surface topography and wetting modifications of PEEK for implant applications. *Lasers Med. Sci.* **2014**, *29*, 1633–1639. [CrossRef] [PubMed]
13. Walsh, W.R.; Pelletier, M.H.; Bertollo, N.; Christou, C.; Tan, C. Does PEEK/HA enhance bone formation compared with Peek in a sheep cervical fusion model? *Clin. Orthop. Relat. Res.* **2016**, *474*, 2364–2372. [CrossRef]
14. Willems, K.; Lauweryns, P.; Verleye, G.; Van Goethem, J. Randomized controlled trial of posterior lumbar interbody fusion with Ti- and CaP-nanocoated polyetheretherketone cages: Comparative study of the 1-year radiological and clinical outcome. *Int. J. Spine Surg.* **2019**, *13*, 575–587. [CrossRef] [PubMed]
15. Buck, E.; Li, H.; Cerruti, M. Surface modification strategies to improve the osseointegration of Poly(etheretherketone) and its composites. *Macromol. Biosci.* **2020**, *20*, e1900271. [CrossRef]
16. Poulsson, A.H.; Eglin, D.; Zeiter, S.; Camenisch, K.; Sprecher, C.; Agarwal, Y.; Nehrbass, D.; Wilson, J.; Richards, R.G. Osseointegration of machined, injection moulded and oxygen plasma modified PEEK implants in a sheep model. *Biomaterials* **2014**, *35*, 3717–3728. [CrossRef]
17. Sunarso; Tsuchiya, A.; Fukuda, N.; Toita, R.; Tsuru, K.; Ishikawa, K. Effect of micro-roughening of poly(ether ether ketone) on bone marrow derived stem cell and macrophage responses, and osseointegration. *J. Biomater. Sci. Polym. Ed.* **2018**, *29*, 1375–1388. [CrossRef]
18. Ouyang, L.; Chen, M.; Wang, D.; Lu, T.; Wang, H.; Meng, F.; Yang, Y.; Ma, J.; Yeung, K.W.K.; Liu, X. Nano textured PEEK surface for enhanced osseointegration. *ACS Biomater. Sci. Eng.* **2019**, *5*, 1279–1289. [CrossRef]

19. Rechendorff, K.; Hovgaard, M.B.; Foss, M.; Zhdanov, V.P.; Besenbacher, F. Enhancement of protein adsorption induced by surface roughness. *Langmuir* **2006**, *22*, 10885–10888. [CrossRef]
20. Kubasiewicz-Ross, P.; Hadzik, J.; Seeliger, J.; Kozak, K.; Jurczyszyn, K.; Gerber, H.; Dominiak, M.; Kunert-Keil, C. New nano-hydroxyapatite in bone defect regeneration: A histological study in rats. *Ann. Anat.* **2017**, *213*, 83–90. [CrossRef]
21. Lukaszewska-Kuska, M.; Krawczyk, P.; Martyla, A.; Hedzelek, W.; Dorocka-Bobkowska, B. Effects of a hydroxyapatite coating on the stability of endosseous implants in rabbit tibiae. *Dent. Med. Probl.* **2019**, *56*, 123–129. [CrossRef]
22. Khaled, H.; Atef, M.; Hakam, M. Maxillary sinus floor elevation using hydroxyapatite nano particles vs tenting technique with simultaneous implant placement: A randomized clinical trial. *Clin. Implant Dent. Relat. Res.* **2019**, *21*, 1241–1252. [CrossRef]
23. Almasi, D.; Izman, S.; Assadian, M.; Ghanbari, M.; Abdul Kadir, M.R. Crystalline ha coating on peek via chemical deposition. *Appl. Surf. Sci.* **2014**, *314*, 1034–1040. [CrossRef]
24. Adam, M.; Ganz, C.; Xu, W.; Sarajian, H.R.; Gotz, W.; Gerber, T. In vivo and in vitro investigations of a nanostructured coating material—A preclinical study. *Int. J. Nanomed.* **2014**, *9*, 975–984. [CrossRef]
25. Rogers, W.J. Steam and Dry Heat Sterilization of Biomaterials and Medical Devices. In *Sterilisation of Biomaterials and Medical Devices*; Lerouge, S., Simmons, A., Eds.; Woodhead Publishing: Shaston, UK, 2012; p. 352.
26. Li, X.; Xue, W.; Cao, Y.; Long, Y.; Xie, M. Effect of lycopene on titanium implant osseointegration in ovariectomized rats. *J. Orthop. Surg. Res.* **2018**, *13*, 237. [CrossRef] [PubMed]
27. NemToi, A.; Trandafir, V.; Pasca, A.S.; Sindilar, E.V.; Dragan, E.; Odri, G.A.; NemToi, A.; Haba, D.; Sapte, E. Osseointegration of chemically modified sandblasted and acid-etched titanium implant surface in diabetic rats: A histological and scanning electron microscopy study. *Rom. J. Morphol. Embryol.* **2017**, *58*, 881–886. [PubMed]
28. Kilkenny, C.; Browne, W.J.; Cuthill, I.C.; Emerson, M.; Altman, D.G. Improving bioscience research reporting: The ARRIVE guidelines for reporting animal research. *J. Pharmacol. Pharmacother.* **2010**, *1*, 94–99. [CrossRef] [PubMed]
29. Donath, K.; Breuner, G. A method for the study of undecalcified bones and teeth with attached soft tissues. The Sage-Schliff (sawing and grinding) technique. *J. Oral Pathol.* **1982**, *11*, 318–326. [CrossRef]
30. Park, P.J.; Lehman, R.A. Optimizing the spinal interbody implant: Current advances in material modification and surface treatment technologies. *Curr. Rev. Musculoskelet. Med.* **2020**, *13*, 688–695. [CrossRef]
31. Gu, X.; Sun, X.; Sun, Y.; Wang, J.; Liu, Y.; Yu, K.; Wang, Y.; Zhou, Y. Bioinspired modifications of PEEK implants for bone tissue engineering. *Front. Bioeng. Biotechnol.* **2020**, *8*, 631616. [CrossRef]
32. Torstrick, F.B.; Lin, A.S.P.; Potter, D.; Safranski, D.L.; Sulchek, T.A.; Gall, K.; Guldberg, R.E. Porous PEEK improves the bone-implant interface compared to plasma-sprayed titanium coating on PEEK. *Biomaterials* **2018**, *185*, 106–116. [CrossRef] [PubMed]
33. Hickey, D.J.; Lorman, B.; Fedder, I.L. Improved response of osteoprogenitor cells to titanium plasma-sprayed PEEK surfaces. *Colloids Surf. B Biointerfaces* **2019**, *175*, 509–516. [CrossRef] [PubMed]
34. Ng, A.M.; Tan, K.K.; Phang, M.Y.; Aziyati, O.; Tan, G.H.; Isa, M.R.; Aminuddin, B.S.; Naseem, M.; Fauziah, O.; Ruszymah, B.H. Differential osteogenic activity of osteoprogenitor cells on HA and TCP/HA scaffold of tissue engineered bone. *J. Biomed. Mater. Res. A* **2008**, *85*, 301–312. [CrossRef] [PubMed]
35. Barkarmo, S.; Wennerberg, A.; Hoffman, M.; Kjellin, P.; Breding, K.; Handa, P.; Stenport, V. Nano-hydroxyapatite-coated PEEK implants: A pilot study in rabbit bone. *J. Biomed. Mater. Res. A* **2013**, *101*, 465–471. [CrossRef]
36. Johansson, P.; Jimbo, R.; Naito, Y.; Kjellin, P.; Currie, F.; Wennerberg, A. Polyether ether ketone implants achieve increased bone fusion when coated with nano-sized hydroxyapatite: A histomorphometric study in rabbit bone. *Int. J. Nanomed.* **2016**, *11*, 1435–1442. [CrossRef]
37. Suska, F.; Omar, O.; Emanuelsson, L.; Taylor, M.; Gruner, P.; Kinbrum, A.; Hunt, D.; Hunt, T.; Taylor, A.; Palmquist, A. Enhancement of CRF-PEEK osseointegration by plasma-sprayed hydroxyapatite: A rabbit model. *J. Biomater. Appl.* **2014**, *29*, 234–242. [CrossRef] [PubMed]
38. Gerber, T.; Lenz, S.; Holzhüter, G.; Götz, W.; Helms, K.; Harms, C.; Mittlmeier, T. Nanostructured bone grafting substitutes—A pathway to osteoinductivity. *Key Eng. Mater.* **2012**, *493–494*, 147–152. [CrossRef]
39. Gotz, W.; Papageorgiou, S.N. Molecular, cellular and pharmaceutical aspects of synthetic hydroxyapatite bone substitutes for oral and maxillofacial grafting. *Curr. Pharm. Biotechnol.* **2017**, *18*, 95–106. [CrossRef]
40. Punke, C.; Zehlicke, T.; Just, T.; Holzhuter, G.; Gerber, T.; Pau, H.W. Matrix change of bone grafting substitute after implantation into guinea pig bulla. *Folia Morphol.* **2012**, *71*, 109–114.
41. Xu, W.; Holzhuter, G.; Sorg, H.; Wolter, D.; Lenz, S.; Gerber, T.; Vollmar, B. Early matrix change of a nanostructured bone grafting substitute in the rat. *J. Biomed. Mater. Res. B Appl. Biomater.* **2009**, *91*, 692–699. [CrossRef]
42. Rodella, L.F.; Bonazza, V.; Labanca, M.; Lonati, C.; Rezzani, R. A review of the effects of dietary silicon intake on bone homeostasis and regeneration. *J. Nutr. Health Aging* **2014**, *18*, 820–826. [CrossRef] [PubMed]
43. Schwarz, K.; Milne, D.B. Growth-promoting effects of silicon in rats. *Nature* **1972**, *239*, 333–334. [CrossRef]
44. Gittens, R.A.; Olivares-Navarrete, R.; Schwartz, Z.; Boyan, B.D. Implant osseointegration and the role of microroughness and nanostructures: Lessons for spine implants. *Acta Biomater.* **2014**, *10*, 3363–3371. [CrossRef] [PubMed]
45. Sengupta, P. The laboratory rat: Relating its age with human's. *Int. J. Prev. Med.* **2013**, *4*, 624–630. [PubMed]

Article

Multiple Porous Synthetic Bone Graft Comprising EngineeredMicro-Channel for Drug Carrier and Bone Regeneration

Chun-Sik Bae [1], Seung-Hyun Kim [1], Taeho Ahn [1], Yeonji Kim [2], Se-Eun Kim [1], Seong-Soo Kang [1], Jae-Sung Kwon [3], Kwang-Mahn Kim [3], Sahng-Gyoon Kim [4] and Daniel Oh [3],*

1. College of Veterinary Medicine, Chonnam National University, Gwangju 61186, Korea; csbae210@chonnam.ac.kr (C.-S.B.); leicia@naver.com (S.-H.K.); thahn@jun.ac.kr (T.A.); ksevet@gmail.com (S.-E.K.); vetkang@chonnam.ac.kr (S.-S.K.)
2. OsteoGene Bio, 75 Oak Street, Norwood, NJ 07648, USA; yeonji27@gmail.com
3. College of Dentistry, Yonsei University, Seoul 03722, Korea; jkwon@yuhs.ac (J.-S.K.); kmkim@yuhs.ac (K.-M.K.)
4. College of Dental Medicine, Columbia University, New York, NY 10032, USA; drmartinkim@gmail.com
* Correspondence: dso0301@yuhs.ac; Tel.: +1-551-214-7788

Abstract: Due to high demand but limited supply, there has been an increase in the need to replace autologous bone grafts with alternatives that fulfill osteogenic requirements. In this study, two different types of bone grafts were tested for their drug carrying abilities along with their osteogenic properties. Two different types of alendronate-loaded bone grafts, Bio-Oss (bovine bone graft) and InRoad (biphasic synthetic bone graft) were observed to see how different concentrations of alendronate would affect the sustained release to enhance osteogenesis. In this study, defected ovariectomize-induced osteoporotic rat calvarias were observed for 28 days with three different concentrations of alendronate (0 mg, 1 mg, 5 mg) for both Bio-Oss and InRoad. A higher concentration (5 mg) allowed for a more controlled and sustained release throughout the 28-day comparison to those of lower concentrations (0 mg, 1 mg). When comparing Bio-Oss and InRoad through histology and Micro-CT, InRoad showed higher enhancement in osteogenesis. Through this study, it was observed that alendronate not only brings out robust osteogenesis with InRoad bone grafts, but also enhances bone regeneration in an alendronate-concentration-dependent manner. The combination of higher concentration of alendronate and multiple porous bone graft containing internal micro-channel structure of InRoad resulted in higher osteogenesis with a sustained release of alendronate.

Keywords: synthetic bone graft; micro-channel; multiple pores; drug carrier; bone regeneration

1. Introduction

Performed in over two million patients worldwide, the bone grafting procedure is the second most prevalent tissue transplantation [1]. Currently, autologous bone grafts are the preferred method over other procedures because it contains important properties—osteoconduction, osteoinduction, and osteogenesis, etc.—to ensure a successful bone transplantation [2]. Unfortunately, complications such as inferior healing and limited supply compared to the high demand of autologous bone grafts require different alternatives that would suffice osteogenic properties mentioned above [3–5].

Biphasic calcium bone grafts are rising alternatives to stimulate cell growth within the porous structure, whether it is through the seeding or migrating of cells from nearby tissues. In an ideal design, synthetic bone grafts need to be equipped for cell migration, proliferation, attachment, and differentiation. In addition, synthetic bone grafts can qualify as drug carriers that can supply to the site of transplantation. This drug delivery system would bypass the need for cells to be locally seeded in order to increase tissue repair and regeneration [6–11]. Moreover, many other studies focused on bone regeneration utilizing induced osteoporotic animal models were performed in subcritical size defects [12–14]. Even with the established proof of bone regeneration models, critical size defects need to be promoted for the confirmation of translational feasibility and regenerative capacity of synthetic bone graft materials [15].

Previous studies have shown that the wicking property of highly porous, multi-level configurational hydroxyapatite bone void filler (HA-BVF) granules can organically stimulate healing cascade without the addition of exogenous factors or cells [15]. The result concluded that the new bone formation of HA-BVF outperformed that of Bio-Oss, one of the most popular clinically used bone grafting materials in dentistry [15,16]. Also, the uniformly packed granules used in the study are known to support cell migration and extracellular matrix (ECM) growth in the vacant spots within the granules [17]. However, many studies have raised concerns that scaffolds and granules or any type of implants are open to the possibilities of becoming infected, requiring drug release from the implant materials [18]. Therefore, in this study, the dual function of HA-BVF for drug carrying and its bone regeneration ability was explored with the addition of alendronate.

Osteoporosis is defined by decreasing bone mass, deteriorating bone tissue, and degrading bone microarchitecture [19]. Although osteoporosis is widespread, the advancement for the utilization of synthetic bone graft to resolve deteriorating bone density and bone mass has been lacking greatly [20–23]. Bisphosphonates such as alendronate, risedronate, zoledronate, etc., are a class of drugs that prevent the loss of bone density, commonly used to treat osteoporosis and similar diseases. Based on clinical studies, there are pros and cons among the drugs. No significant differences were observed through the studies. Hence, in this study, we decided to use alendronate as a model drug to test preclinical feasibility and capability of InRoad for bone regeneration together with drug carrier. Furthermore, alendronate is the preferred choice of bisphosphate because of their efficacy, administration method, cost efficiency, and the availability of long-term safety data [24–28]. The effect of alendronate was proven through numerous and extensive studies to significantly impact osteogenesis by decreasing bone turnover through bone resorption inhibition [29,30]. In addition, the specific type of granule used for this study has micro-channels and submicron holes that would advance drug loading and releasing behavior [31].

Hypothesizing that the addition of alendronate will enhance bone regeneration using the localized drug carrying ability, HA-BVF named InRoad synthetic bone graft was observed and compared with Bio-Oss, a popularly used allograft. In this study, in vivo calvaria bone regeneration was observed using ovariectomize-induced osteoporotic rat models. Calvaria critical size (5 mm in diameter) defect models were created to test the hypothesis. For a deeper understanding of bone regeneration impact associated with Aln, three different concentrations of Aln (0 mg, 1 mg, and 5 mg) were examined for 4, 8, and 12 weeks using both Bio-Oss (bovine bone graft) and InRoad (biphasic synthetic bone graft; $90 \pm 5\%$ hydroxyapatite (HA) and $10 \pm 5\%$ β-tricalcium phosphate (β-TCP)). Through the comparison between popular allograft (Bio-Oss) and synthetic HA-BVF (InRoad), the osteogenic properties from InRoad compared to that of Bio-Oss was also observed. All the results were analyzed through a scanning electron microscope, microtomographic, histomorphometric, and histological evaluation.

2. Materials and Methods

This study was carried out to evaluate the feasibility and capability for bone regeneration together with drug carrier in comparison between commercial xenograft (Bio-Oss) and newly developed synthetic calcium phosphate bone graft (InRoad) (Figure 1).

Figure 1. Graphical abstract of the study.

2.1. Preparation of Alendronate (Aln) Loaded Bone Grafts

InRoad bone grafts ranging 0.3–1.0 mm in diameter were fabricated following the procedure, identical to the previous fabrication [15]. Briefly, 200 to 400 nm sized biphasic powder was mixed with mixing solution with a 1.7–1.8 powder/solution ratio. The mixing solution was made of 1.0 wt% poly(vinyl alcohol) (Sigma-Aldrich, Milwaukee, WI, USA), 0.1 wt% carboxymethyl cellulose (Sigma-Aldrich, Milwaukee, WI, USA) as binders, and 1.0 wt% ammonium polyacrylate (R.T. Vanderbilt Company, Norwalk, CT, USA) as an anionic dispersant. The prepared biphasic paste was coated onto a polyurethane sponge template and sintered at 1230 °C for 3 h. Then, the sintered scaffold blocks were granulated using a mortar and pestle and sieved with correlating experimental size. The obtained granule should have a porous cancellous-bonelike structure containing internal microchannel. And 0.25–1 mm range in diameter Bio-Oss was used in this study as a comparison. In order to prepare the Aln-loaded bone grafts, 1 mg and 5 mg of Aln was dissolved in 0.1 M MES buffer (pH 5.6). Bone grafts were then immersed in solution and underwent gentle shaking to react for 24 h. Aln-loaded bone grafts were collected after shaking, washed with distilled water, and vacuum-dried for one day.

2.2. Characterization of Aln-Loaded Bone Grafts (BGs)

The surface morphologies, overall and internal structures of bone grafts were investigated using scanning electron microscopy (SEM, JEOL 5700, Tokyo, Japan) and micro-computed tomography (micro-CT: SKYSCAN 1727, Billerica, MA, USA) was performed. Elemental analysis of Aln (1 mg)/BGs and Aln (5 mg)/BGs was assessed using X-ray diffraction.

2.3. Release Kinetic of Aln from Bone Grafts

For the in vitro kinetic evaluation of Aln release from Aln (1 mg)/InRoad or Bio-Oss and Aln (5 mg)/InRoad or Bio-Oss, each sample was immersed in 1 mL of PBS buffer (pH 7.4) with gentle shaking (100 rpm) at 37 °C. At predetermined time periods of 1, 3, 5, and 10 h, and 1, 3, 5, 7, 14, 21, and 28 days, the supernatants of the specimens were collected and replaced with an equal volume of fresh PBS solution. To record the absorbance, the wavelength was set at 293 nm with a complex of Aln and standard iron (III) chloride solution, using a Flash Multimode Reader (Varioskan™, Thermo Scientific, Waltham, MA, USA).

2.4. Animals

All procedures were authorized under the Institutional Animal Care and Use Committee of Chonnam National University (CNU IACUC-YB-2019-30). Animal care protocols were under abidance with the Guidelines for Animal Experiments of Chonnam National University. General anesthesia was used for all surgical procedures and tramadol was used for postoperative analgesic care in effort to lessen animal distress. A total of 90 Sprague–Dawley rats (Female, 11 weeks old, 251.9 ± 10.15 g; Samtaco, Osan, Korea) were operated and observed. All 90 rats were distributed appropriately in a random manner into 6 different groups (n = 5). The bone regeneration process of each group was observed and assessed at 4, 8, and 12 weeks following the bone grafts implantation. For each observation date, a control group was set up for Bio-Oss (Geistlich, Wolhusen, Switzerland; bovine bone graft small particles 0.25–1.0 mm) and experimental group for InRoad (OsteoGene Tech, Norwood, NJ, USA; dental synthetic bone graft small particles 0.3–1.0 mm) [15]. Then, the rats were further divided into three groups with different concentrations of Aln, leading to total of six different groups:

- Group I: Aln (0 mg)/Bio-Oss;
- Group II: Aln (1 mg)/Bio-Oss;
- Group III: Aln (5 mg)/Bio-Oss;
- Group IV: Aln (0 mg)/InRoad;
- Group V: Aln (1 mg)/InRoad;
- Group VI: Aln (5 mg)/InRoad.

2.5. Ovariectomize (Ovx) Operation and Defect Surgery

Preoperatively, the rats were first fasted for 12 h then weighed. Rats were also under subcutaneous treatments with atropine (0.1 mg/kg; Jeil Pharmaceutical, Daegu, Korea) and enrofloxacine (2.5 mg/kg; Bayerkorea, Seoul, Korea). Using intraperitoneal injection, rats were placed under anesthesia with a mixture of xylazine (10 mg/kg; Bayerkorea, Seoul, Korea) and ketamine (40 mg/kg; Yuhan Co., Seoul, Korea). Under a sterile environment, two separate flank incisions were made through the epidermis and muscle. Once the ovaries became visible, ovariectomy was carried out by first placing a hemostat to hold the uterine horns after pulling the ovary gently. Then, a 4–0 silk ligature was placed below the hemostat. The ovary was extracted, and the uterus was placed back to the abdomen. Operated abdominal muscle layer was sutured back with absorbable 4–0 suture (Surgisorb, Samyang Co., Seoul, Korea). Epidermis was sutured back with non-absorbable 3–0 suture (Silk, Ailee Co., Busan, Korea) and treated subcutaneously with 3 mL of normal warmed saline. After 8 weeks post operation, bone graft was performed using the exact same anesthesia protocol to that of ovariectomy. Then, to prepare for the skin incision, the rat head was shaved and disinfected. After the epidermis was cut, an L-shaped incision was made to the exposed periosteum. The incision area was separated from the skull with blunt scraping. Then, using a trepan bur (5 mm in diameter), a circular critical size bone defect was made. The void of the defected area was filled with either Bio-Oss or InRoad (Figure 2) with appropriate Aln concentration. After filling the void, the periosteum was sutured back with absorbable 4–0 suture. Epidermis was sutured back with non-absorbable 3–0 suture and treated subcutaneously with tramadol (10 mg/kg) at 12, 24, and 36 h for continued postoperative analgesia.

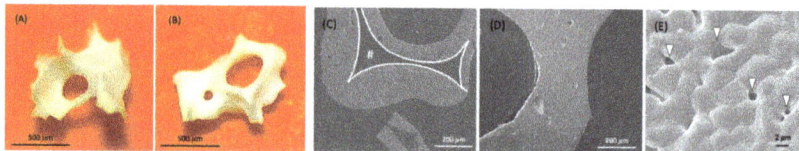

Figure 2. Digital and SEM comparison images between InRoad (**A**,**C**,**E**) and Bio-Oss (**B**,**D**). Both graft types similar under stereo microscope (**A**,**B**). Cross-section image under SEM showed distinguishable difference in internal structure; InRoad showed an embedded micro-channel (**C**) and Bio-Oss did not (**D**). Micro-sized holes (arrowhead) were observed on the InRoad surface (**E**).

2.6. Analysis of Bone Formation

Calvaria harvest was performed at 4, 8, and 12 weeks post bone graft implantation. The manually selected region of interest (ROI) comprised of the defected area. Harvest calvarias were stored in 10% buffered formalin for radiographic, microtomographic, and histological analysis (n = 5). Radiographic analysis was carried out using a digital dental

X-ray (Elytis, Trophy, MARNE LA VALLEE CEDEX 2, France at 60 kVp, 4 mA, film-focus distance of 20 cm and exposure time of 0.344 s). Radiographic analysis for micro-CT was carried out using X-ray tube voltage of 70 kVp with current intensity of 220 μA and integration time of 500 ms. Additionally, bone volume quantification was carried out using CTAn software (BRUKER, Billerica, MA, USA). For visual and display analysis, 3D images were also obtained. Histomorphometric analysis was carried out using Bioquant image analysis software (BIOQUANT Image Analysis Corporation, Nashville, TN, USA) to manually define the total bone area, newly formed bone area, and bone graft area.

2.7. Histological Evaluation

Fixed calvaria samples were processed for undecalcified histological preparation. Using EXAKT Grinding System (EXAKT Technologies, Norderstedt, Germany). Ground sections with 10–20 μm thickness were generated. All ground sections were applied with Masson–Goldner staining. Of the prepared ground sections, the most central section of the defect was selected. The central section was characterized by the ones displaying the widest extension. After the central sections of each defect area were identified, they were subjected to histologic and histomorphometric analysis. The defected area images are under 50× magnification and 200× magnification.

2.8. Statistical Analysis

Quantitative data calculation as means ± standard deviation and comparisons were carried out using one-way ANOVA (Systat Software Inc., Chicago, IL, USA). * $p < 0.05$ was considered statistically significant.

3. Results

3.1. Clinical Observation

During the 28 days post operation of the implant, no complications were observed from all groups. No significant pathological variations or presence of abnormal fluids were identified in any test specimens.

3.2. Characterization of Aln-Loaded Bone Graft

There were no observed changes of the appearance and structure of both InRoad and Bio-Oss bone grafts associated with Aln loading. External morphology of both grafts was similar including porous cancellous structure (Figure 2A,B). Internal morphology of InRoad was clearly distinguishable compared to that of Bio-Oss with its micro-channel (#) structure inside trabecular septum, resulting in superior wicking property and enlarged surface area (Figure 2C,D). Lots of sub-micron-sized holes on the InRoad surface were observed to encourage cells to anchor (Figure 2E). Exceptional wicking property was demonstrated by dropping the bone graft onto fresh rat blood from both bone grafts (Figure 3). By dropping InRoad and Bio-Oss on the top of rat blood, InRoad was saturated with blood immediately without interruption (Figure 3A,B) while the Bio-Oss was floating on top of blood even after 7–10 min (Figure 3C,D).

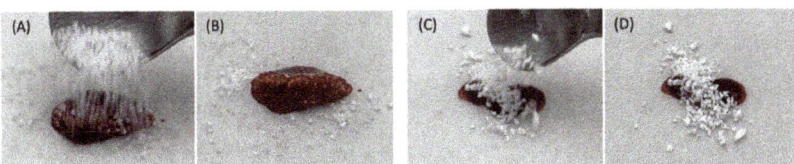

Figure 3. Comparison of wicking property between InRoad (**A,B**) and Bio-Oss (**C,D**) using fresh rat blood. Superior wicking property was demonstrated on InRoad compare to that of Bio-Oss.

The average loading amount of Aln on InRoad and Bio-Oss was 825.72 ± 8.03 μg and 547.60 ± 7.86 μg in Aln (1 mg), and 3601.05 ± 7.86 μg and 2752.35 ± 7.00 μg in Aln (5 mg), respectively. Their loading efficiency was 82.57 ± 0.80%, 54.76 ± 0.79%, 72.02 ± 0.16%, and 55.05 ± 0.14%, respectively (Table 1).

Table 1. Alendronate loading amount and efficiency in both InRoad and Bio-Oss in 1 mg and 5 mg concentration.

Samples	Loading Amount (µg)	Loading Efficiency (%)
Aln (1 mg)/InRoad	825.72 ± 8.03	82.57 ± 0.80
Aln (1 mg)/Bio-Oss	547.60 ± 7.86	54.76 ± 0.79
Aln (5 mg)/InRoad	3601.05 ± 7.86	72.02 ± 0.16
Aln (5 mg)/Bio-Oss	2752.35 ± 7.00	55.05 ± 0.14

InRoad (90 ± 5% HA and 10 ± 5% β-TCP) was synthesized by the reaction of calcium hydroxide and phosphoric acid. The powder diffraction files (PDF) used for analyses were 72-1243 (HA), 09-0169 (β-TCP), 09-0348 (α-TCP), 04-0777 (CaO), and 25-1137 (TTCP), in accordance with the International Centre for Diffraction Data (JCPDS-ICDD). Crystallinity was determined by calculating the percentage of pattern that was contained by any amorphous characteristic, such as the broad amorphous hump observed in the scraped samples. For the Ca:P ratio, the nominal peaks chosen for calibration were HA at 31.7° 2θ, α-TCP at 30.7° 2θ, β-TCP at 31.01° 2θ, and CaO at 37.4° 2θ. The ratio for Ca/P was 1.66, calculated by measuring their respective peak intensities using XRD pattern. The intense high and narrow crystallization peak are visible (Figure 4). The absence of decomposition phase, such as α-TCP or TTCP in HA and β-TCP, were confirmed.

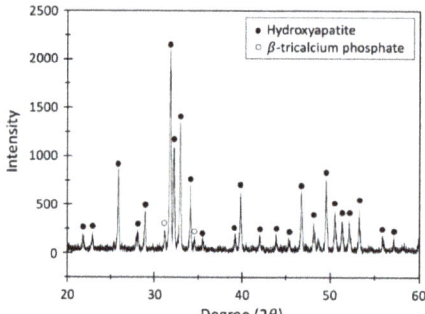

Property	InRoad granule (0.3-1.0 mm)
wt. % HA	93.7 ± 0.8
wt. % β-TCP	6.3 ± 0.8
wt. % α-TCP	-
wt. % CaO	-
wt. % TTCP	-
Crystallinity [%]	90 ± 4
Ca:P ratio by XRD	1.66 ± 0.00

Figure 4. XRD for InRoad composed of 93.7% hydroxyapatite (HA) and 6.3% β-tricalcium phosphate (β-TCP).

3.3. Release Kinetic of Aln from Bone Grafts

Figure 5A showed the release profile of Aln from Aln (1 mg) and Aln (5 mg) on InRoad and Bio-Oss bone grafts. Sustained release of Aln up to 28 days was observed from all samples. On the first day, 248.54 ± 3.21 µg from Aln (1 mg)/InRoad, 263.29 ± 3.37 µg from Aln (1 mg)/Bio-Oss, 386.60 ± 3.39 µg from Aln (5 mg)/InRoad, and 429.02 ± 4.87 µg from Aln (5 mg)/Bio-Oss were released. A total of 577.27 ± 4.56 µg from Aln (1 mg)/InRoad, 461.40 ± 7.09 µg from Aln (1 mg)/Bio-Oss, 994.98 ± 3.38 µg from Aln (5 mg)/InRoad, and 791.38 ± 7.59 µg from Aln (5 mg)/Bio-Oss were released. The observed Aln release proportion differed depending on the Aln concentration for both InRoad and Bio-Oss. It was noted that higher concentration of Aln (5 mg) showed better sustension of Aln release. For 28 days, it was observed that 80% of Aln was released from Aln (1 mg)/Bio-Oss and less than 30% of Aln was released from Aln (5 mg)/Bio-Oss and Aln (5 mg)/InRoad. However, less than 70% (68.73 ± 1.15%) of Aln was released from Aln (1 mg)/InRoad (Figure 5B).

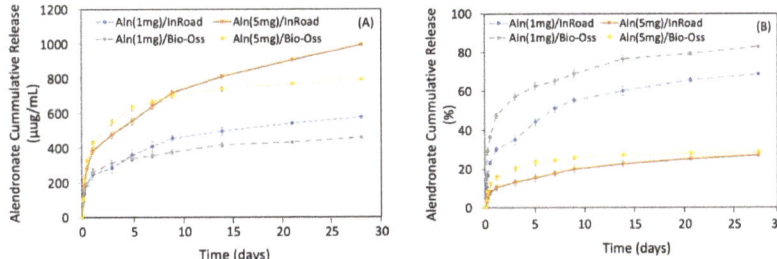

Figure 5. (**A**) Cumulative in vitro release profile of Aln from Aln (1 mg)/InRoad, Aln (1 mg)/Bio-Oss, Aln (5 mg)/InRoad, and Aln (5 mg)/Bio-Oss. Released Aln from both InRoad and Bio-Oss were similar within each concentration. (**B**) Percentage cumulative in vitro release profile of Aln showed that Aln concentration affected Aln releasing pattern. On the first day, 30.21 ± 1.09% and 47.62 ± 1.18% of Aln was released from Aln (1 mg)/InRoad and Aln (1 mg)/Bio-Oss, respectively. While 10.41 ± 0.99% and 15.71 ± 1.26% of Aln was released from Aln (5 mg)/InRoad and Aln (5 mg)/Bio-Oss, respectively. On the 28th day, 68.73 ± 1.15% and 83.09 ± 0.89% of Aln was released from Aln (1 mg)/InRoad and Aln (1 mg)/Bio-Oss, whereas 27.02 ± 0.67% and 28.70 ± 0.63% of Aln was released from Aln (5 mg)/InRoad and Aln (5 mg)/Bio-Oss.

3.4. Analysis of Bone Formation

Using micro-CT, bone formation at 4, 8, and 12 weeks after implantation were evaluated. Figure 6 shows the disappearance of the sharp margin of implantation sites with a lapse of time. For some specimens, blunt margins were still observed until 8 weeks along with higher bone formation and radio-opaque consolidation of the defected area for both InRoad and Bio-Oss. Micro-CT images of InRoad specimens showed relatively consolidated new bone formation compared to Bio-Oss specimens throughout the study period, regardless of any concentrations of Aln. The bone mineral density and bone formation volume increased over time in an Aln-concentration dependent manner (Figure 7). Bone mineral density increased significantly at 8 and 12 weeks in Aln (5 mg) compared to that of Aln (1 mg) for both InRoad and Bio-Oss. Bone formation volume (%BV) was calculated as the percentage of new bone area in the surgical defect area. For the defect area, all tissues of the newly formed bone were included.

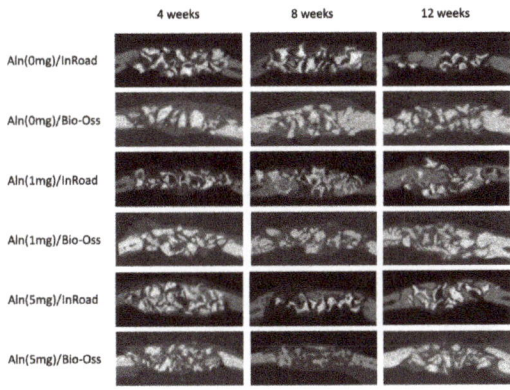

Figure 6. Micro-computed tomography (CT) analysis used to observe the amount of bone formation at the 4, 8, and 12 weeks after implantation. The amount of bone formation was observed using the calculated bone mineral density and bone formation volume (%BV). For the defect area, all tissues of the newly formed bone were included.

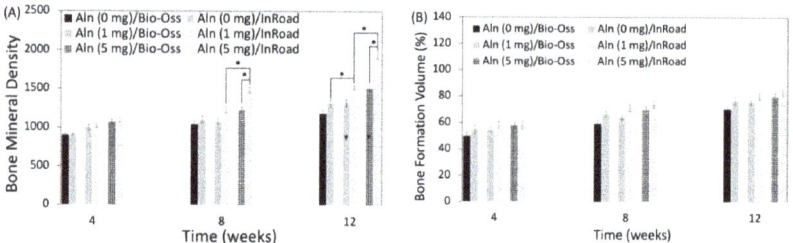

Figure 7. (**A**) Bone mineral density and (**B**) bone formation volume (%) at 4, 8, and 12 weeks after implantation. The error bars represent mean ± SD (n = 5). (* $p < 0.05$).

3.5. Histological Evaluation

Histology analysis at 50× and 200× magnification confirmed that Aln/InRoad bone grafts improved new bone formation. Masson–Goldner staining showed active new bone formation evidenced in all groups in an Aln-concentration-dependent manner. After 8 weeks of implantation, mature bone was distinctly observed around the residual materials in both InRoad and Bio-Oss in accordance with Aln release. After implantation for 12 weeks, the greatest extent of regenerated bone and blood vessels (arrowhead) were observed in Aln (5 mg)/InRoad group.

4. Discussion

Cell adhesion, proliferation, and mineralization are critical to be supported from bone graft materials in bone tissue engineering. Therefore, osteoconductive bone substitutes such as HA and β-TCP are usually considered. Although osteoconductive materials provide a framework for vascular and cellular infiltration, they lack the ability to stimulate the differentiation of mesenchymal stem cells or osteoblast-like cells. Therefore, synthetic bone grafts bring out more effective bone regeneration with the addition of osteoinductive materials such as Aln. The dominant role of Aln is to inhibit osteoclast function through the mevalonate pathway involving cholesterol synthesis inhibition [32]. However, recent studies revealed that local delivery of Aln using calcium phosphate scaffolds can also promote osteoblast differentiation and mineralization in vitro [33,34]. In addition, Aln is known to increase osteocalcin expression, mineralization, and unprenylated Rap1 in human mesenchymal stem cells [35]. Enhancement of proliferation and differentiation in bone-forming cells adjacent to the bone surface in vivo is also possible with the local treatment of Aln [36]. Local delivery of Aln can be explained as a two-step process, (1) preparing InRoad or Bio-Oss bone grafts and (2) the Aln loading on InRoad or Bio-Oss bone grafts. There are concerns of the limitations for the Aln loading process: heterogeneous distribution of Aln within the grafts and irregular release kinetics [37,38]. In this study, the characteristics of InRoad and Bio-Oss containing Aln were investigated to see whether Aln loaded InRoad or Bio-Oss enhanced new bone formation compared to that of InRoad or Bio-Oss without Aln using ovariectomize-induced osteoporotic rat calvaria defect model. In addition, by comparing the popular allograft with a synthetic bone graft, it was observed how much osteogenic property of the allograft could be supported with a synthetic bone graft.

The SEM images proved the porous structures even with the addition of Aln on both InRoad and Bio-Oss bone grafts. The InRoad bone graft is unique with its internal micro-channel structure with macro-pores and micro-holes on its surface. This unique physical characteristic provides larger surface area for cells to attach, encouraging cells to anchor and promote wicking property to enhance cells migration with blood (Figure 2). Simultaneously, this structural advantage is also beneficial for efficiency in Aln loading and releasing kinetics. In addition, macro-pores enable easy neovascularization and new bone ingrowth. Therefore, engineered micro-channels like that of InRoad demonstrates multiple functions. It provides double surface unlike any other bone grafts, resulting in a larger amount of Aln holding capacity and sustainable release capability, along with a larger space for new bone ingrowth. The larger space provided from InRoad achieving larger amount of new bone formation helps seamless integration between bone grafts and new bone after completion of defect healing. As shown in Figure 3, strong wicking property by the combinatorial structure of InRoad may have contributed to the active recruitment of heterogeneous cells from the host body with a surplus amount of blood. Essentially, without active cell migration from the surplus amount of blood in accordance with bone graft implantation, meaningful bone regeneration may be difficult. Hence, the wicking capability in any bone grafts is a crucial property for bone regeneration purposes. In this regard, InRoad may be considered as an innovative and beneficial approach to replace auto-, allo-, and xeno-grafts as a synthetic bone graft. The test results have been compared to the parameters specified by the standards and are summarized in Figure 4. InRoad granules exhibited Ca:P ratios that meet the ISO 13779 specifications.

A relatively linear Aln release kinetic was preserved over the 28-day period. A burst release was observed in the first 24 h (Figure 5A,B). It is hypothesized that the burst may have been due to the Aln bound on the surface of the bone grafts. The sustained and slower release of Aln from InRoad than that of Bio-Oss might be ascribed from the biphasic characteristics of InRoad ($90 \pm 5\%$ HA and $10 \pm 5\%$ β-TCP) by dissolution of the calcium phosphate mineral [17,28–30]. Moreover, a naturally complicated structure like InRoad, composed of inner micro-channel structure and micro-holes, may consider the structural complication a favorable characteristic for better release kinetics than that of Bio-Oss. Considering the sustained release kinetics observed from this study, the Aln/InRoad bone graft may be capable of releasing Aln over several months. Therefore, Aln/InRoad bone graft shows a possibility to be used as a local long-term Aln delivery system for bone defect model in vivo.

The analysis from X-ray and micro-CT of Aln-loaded InRoad and Bio-Oss showed that bone growth was proportional to the amount of Aln content (Figure 6). No significance was observed when comparing bone mineral density and bone formation volume between 1 mg Aln loaded Bio-Oss and InRoad bone graft group and the control group (Aln-unloaded group) except at 12 weeks in InRoad group. However, significant differences in bone mineral density InRoad group at 8 and 12 weeks after implantation with increased Aln content (Figure 7) were observed. Whether increase in bone mineral density is due from the high doses of Aln or long-term exposure of Aln is difficult to determine. Through the histological analysis, it can be observed that Aln (5 mg)/InRoad bone graft showed woven bone formation at 4 weeks after implantation which most converted into mature bone at 8 and 12 weeks (Figure 8). These observations show Aln/InRoad bone graft has good potential for osteoinduction, osteogenesis enhancement, and bone regeneration. The mechanical strength of regenerated bone and patterns of biodegradation was unable to be determined.

Even with the well-known advantages of Aln, there are some concerns due to its uncoupling effect [39–43]. The over-suppression of bone turnover and osteogenesis inhibition are likely to be due to prolonged exposure and systemic administration of Aln [43,44]. All rats were observed up to 12 weeks after implantation and no adverse side effects were detected through the local Aln delivery through Bio-Oss and InRoad. Moreover, the Aln/InRoad bone grafts from this study showed a significant difference in bone formation over 12 weeks period with higher dose of Aln (5 mg) suggesting in vivo applications of Aln/InRoad synthetic bone graft (biphasic calcium phosphate: $90 \pm 5\%$ HA and $10 \pm 5\%$ β-TCP) is effective for bone regeneration.

There are limitations that are yet to be observed from the study. It is suggested to investigate the treated species for a longer period (more than 12 weeks after implantation) to observe the cumulative effect of Aln. With the elongated observation period, it may be possible to study how Aln affects the newly formed bone as well as the absorption of osteoclast-related material. Further evaluation with micro-CT is needed to test the length of bone consolidation, remodeling, and material absorption. In addition, biomechanical study of the regenerated bone is suggested for a deeper understanding of the structural stability and the quality of the regenerated area.

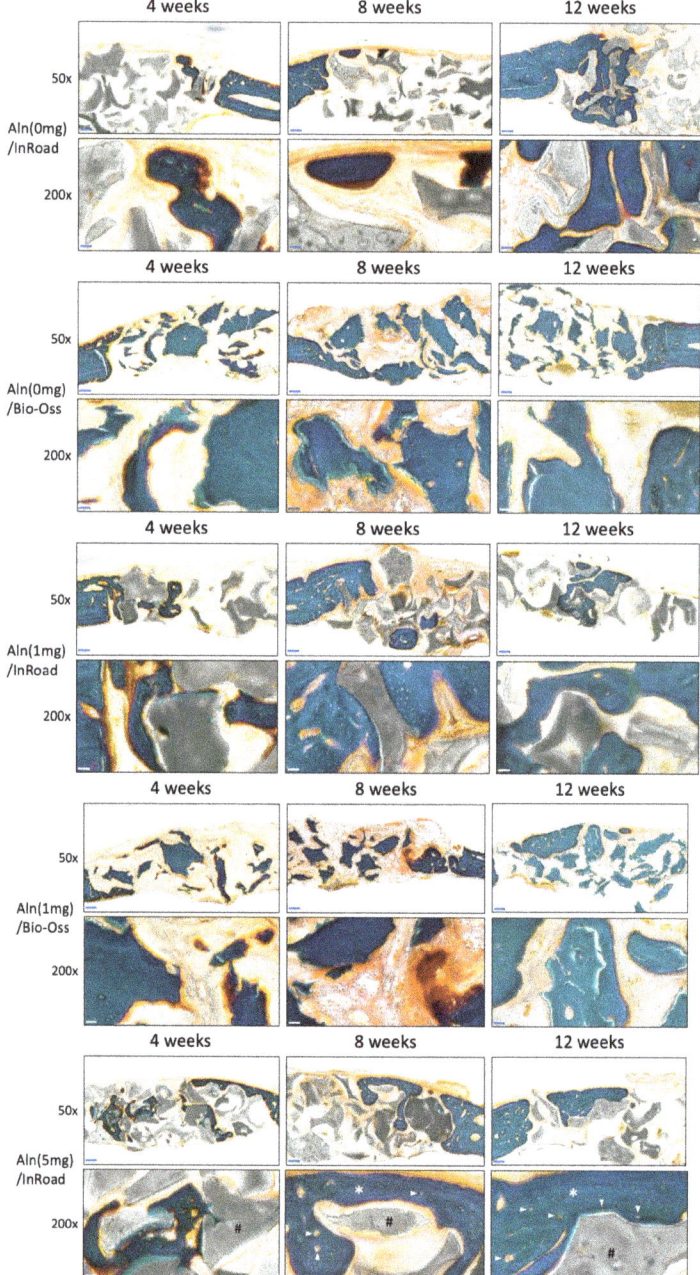

Figure 8. *Cont.*

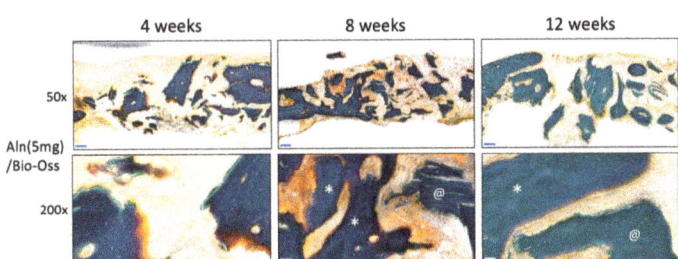

Figure 8. Representative sections of Masson–Goldner stains of 4, 8, and 12 weeks after implantation (50× magnification, 200 µm scale bar). Abundant surrounding fibrous tissue formation and woven bone formation were visible in the most of both samples, especially at 8- and 12-week point. Similar findings were observed on high magnification field (200× magnification, 50 µm scale bar). Blood vessels (arrowhead). InRoad bone graft (#). Bio-Oss bone graft (@). New bone formation (*).

5. Conclusions

In this study, the bone regeneration feasibility and capability of granule type synthetic bone grafts with drug-carting ability were studied using osteoporotic rat model. The sustained release of Alendronate was demonstrated from both Aln/Bio-Oss and Aln/InRoad bone grafts. However, the slower and higher cumulating release of Aln was noticed from Aln/InRoad, along with enhanced bone formation and mineralization. The result was also in respect to the complicated macro–micro–submicron structure and superior wicking property of InRoad. In addition, InRoad synthetic bone grafts significantly enhanced the osteogenetic effect in an ovariectomize-induced osteoporotic rat calvaria defect model in vivo when compared with that of Bio-Oss. This presents great potential as a bone graft material with drug carrying ability for large bone defects even in osteoporotic clinical conditions.

Author Contributions: Conceptualization, D.O. and C.-S.B.; methodology, C.-S.B., S.-H.K., S.-S.K., J.-S.K., and K.-M.K.; software, Y.K.; validation, S.-H.K., T.A., S.-E.K. and S.-G.K.; investigation, S.-H.K., T.A., S.-S.K., S.-E.K., and S.-G.K.; data curation, Y.K.; writing—original draft preparation, D.O. and Y.K.; writing—review and editing, D.O. and C.-S.B.; supervision, D.O. and C.-S.B. All authors have read and agreed to the published version of the manuscript.

Funding: This research was funded by Chonnam National University, 2016 and the Basic Science Research Program through the National Research Foundation of Korea (NRF) funded by Ministry of Education (2017R1D1A1B03034829).

Institutional Review Board Statement: Animal experimental procedures were approved by the Institutional Animal Care and Use Committee of Chonnam National University (CNU IACUC-YB-2019-30) and the animals were cared for in accordance with the Guidelines for Animal Experiments of Chonnam National University.

Informed Consent Statement: Not applicable.

Data Availability Statement: The data presented in this study are available on request from the corresponding author. The data are not publicly available due to intellectual property issues.

Acknowledgments: This study was partially supported by Chonnam National University, 2016 and Basic Science Research Program through the National Research Foundation of Korea (NRF) funded by Ministry of Education (2017R1D1A1B03034829). Compliance with ethical standards.

Conflicts of Interest: The authors declare no conflict of interest.

References

1. Campana, V.; Milano, G.; Pagano, E.; Barba, M.; Cicione, C.; Salonna, G.; Lattanzi, W.; Logroscino, G. Bone substitutes in orthopedic surgery: From basic science to clinical practice. *J. Mater. Sci. Mater. Med.* **2014**, *25*, 2445–2461. [CrossRef] [PubMed]
2. Bauer, T.W.; Muschler, G.F. Bone graft materials: An overview of the basic science. *Clin. Orthop. Relat. Res.* **2000**, *371*, 10–27. [CrossRef]

3. Fillingham, Y.; Jacobs, J. Bone grafts and their substitutes. *Bone Jt. J.* **2016**, *98*, 6–9. [CrossRef]
4. Centers for Disease Control (CDC). Transmission of HIV through bone transplantation: Case report and public health recommendations. *MMWR Morb. Mortal. Wkly. Rep.* **1988**, *37*, 597.
5. Stevenson, S.; Horowitz, M. The response to bone allografts. *J. Bone Jt. Surg.* **1992**, *74*, 939–950. [CrossRef]
6. Garg, T.; Singh, O.; Arora, S.; Murthy, R.S.R. Scaffold: A novel carrier for cell and drug delivery. *Crit. Rev. Ther. Drug Carr. Syst.* **2017**, *29*, 1–63. [CrossRef] [PubMed]
7. Papkov, M.S.; Agashi, K.; Olaye, A.; Shakesheff, K.; Domb, A.J. Polymer carriers for drug delivery in tissue engineering. *Adv. Drug. Deliv. Rev.* **2007**, *59*, 187–206. [CrossRef]
8. Khang, G.; Lee, S.J.; Kim, M.S.; Lee, H.B. Biomaterials: Tissue engineering and scaffold. In *Encyclopedia of Medical Devices and Instrumentation*; Webster, J., Ed.; Wiley: Hoboken, NJ, USA, 2006; Volume 2, pp. 366–383.
9. Mandal, B.B.; Kundu, S.C. Non-bioengineered high strength three-dimensional gland fibroin scaffolds from tropical non-mulberry silkworm for potential tissue engineering applications. *Macromol. Biosci.* **2008**, *8*, 807–818. [CrossRef] [PubMed]
10. Mandal, B.B.; Kundu, S.C. Cell proliferation and migration in silk fibroin 3D scaffolds. *Biomaterials* **2009**, *30*, 2956–2965. [CrossRef]
11. Mandal, B.B.; Kundu, S.C. Osteogenic and adipogenic differentiation of rat bone marrow cells on nonmulberry and mulberry silk gland fibroin 3D scaffolds. *Biomaterials* **2009**, *30*, 5019–5030. [CrossRef] [PubMed]
12. Teofilo, J.M.; Brentegani, L.G.; Lamano-Carvalho, T.L. Bone healing in osteoporotic female rats following intra-alveolar grafting of bioactive glass. *Arch. Oral. Biol.* **2004**, *49*, 755–762. [CrossRef]
13. Okazaki, A.; Koshino, T.; Saito, T.; Takagi, T. Osseous tissue reaction around hydroxyapatite block implanted into proximal metaphysis of tibia of rat with collagen-induced arthritis. *Biomaterials* **2000**, *21*, 483–487. [CrossRef]
14. Tami, A.E.; Leitner, M.M.; Baucke, M.G.; Mueller, T.L.; van Lenthe, G.H.; Muller, R.; Ito, K. Hydroxyapatite particles maintain peri-implant bone mantle during osseointegration in osteoporoticbone. *Bone* **2009**, *45*, 1117–1124. [CrossRef]
15. Kim, S.; Ahn, T.; Han, M.H.; Bae, C.; Oh, D. Wicking Property of Graft Material Enhanced Bone Regenerationin the Ovariectomized Rat Model. *Tissue Eng. Regen. Med.* **2018**, *15*, 503–510. [CrossRef]
16. Xuan, F.; Lee, C.U.; Son, J.S.; Jeong, S.M.; Choi, B.H. A comparative study of the regenerative effect of sinus bone grafting with platelet-rich fibrin-mixed Bio-Oss and commercial fibrin-mixed Bio-Oss: An experimental study. *J. Cranio-Maxillofacial Surg.* **2014**, *42*, 47–50. [CrossRef]
17. Ribeiro, C.C.; Barrias, C.C.; Barbosa, M.A. Preparation and characterisation of calcium-phosphate porous microspheres with a uniform size for biomedical applications. *J. Mater. Sci. Mater. Med.* **2006**, *17*, 455–463. [CrossRef]
18. Schlapp, M.; Friess, W. Collagen/PLGA microparticle composites for local controlled delivery of gentamicin. *J. Pharm. Sci.* **2003**, *92*, 2145–2151. [CrossRef] [PubMed]
19. NIH Consensus development panel on osteoporosis prevention, diagnosis, and therapy. *JAMA* **2001**, *285*, 785–795. [CrossRef]
20. Poole, K.E.; Treece, G.M.; Ridgway, G.R.; Mayhew, P.M.; Borggrefe, J.; Gee, A.H. Targeted regeneration of bone in the osteoporotichuman femur. *PLoS ONE* **2011**, *6*, e16190. [CrossRef] [PubMed]
21. Leppanen, O.V.; Sievanen, H.; Jokihaara, J.; Pajamaki, I.; Kannus, P.; Jarvinen, T.L. Pathogenesis of age-related osteoporosis: Impaired mechano-responsiveness of bone is not the culprit. *PLoS ONE* **2008**, *3*, 2540. [CrossRef]
22. Khosla, S.; Westendorf, J.J.; Oursler, M.J. Building bone to reverseosteoporosis and repair fractures. *J. Clin. Investig.* **2008**, *118*, 421–428. [CrossRef] [PubMed]
23. Dominguez, L.J.; Scalisi, R.; Barbagallo, M. Therapeutic options in osteoporosis. *Acta Biomed.* **2010**, *81*, 55–65. [PubMed]
24. Black, D.M.; Schwartz, A.V.; Ensrud, K.E.; Cauley, J.A.; Levis, S.; Quandt, S.A.; Satterfield, S.; Wallace, R.B.; Bauer, D.C.; Palermo, L.; et al. Effects of continuing or stopping alendronate after 5 years of treatment: The Fracture Intervention Trial Long-term Extension (FLEX): A randomized trial. *JAMA* **2006**, *296*, 2927. [CrossRef] [PubMed]
25. Yates, J. A meta-analysis characterizing the dose-response relationships for three oral nitrogen-containing bisphosphonates in postmenopausal women. *Osteoporos. Int.* **2013**, *24*, 253. [CrossRef] [PubMed]
26. Zhang, J.; Wang, R.; Zhao, Y.L.; Sun, X.H.; Zhao, H.X.; Tan, L.; Chen, D.C.; Xu, B.H. Efficacy of intravenous zoledronic acid in the prevention and treatment of osteoporosis: A meta-analysis. *Asian Pac. J. Trop. Med.* **2012**, *5*, 743. [CrossRef]
27. Crandall, C.J.; Newberry, S.J.; Diamant, A.; Lim, Y.W.; Gellad, W.F.; Booth, M.J.; Motala, A.; Shekelle, P.G. Comparative effectiveness of pharmacologic treatments to prevent fractures: An updated systematic review. *Ann. Intern. Med.* **2014**, *161*, 711. [CrossRef]
28. Liberman, U.A.; Weiss, S.R.; Bröll, J.; Minne, H.W.; Quan, H.; Bell, N.H.; Rodriguez-Portales, J.; Downs, R.W.; Dequeker, J.; Favus, M. Effect of oral alendronate on bone mineral density and the incidence of fractures in postmenopausal osteoporosis. The Alendronate Phase III Osteoporosis Treatment Study Group. *N. Engl. J. Med.* **1995**, *333*, 1437. [CrossRef] [PubMed]
29. Ward, L.M.; Rauch, F.; Whyte, M.P.; D'Astous, J.; Gates, P.E.; Grogan, D.; Lester, E.L.; McCall, R.E.; Pressly, T.A.; Sanders, J.O.; et al. Alendronate for the Treatment of Pediatric Osteogenesis Imperfecta: A Randomized Placebo-Controlled Study. *J. Clin. Endocrinol. Metab.* **2011**, *96*, 355–364. [CrossRef]
30. Tsuchimoto, M.; Azuma, Y.; Higuchi, O.; Sugimoto, I.; Hirata, N.; Kiyoki, M.; Yamamoto, I. Alendronate Modulates osteogenesis of Human Osteoglastic Cells in Vitro. *Jpn. J. Pharmacol.* **1994**, *66*, 25–33. [CrossRef] [PubMed]
31. Hong, M.H.; Son, J.S.; Kim, K.M.; Han, M.; Oh, D.S.; Lee, Y.K. Drug-loaded porous spherical hydroxyapatite granules for bone regeneration. *J. Mater. Sci. Mater. Med.* **2011**, *22*, 349–355. [CrossRef]
32. Rodan, G.A.; Martin, T.J. Therapeutic approaches to bone diseases. *Science* **2000**, *289*, 1508–1514. [CrossRef] [PubMed]

33. Chen, J.; Luo, Y.; Hong, L.; Ling, Y.; Pang, J.; Fang, Y.; Wei, K.; Gao, X. Synthesis, characterization and osteoconductivity properties of bone fillers based on alendronate-loaded poly(epsilon-caprolactone)/hydroxyapatite microspheres. *J. Mater. Sci. Mater. Med.* **2011**, *22*, 547–555. [CrossRef] [PubMed]
34. Boanini, E.; Torricelli, P.; Gazzano, M.; Fini, M.; Bigi, A. The effect of alendronate doped calcium phosphates on bone cells activity. *Bone* **2012**, *51*, 944–952. [CrossRef] [PubMed]
35. Duque, G.; Vidal, C.; Rivas, D. Protein isoprenylation regulates osteogenic differentiation of mesenchymal stem cells: Effect of alendronate, and farnesyl and geranylgeranyl transferase inhibitors. *Br. J. Pharmacol.* **2011**, *162*, 1109–1118. [CrossRef]
36. Komatsu, K.; Shimada, A.; Shibata, T.; Wada, S.; Ideno, H.; Nakashima, K.; Amizuka, N.; Noda, M.; Nifuji, A. Alendronate promotes bone formation by inhibiting protein prenylation in osteoblasts in rat tooth replantation model. *J. Endocrinol.* **2013**, *219*, 145–158. [CrossRef]
37. Peter, B.; Pioletti, D.P.; Laib, S.; Bujoli, B.; Pilet, P.; Janvier, P.; Guicheux, J.; Zambelli, P.Y.; Bouler, J.M.; Gauthier, O. Calcium phosphate drug delivery system: Influence of local zoledronate release on bone implant osteointegration. *Bone* **2005**, *36*, 52–60. [CrossRef] [PubMed]
38. Josse, S.; Faucheux, C.; Soueidan, A.; Grimandi, G.; Massiot, D.; Alonso, B.; Janvier, P.; Laib, S.; Pilet, P.; Gauthier, O.; et al. Novel biomaterials for bisphosphonate delivery. *Biomaterials* **2005**, *26*, 2073–2080. [CrossRef]
39. Ralston, S.H.; Hacking, L.; Willocks, L.; Bruce, F.; Pitkeathly, D.A. Clinical, biochemical, and radiographic effects of aminohydroxypropylidene bisphosphonate treatment in rheumatoid arthritis. *Ann. Rheum. Dis.* **1989**, *48*, 396–399. [CrossRef]
40. Eggelmeijer, F.; Papapoulos, S.E.; van Paassen, H.C.; Dijkmans, B.A.; Breedveld, F.C. Clinical and biochemical response to single infusion of pamidronate in patients with active rheumatoid arthritis: A double blind placebo controlled study. *J. Rheumatol.* **1994**, *21*, 2016–2020.
41. Mashiba, T.; Hirano, T.; Turner, C.H.; Forwood, M.R.; Johnston, C.C.; Burr, D.B. Suppressed bone turnover by bisphosphonates increases microdamage accumulation and reduces some biomechanical properties in dog rib. *J. Bone Miner. Res.* **2000**, *15*, 613–620. [CrossRef]
42. Sama, A.A.; Khan, S.N.; Myers, E.R.; Huang, R.C.; Cammisa, F.P., Jr.; Sandhu, H.S.; Lane, J.M. High-dose alendronate uncouples osteoclast and osteoblast function: A study in a rat spine pseudarthrosis model. *Clin. Orthop. Related Res.* **2004**, *425*, 135–142. [CrossRef] [PubMed]
43. Odvina, C.V.; Zerwekh, J.E.; Rao, D.S.; Maalouf, N.; Gottschalk, F.A.; Pak, C.Y. Severely suppressed bone turnover: A potential complication of alendronate therapy. *J. Clin. Endocrinol. Metab.* **2005**, *90*, 1294–1301. [CrossRef] [PubMed]
44. Iwata, K.; Li, J.; Follet, H.; Phipps, R.J.; Burr, D.B. Bisphosphonates suppress periosteal osteoblast activity independently of resorption in rat femur and tibia. *Bone* **2006**, *39*, 1053–1058. [CrossRef] [PubMed]

MDPI
St. Alban-Anlage 66
4052 Basel
Switzerland
Tel. +41 61 683 77 34
Fax +41 61 302 89 18
www.mdpi.com

Materials Editorial Office
E-mail: materials@mdpi.com
www.mdpi.com/journal/materials